CISM COURSES AND LECTURES

The series presents lecture notes, monographs, edited works and proceedings in the field of Mechanics, Engineering, Computer Science and Applied Mathematics.
Purpose of the series is to make known in the international scientific and technical community results obtained in some of the activities organized by CISM, the International Centre for Mechanical Sciences.

INTERNATIONAL CENTRE FOR MECHANICAL SCIENCES

COURSES AND LECTURES - No. 447

MULTISCALE MODELING IN CONTINUUM MECHANICS AND STRUCTURED DEFORMATIONS

EDITED BY

GIANPETRO DEL PIERO
UNIVERSITÁ DI FERRARA

DAVID R. OWEN
CARNEGIE MELLON UNIVERSITY, PITTSBURGH

Springer-Verlag Wien GmbH

This volume contains 61 illustrations

Originally published by Springer-Verlag Wien New York in 2004
SPIN 11300984

In order to make this volume available as economically and as rapidly as possible the authors' typescripts have been reproduced in their original forms. This method unfortunately has its typographical limitations but it is hoped that they in no way distract the reader.

ISBN 978-3-211-22425-0 ISBN 978-3-7091-2770-4 (eBook)
DOI 10.1007/978-3-7091-2770-4

PREFACE

The course originated from recent progress in two active and related areas of continuum mechanics, the mathematical theory of fracture and the theory of structured deformations. The course was addressed to researchers in the field of continuum mechanics interested in methods for modeling material response, theoretical approaches to fracture mechanics, unified theories for different aspects of material behavior, and problems involving materials with microstructure. In fact, participants came both from industry and academia, and this diversity gave rise to many interesting discussions during and following the lectures.

The first chapter of this book is a general introduction to energy minimization for isotropic nonlinear elastic bodies; we consider this introduction to be a conceptual reference frame both for the theories treated in this volume and for future developments. The second chapter deals with a variational formulation for nonlinear crack problems. In it, conditions for crack equilibrium or propagation are determined by the asymptotic behavior of the strain and stress fields near the tip of the crack. An asymptotic analysis is also provided for the two-dimensional non-linear dynamic problem of a cracked elastic body undergoing anti-plane shear motion. The third chapter describes an evolutionary procedure within Griffith's model for brittle fracture. It is accompanied by examples solved either by asymptotic methods or by an accurate finite element scheme, both based on the mathematical theory of Γ*-convergence.*

The remaining chapters form a distinct part of the book, dealing with the theory of structured deformations. Unlike the classical deformations of continuum mechanics, structured deformations provide a multiscale geometrical setting that can be used to describe smooth and non-smooth geometrical changes at different length scales, such as macro- and microfracture as well as macro- and microslip in crystals. This part starts with a self-contained presentation of the foundations of the theory of structured deformations covering geometrical and energetic aspects, with emphasis on motivation and illustration through examples. A second chapter on foundations deals with structured deformations involving higher-order gradients. These deformations are intended to provide a basis for the study of multiscale motions and of mesolevel phase transitions. The next chapter contains applications to single crystals undergoing single or multiple slip, including a model for hardening based on a specific assumption on mesolevel geometrical changes. It also includes a purely geometrical analysis of multiple slip that yields the fundamental Schmid representation of plastic stretching assumed in standard treatments of crystalline plasticity. The final chapter of the book deals with the formulation of field theories for bodies undergoing deformations at different scales, both in statical and dynamical context. The resulting field equations provide a generalization of non-linear elastodynamics that accounts for dissipation in smooth dynamical processes.

The subjects treated in the book are evolving rapidly; most of the results reported are recently published, and some are published here for the first time. In spite of the many ongoing developments, both in the direction of refining the underlying theories and in the direction of broadening the range of applications, we believe that this book provides an up-to-date account of these subjects.

Gianpietro Del Piero and David R. Owen

CONTENTS

CONTENTS

Energy Minimization for Isotropic Nonlinear Elastic Bodies

M. Šilhavý*†‡

† Mathematical Institute AS CR, Prague, Czech Republic

‡ Department of Mathematics, University of Pisa, Pisa, Italy

1 Nonelliptic Energies, Semiconvexity, and Effective Response

The microstructures accompanying phase transitions in solids are governed by the minimum energy principle. Basic properties of the energy functional (the existence/nonexistence of minima, occurrence/absence of a microstructure) are reflected by the semiconvexity properties of the energy: convexity, rank 1 convexity, polyconvexity, and quasiconvexity. The occurrence of microstructures is connected with the incompatibility of individual solid phases and with the nonexistence of minimizers of energy. In their presence, one is concerned with effective properties. The procedure is the relaxation.

The course treats the semiconvexity properties and relaxation for isotropic bodies. Conditions for the semiconvexity are presented and a procedure for finding the rank 1 convex hull described, with an emphasis on the Baker–Ericksen inequalities. Classes of explicitly calculable models are determined for which the mechanisms of relaxation can be tested. To these calculable models belong nematic elastomers, isotropic rubber-like solids with nonelliptic energies, treated in a separate chapter.

This introduction first presents some examples and discusses preliminarily the relaxation. The minimum energy sets are described for the cubic to tetragonal phase transformation in shape memory alloys, and for the isotropic to nematic transformation in elastomers. The relaxation needs an apparatus of semiconvexity properties introduced systematically in Section 1.3. Of these, the quasiconvexity and rank 1 convexity have a direct physical meaning, the convexity and polyconvexity are but convenient tools. One has

$$\text{convexity} \Rightarrow \text{polyconvexity} \Rightarrow \text{quasiconvexity} \Rightarrow \text{rank 1 convexity},$$

with a “large” gap between the convexity and polyconvexity and “narrow” gaps among the remaining notions. Section 1.4 provides the motivation for quasiconvexity and discusses a consequence of the rank 1 convexity, Maxwell’s rule, i.e., the continuity of the

*Supported by Grant 201/00/1516 of the Grant Agency of the Czech Republic. The author thanks Gianpietro Del Piero for valuable comments on the previous draft of the manuscript.

normal component of Eshelby's tensor across the phase interface. Then the relaxed energies or sets are reconsidered in Section 1.5 using the semiconvexity notions. Besides the quasiconvex hull (relaxation), also the convex, polyconvex, and rank 1 convex hulls are defined for functions an sets.

1.1 Effective Energy

Convex and nonconvex energies in dimension 1 Consider a one–dimensional elastic solid occupying thc interval $[0, 1]$ in its reference configuration. The placement is described by a Lipschitz continuous function $u : [0, 1] \to \mathbb{R}$ with the deformation gradient $F = u_x$. The body is subjected to Dirichlet boundary conditions $u(0) = 0, u(1) = A$ where A is a prescribed value. In the absence of body forces, the energy of the body is given by

$$I(u) = \int_0^1 f(u_x)\,dx$$

where $f : \mathbb{R} \to \mathbb{R}$ is the stored energy. We neglect the invertibility constraint $F > 0$. The goal is to determine states of minimum energy in the Dirichlet class

$$\mathcal{A} = \{u \in W^{1,\infty}(0,1) : u(0) = 0, u(1) = A\},$$

that is, a $u_0 \in \mathcal{A}$ such that

$$I(u_0) \le I(u)$$

for all $u \in \mathcal{A}$, and the minimum energy

$$f^{qc}(A) = \inf\{I(u) : u \in \mathcal{A}\}.$$

The character of the minimum problem depends on whether f is convex or not.

Example 1.1. Let f be strictly convex in the sense that

$$f(G) > f(F) + f'(F)(G - F) \tag{1}$$

for all $F, G \in \mathbb{R}, F \ne G$ where we assume that f is differentiable with the derivative denoted by f'. Under this assumption,
(i) the minimizer is unique and realized by a homogeneous state: $u_0(x) = Ax$;
(ii) the minimum energy $f^{qc}(A)$ coincides with the original energy

$$f^{qc}(A) = f(A).$$

Proof Let $u \in \mathcal{A}$ be an arbitrary state. Inserting into (1), we obtain

$$f(u_x(x)) \ge f(A) + f'(A)(u_x(x) - A). \tag{2}$$

The integration gives

$$I(u) = \int_0^1 f(u_x(x))\,dx \ge f(A) + f'(A)\int_0^1 (u_x(x) - A)\,dx = I(u_0)$$

where we have used the boundary condition $u(0) = u_0(0) = 0, u(1) = u_0(1) = A$. This proves that u_0 is a minimizer. If $u \neq u_0$ then (2) holds with the strict inequality sign on a set of nonzero measure by the strict convexity and thus we obtain $I(u) > I(u_0)$; the minimizer is unique. Also it is clear that (ii) holds. □

Example 1.2. Let f be a double well potential

$$f(F) = (F^2 - 1)^2$$

with minima at $F = \pm 1$. For this potential,
(i) the problem has a minimizer for each $A \in \mathbb{R}$;
(ii) for $A \in \mathbb{R} \setminus (-1, 1)$ the minimizer is unique and homogeneous, $u_0(x) = Ax$; for $A \in (-1, 1)$ there are infinitely many minimizers: the set of all minimizers consists of all $u \in \mathcal{A}$ such that

$$u_x(x) \in \{-1, 1\} \quad \text{a.e. on} \quad [0, 1]. \tag{3}$$

(iii) the minimum energy is given by

$$f^{qc}(A) = \begin{cases} (A^2 - 1)^2 & \text{if} \quad A \in \mathbb{R} \setminus (-1, 1) \\ 0 & \text{if} \quad A \in (-1, 1); \end{cases}$$

thus f^{qc} is the convexification of f in this case.

Remarks 1.3.
- In (3) we encounter deformations with discontinuous gradient ± 1. Such deformations are interpreted as states with two or several phases. The interpretation is analogous to that of Maxwell and Gibbs for phase transitions in fluids.
- This one–dimensional example does not reveal the right convexity notion: in higher dimensions we need the semiconvexity notions: quasiconvexity, rank 1 convexity, polyconvexity.
- The *relaxed energy* f^{qc} is not in conflict with the balance of energy: f^{qc} does not occur in the standard equation of balance of energy of elastodynamics.

Problems in arbitrary dimension Denote by $\mathbb{M}^{m \times n}$ the set of all m by n real matrices with the scalar product $\mathbf{A} \cdot \mathbf{B} = \operatorname{tr}(\mathbf{A}\mathbf{B}^T)$ and norm $|\mathbf{A}| = \sqrt{\operatorname{tr}(\mathbf{A}\mathbf{A}^T)}$. Consider vector-valued dependent variables $\mathbf{u} = (u_1, \ldots, u_m) \in \mathbb{R}^m$ as functions of $\mathbf{x} = (x_1, \ldots, x_n) \in \Omega$ where Ω is an open bounded subset of $\mathbb{R}^n$. Furthermore, let f be a continuous function on $\mathbb{M}^{m \times n}$,[1] bounded from below, and define

$$I(\mathbf{u}) = \int_\Omega f(D\mathbf{u})\, d\mathbf{x}, \quad \mathbf{u} \in W^{1,\infty}(\Omega, \mathbb{R}^m). \tag{4}$$

Here $W^{1,\infty}(\Omega, \mathbb{R}^m)$ is the Sobolev space of all (classes of equivalences of) measurable functions $\mathbf{u} : \Omega \to \mathbb{R}^m$ whose generalized derivatives $D\mathbf{u} : \Omega \to \mathbb{M}^{m \times n}$ are (essentially) bounded on Ω. Alternatively, $W^{1,\infty}(\Omega, \mathbb{R}^m)$ is the set of all locally Lipschitz

[1] By a function on $\mathbb{M}^{m \times n}$ is meant a function with values in $\mathbb{R} \cup \{\infty\}$. Functions with values in $\mathbb{R}$ are termed finite–valued functions. If f is referred to as continuous, it is understood that it is finite–valued.

continuous functions $\mathbf{u}$ on Ω whose total differential $D\mathbf{u}$,[2] which exists a.e. on Ω, is essentially bounded. If Ω is a connected bounded open set with Lipschitz boundary, then $W^{1,\infty}(\Omega, \mathbb{R}^m)$ coincides with the set of all Lipschitz continuous functions on Ω. If the target space $\mathbb{R}^m$ is clear from the context, we write $W^{1,\infty}(\Omega)$ for $W^{1,\infty}(\Omega, \mathbb{R}^m)$. Furthermore, for any bounded open set $\Omega \subset \mathbb{R}^n$, $W_0^{1,\infty}(\Omega, \mathbb{R}^m)$ denotes the set of all $\mathbf{u} \in W^{1,\infty}(\Omega, \mathbb{R}^m)$ which vanish on the boundary $\partial\Omega$ in the sense that $\mathbf{u}$ can be extended to a globally Lipschitz continuous function $\mathbf{v} : \mathbb{R}^n \to \mathbb{R}^m$ which vanishes outside Ω. If $\mathbf{u}_1, \mathbf{u}_2 \in W^{1,\infty}(\Omega, \mathbb{R}^m)$, we write $\mathbf{u}_1(\mathbf{x}) = \mathbf{u}_2(\mathbf{x})$ on $\partial\Omega$ if $\mathbf{u}_1 - \mathbf{u}_2 \in W_0^{1,\infty}(\Omega, \mathbb{R}^m)$.

Problem (4) includes nonlinear hyperelasticity (at large deformations). In this case $m = n$, $\mathbf{u}$ denotes the placement giving the actual position $\mathbf{y} = \mathbf{u}(\mathbf{x})$ in terms of the reference position $\mathbf{x} \in \Omega$, $\mathbf{F} = D\mathbf{u}$ is the deformation gradient and we have the constitutive equations

$$f = f(\mathbf{F}), \quad \mathbf{S} = \hat{\mathbf{S}}(\mathbf{F}) = Df,$$

giving the stored energy f and Piola Kirchhoff stress $\mathbf{S}$ in terms of $\mathbf{F}$. The constitutive functions are defined on $U = \mathbb{M}_+^{n\times n}$, the set of all n by n matrices with positive determinant. However, if one extends f by ∞ outside of $\mathbb{M}_+^{n\times n}$, one obtains a function on $\mathbb{M}^{n\times n}$. Thus one can assume that f is defined on $\mathbb{M}^{n\times n}$ even in the case of nonlinear elasticity. The frame indifference requires that

$$f(\mathbf{QF}) = f(\mathbf{F}), \quad \mathbf{Q} \in SO(n), \tag{5}$$

where $SO(n)$ is the proper orthogonal group. The symmetry group of the material is defined by

$$\mathcal{G} := \{\mathbf{H} \in \mathbb{M}_+^{n\times n} : f(\mathbf{F}) = f(\mathbf{FH}) \text{ for all } \mathbf{F} \in U\}.$$

For isotropic materials, $\mathcal{G} = SO(n)$, for crystals, $\mathcal{G} \subset SO(n)$ is one of the point groups of crystal lattices.

Effective energy in arbitrary dimension A direct generalization of Examples 1.1 and 1.2 is the following relaxation of functions: for each $\mathbf{A} \in \mathbb{M}^{m\times n}$, find

$$f^{qc}(\mathbf{A}) := \frac{1}{|\Omega|} \inf\{I(\mathbf{u}) : \mathbf{u} \in W^{1,\infty}(\Omega, \mathbb{R}^m), \mathbf{u}(\mathbf{x}) = \mathbf{A}\mathbf{x} \text{ on } \partial\Omega\}. \tag{6}$$

Here $|\Omega|$ is the Lebesgue measure of Ω. The function f^{qc} is called the relaxation of f, the relaxed energy, or the quasiconvex hull of f. It turns out that f^{qc} is independent of the choice of Ω. In the context of continuum mechanics, the laboratory experiments measure the mean deformation gradient versus minimum energy and hence f^{qc}.

In the one dimensional case 1.2 the calculation of the relaxed energy reduces to the convexification f^c of f. The states of minimum energy in the nonconvex interval $\in (-1, 1)$ were obtained by mixing the phases at $u_x = -1, u_x = 1$, with appropriate weights. In higher dimensions generally

$$f^{qc} > f^c,$$

[2]For any function f whatsoever, Df denotes the derivative of f whenever it exists.

because a free mixing is impossible since the phases must be gradients of $\mathbf{u}$. This imposes a strong restriction on the interfaces, Hadamard's compatibility condition (11). The higher-dimensional analogs of mixtures of phases are laminates (of order $r = 1, 2, \ldots$) and will be discussed in Example 1.6 and Remark 1.7.

If $f^{qc}(\mathbf{A}) = f(\mathbf{A})$ then the homogeneous state $\mathbf{u}(\mathbf{x}) = \mathbf{A}\mathbf{x}$ realizes the infimum in (6) and the material displays no instabilities. If

$$f^{qc}(\mathbf{A}) < f(\mathbf{A}), \tag{7}$$

then $\mathbf{A}$ is unstable and in the process of minimizing its energy, the body exhibits microstructure. The set $\mathcal{T} \subset \mathbb{M}^{m\times n}$ of all $\mathbf{A}$ where the inequality (7) holds is thus the phase transition zone (cf. Freidin, Vilchevskaya and Sharipova (2002)).

The argument in Example 1.1 still holds: if f is convex at $\mathbf{A}$ in the sense that

$$f(\mathbf{B}) \geq f(\mathbf{A}) + Df(\mathbf{A}) \cdot (\mathbf{B} - \mathbf{A})$$

for all $\mathbf{B} \in \mathbb{M}^{m\times n}$ then

$$f^{qc}(\mathbf{A}) = f(\mathbf{A}). \tag{8}$$

However, it must be emphasized that the convexity of f at $\mathbf{A}$ is not necessary: Suppose that $m = n = 2$ and that for a given $\mathbf{A} \in \mathbb{M}^{2\times 2}$

$$f(\mathbf{B}) \geq f(\mathbf{A}) + \mathbf{C} \cdot (\mathbf{B} - \mathbf{A}) + c(\det \mathbf{B} - \det \mathbf{A}) \tag{9}$$

for all $\mathbf{B} \in \mathbb{M}^{2\times 2}$ where $\mathbf{C} \in \mathbb{M}^{2\times 2}$ is a fixed matrix constant and $c \in \mathbb{R}$ is a constant. Then we still have (8). For $m = n = 2$, inequality (9) is a special case of polyconvexity, a condition much weaker than the convexity.

Exercise 1.4. Prove (8) under (9). Hint: notice that for $\mathbf{u}$ satisfying the boundary condition in (6) the Gauss–Green theorem and the change of variables formula for integrals give

$$\int_\Omega D\mathbf{u}\, d\mathbf{x} = |\Omega|\mathbf{A}, \quad \int_\Omega \det D\mathbf{u}\, d\mathbf{x} = |\Omega| \det \mathbf{A}. \tag{10}$$

Hadamard's lemma Let $\mathbf{u} : \Omega \to \mathbb{R}^m, \Omega \subset \mathbb{R}^n$, be a continuous and piecewise continuously differentiable placement with a jump discontinuity of $\mathbf{F} = D\mathbf{u}$ across an oriented surface $\mathcal{S} \subset \Omega$ with a unit normal $\mathbf{n} \in \mathbb{R}^n$. Denote by $\mathbf{F}^\pm$ the limits of $\mathbf{F}$ on $\mathcal{S}$. The Hadamard lemma asserts the existence of a vector-valued function $\mathbf{a} : \mathcal{S} \to \mathbb{R}^m$ such that

$$[\mathbf{F}] = \mathbf{a} \otimes \mathbf{n} \tag{11}$$

where $[\mathbf{F}] := \mathbf{F}^+ - \mathbf{F}^-$. The class of continuous and piecewise continuously differentiable placements (and more generally of those from $W^{1,\infty}(\Omega)$) models coherent phase boundaries (no dislocations), Figure 1.

Exercise 1.5. Prove (11) by using the continuity of $\mathbf{u}^\pm$ on $\mathcal{S}$: Each component f of the function $\mathbf{u}^+ - \mathbf{u}^-$ is zero on $\mathcal{S}$ and thus each $\mathbf{x} \in \mathcal{S}$ is a point of local extreme of f under the constraint $\mathbf{x} \in \mathcal{S}$. The theorem on Lagrange multipliers gives $Df = \lambda\mathbf{n}$ which leads to (11).

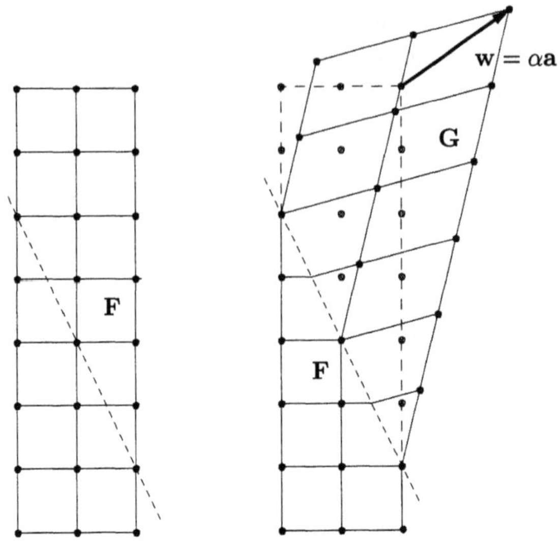

(a) $\mathbf{F} = \mathbf{1}$ (b) $\mathbf{F} = \mathbf{1}, \mathbf{G} = \mathbf{1} + \mathbf{a} \otimes \mathbf{n}$

Figure 1. Compatibility of lattices

Two tensors (= matrices) $\mathbf{F}, \mathbf{G} \in \mathbb{M}^{m \times n}$ are said to be rank 1 connected (equivalently, $\mathbf{F}, \mathbf{G}$ are said to be compatible, and $\mathbf{G}$ is said to be a rank 1 perturbation of $\mathbf{F}$) if $\mathbf{G} - \mathbf{F} = \mathbf{a} \otimes \mathbf{b}$ for some $\mathbf{a} \in \mathbb{R}^m, \mathbf{b} \in \mathbb{R}^n$.

Example 1.6. Let $m = n = 2$, let $\mathbf{B}, \mathbf{C} \in \mathbb{M}^{2 \times 2}$ be given and consider the energy

$$f(\mathbf{A}) = \min \{ |\mathbf{A} - \mathbf{B}|^2, |\mathbf{A} - \mathbf{C}|^2 \}, \tag{12}$$

$\mathbf{A} \in \mathbb{M}^{2 \times 2}$. The function f vanishes at $\mathbf{B}, \mathbf{C}$ and otherwise is positive. It can be interpreted as the energy of a material in which $\mathbf{B}, \mathbf{C}$ represent stable phases of equal energy.[3] Let $\mathcal{L} := \{ \mathbf{A} = t\mathbf{B} + (1-t)\mathbf{C} : 0 < t < 1 \}$ be line segment with endpoints $\mathbf{B}, \mathbf{C}$ and prove that *if* $\mathbf{B}, \mathbf{C}$ *are compatible then* $f^{qc}(\mathbf{A}) = 0$ *for each* $\mathbf{A} \in \mathcal{L}$ *while if* $\mathbf{B}, \mathbf{C}$ *are not compatible then* $f^{qc}(\mathbf{A}) > 0$ *for each* $\mathbf{A} \in \mathcal{L}$.

Proof (Outline) Let $\mathbf{A} = t\mathbf{B} + (1-t)\mathbf{C} \in \mathcal{L}$ where $0 < t < 1$. Assume that $\mathbf{B}, \mathbf{C}$ are compatible and prove $f^{qc}(\mathbf{A}) = 0$. Choose any bounded region $\Omega \subset \mathbb{R}^2$. By the definition (6) we have to find a sequence $\mathbf{u}_j, j = 1, \ldots,$ of deformations of Ω satisfying

$$\left.\begin{aligned} I(\mathbf{u}_j) &\to 0 \quad \text{as} \quad j \to \infty, \\ \mathbf{u}_j(\mathbf{x}) &= \mathbf{A}\mathbf{x} \quad \text{on} \quad \partial\Omega. \end{aligned}\right\} \tag{13}$$

[3] We neglect the objectivity requirement (5) and the invertibility constraint $\det \mathbf{A} > 0$.

Assume that $|\Omega| = 1$ for simplicity of notation. Since $\mathbf{B}, \mathbf{C}$ are compatible, we have $\mathbf{C} = \mathbf{B} + \mathbf{a} \otimes \mathbf{n}$ for some $\mathbf{a}, \mathbf{n} \in \mathbb{R}^2$ with $|\mathbf{n}| = 1$. Let $h : [0,1] \to \mathbb{R}$ be defined by

$$h(s) = \begin{cases} -(1-t)s & \text{if} \quad 0 \le s \le t, \\ t(s-1) & \text{if} \quad t \le s \le 1; \end{cases} \tag{14}$$

notice that $h(0) = h(1) = 0$ and extend h into a 1-periodic function on $\mathbb{R}$, again denoted by h. For each $j = 1, \dots,$ let $\tilde{\mathbf{u}}_j : \Omega \to \mathbb{R}^2$ be defined by

$$\mathbf{u}_j(\mathbf{x}) = \mathbf{A}\mathbf{x} + \frac{1}{j}\mathbf{a}h(j\mathbf{n}\cdot\mathbf{x}),$$

$\mathbf{x} \in \Omega$. One finds that for any $\mathbf{x} \in \Omega$,

$$D\tilde{\mathbf{u}}_j(\mathbf{x}) = \begin{cases} \mathbf{B} & \text{if} \quad \mathbf{x} \in \Omega_j^{\mathbf{B}}, \\ \mathbf{C} & \text{if} \quad \mathbf{x} \in \Omega_j^{\mathbf{C}}, \end{cases}$$

where

$$\Omega_j^{\mathbf{B}} := \left\{ \mathbf{x} \in \Omega : \frac{k-1}{j} < \mathbf{n}\cdot\mathbf{x} < \frac{tk}{j} \text{ for some integer } k \right\},$$
$$\Omega_j^{\mathbf{C}} := \left\{ \mathbf{x} \in \Omega : \frac{tk}{j} < \mathbf{n}\cdot\mathbf{x} < \frac{k}{j} \text{ for some integer } k \right\}.$$

The region Ω thus decomposes into fine layers with alternating phases $\mathbf{B}, \mathbf{C}$. Recalling that $|\Omega| = 1$ one finds that $|\Omega_j^{\mathbf{B}}| \to t$, $|\Omega_j^{\mathbf{C}}| \to 1 - t$ and hence

$$I(\tilde{\mathbf{u}}_j) \to tf(\mathbf{B}) + (1-t)f(\mathbf{C}) = 0$$

as $j \to \infty$. However, $\tilde{\mathbf{u}}_j$ does not satisfy the boundary condition $(13)_2$. The difference is $\tilde{\mathbf{u}}_j(\mathbf{x}) - \mathbf{A}\mathbf{x} = \mathbf{a}h(j\mathbf{n}\cdot\mathbf{x})/j$ and thus $|\tilde{\mathbf{u}}_j(\mathbf{x}) - \mathbf{A}\mathbf{x}| \le 1/j \to 0$ as $j \to \infty$. One can modify $\tilde{\mathbf{u}}_j$ near the boundary $\partial\Omega$ by a cutoff function

$$d_j(\mathbf{x}) = \varphi(j \operatorname{dist}(\mathbf{x}, \partial\Omega)), \quad \mathbf{x} \in \Omega,$$

where φ is a smooth function such that $\varphi(s) = 0$ for $s \le 0$ and $\varphi(s) = 1$ for $s \ge 1$. We thus define a sequence $\mathbf{u}_j : \Omega \to \mathbb{R}^2$ by

$$\mathbf{u}_j(\mathbf{x}) = \mathbf{A}\mathbf{x} + \frac{d_j(\mathbf{x})}{j}\mathbf{a}h(j\mathbf{n}\cdot\mathbf{x}),$$

$\mathbf{x} \in \Omega$. Then the boundary condition $(13)_2$ is satisfied and since $\mathbf{u}_j$ differs from $\tilde{\mathbf{u}}_j$ only on a boundary layer $L_j = \{\mathbf{x} \in \Omega : \operatorname{dist}(\mathbf{x}, \partial\Omega) < 1/j\}$ on which the deformation gradient remains bounded independently of j, one has $|I(\mathbf{u}_j) - I(\tilde{\mathbf{u}}_j)| \le K/j$ where $K < \infty$ is some constant. Thus $I(\mathbf{u}_j) \to 0$ as $j \to \infty$ and $f^{qc}(\mathbf{A}) = 0$.

To prove that $f^{qc}(\mathbf{A}) > 0$ if $\mathbf{B}, \mathbf{C}$ are not compatible, note first that the above construction of $\mathbf{u}_j$ is no longer possible. But this of course does not mean that there is not some other sequence satisfying (13). The proof that there is no such a sequence is obtained by lifting the segment $\mathcal{L}$ into the space $\mathbb{M}^{2\times 2}\times\mathbb{R}$ of pairs $(\mathbf{A},\delta)$ where $\mathbf{A}\in\mathbb{M}^{2\times 2},\delta\in\mathbb{R}$ and employing the convexity notions in the space of lifted objects. Thus let $\tilde{\mathcal{L}}\subset\mathbb{M}^{2\times 2}\times\mathbb{R}$ be line segment with endpoints $(\mathbf{B},\det\mathbf{B})$ and $(\mathbf{C},\det\mathbf{C})$, i.e.,

$$\tilde{\mathcal{L}}=\{t(\mathbf{B},\det\mathbf{B})+(1-t)(\mathbf{C},\det\mathbf{C}):0<t<1\}.$$

Let $\operatorname{dist}(\cdot,\tilde{\mathcal{L}}):\mathbb{M}^{2\times 2}\times\mathbb{R}\to\mathbb{R}$ be the distance from $\tilde{\mathcal{L}}$ and note that since $\tilde{\mathcal{L}}$ is convex, so also is the function $\operatorname{dist}(\cdot,\tilde{\mathcal{L}})$. Independently of whether or not are $\mathbf{B},\mathbf{C}$ compatible, one can prove that there exists a strictly increasing convex function $\omega:[0,\infty)\to[0,\infty)$ such that

$$f(\mathbf{A})\geq\omega(\operatorname{dist}(\mathbf{A},\det\mathbf{A};\tilde{\mathcal{L}}))$$

for every $\mathbf{A}\in\mathbb{M}^{2\times 2}$. Then the function $g:\mathbb{M}^{2\times 2}\times\mathbb{R}\to\mathbb{R}$ defined by $g(\mathbf{A},\delta)=\omega(\operatorname{dist}(\mathbf{A},\delta;\tilde{\mathcal{L}})),(\mathbf{A},\delta)\in\mathbb{M}^{2\times 2}\times\mathbb{R}$, is convex. If $\mathbf{u}:\Omega\to\mathbb{R}^2$ satisfies $\mathbf{u}(\mathbf{x})=\mathbf{A}\mathbf{x}$ on $\partial\Omega$, then we have (10). Hence Jensen's inequality says that

$$\int_\Omega f(D\mathbf{u})\,d\mathbf{x}\geq\int_\Omega g(D\mathbf{u},\det D\mathbf{u})\,d\mathbf{x}\geq g(\mathbf{A},\det\mathbf{A}).$$

The proof will now be completed by showing that if $\mathbf{B},\mathbf{C}$ are not compatible, then $g(\mathbf{A},\det\mathbf{A})>0$. Indeed, the equality $g(\mathbf{A},\det\mathbf{A})=0$ means that $(\mathbf{A},\det\mathbf{A})\in\tilde{\mathcal{L}}$, which in turn means that

$$(\mathbf{A},\det\mathbf{A})=s(\mathbf{B},\det\mathbf{B})+(1-s)(\mathbf{C},\det\mathbf{C})\tag{15}$$

for some $s, 0<s<1$. However, we have $\mathbf{A}=t\mathbf{B}+(1-t)\mathbf{C}$; thus the first component of (15) implies $s=t$ and then the second reads

$$\det(t\mathbf{B}+(1-t)\mathbf{C})=t\det\mathbf{B}+(1-t)\det\mathbf{C}.\tag{16}$$

Using the formula $\det(\mathbf{P}+\mathbf{Q})=\det\mathbf{P}+\mathbf{P}\cdot\operatorname{cof}\mathbf{Q}+\det\mathbf{Q}$ twice, Equation (16) is rearranged to the form $t(1-t)\det(\mathbf{B}-\mathbf{C})=0$ and thus $0<t<1$ implies $\det(\mathbf{B}-\mathbf{C})=0$. Since $m=n=2$, the last equation implies that $\mathbf{B}-\mathbf{C}$ is a rank 1 matrix, in contradiction with the assumption that $\mathbf{B},\mathbf{C}$ are incompatible. □

Remarks 1.7.

- The energy (12) admits an explicit relaxation, see Lurie and Cherkaev (1987), Pipkin (1991), Kohn (1991) for a more general class.
- The construction of $\tilde{\mathbf{u}}_j$ is frequently used; the procedure is called lamination and the deformations $\tilde{\mathbf{u}}_j$ (simple) laminates. If is often necessary to refine the construction to form layers within the layers of an already constructed laminate. Such deformations are called laminates of higher order.
- If $\bar{\mathbf{F}}_j$ is the mean deformation

$$\bar{\mathbf{F}}_j:=\frac{1}{|\Omega|}\int_\Omega D\mathbf{u}_j\,d\mathbf{x}$$

then

$$\bar{\mathbf{F}}_j \to \mathbf{A}, \quad \text{and} \quad \mathbf{u}_j(\mathbf{x}) \to \mathbf{A}\mathbf{x} \ \text{ uniformly on } \ \Omega,$$

but the energy $I(\mathbf{u}_j)$ does not converge to that of the homogeneous deformation of gradient $\mathbf{A}$. Thus if one views $\mathbf{A}$ as the macroscopic deformation, the correct macroscopic energy is not $f(\mathbf{A})$ but $f^{qc}(\mathbf{A})$.

• In crystal structures, laminates of orders 2 (layers within layers) are frequently observed, but rarely laminates of order ≥ 3.

1.2 Minima of Effective Energy

The calculation of f^{qc} is generally difficult; an easier problem is to find the set L of all $\mathbf{A}$ for which $f^{qc}(\mathbf{A})$ takes the minimum value. It turns out that L is completely determined by the set K of minima of f; L is called the relaxation of K and denoted by K^{qc}.

For compact sets $K \subset \mathbb{M}^{m\times n}$ one can define K^{qc} as the set of all $\mathbf{A} \in \mathbb{M}^{m\times n}$ for which there exists a sequence $\mathbf{u}_j \in W^{1,\infty}(\Omega)$ such that

$$\left.\begin{array}{lll} \mathbf{u}_j(\mathbf{x}) = \mathbf{A}\mathbf{x} & \text{on} & \partial\Omega, \\ \operatorname{dist}(D\mathbf{u}_j, K) \to 0 & \text{a.e. on} & \Omega, \\ |D\mathbf{u}_j| \leq C < \infty & \text{a.e. on} & \Omega. \end{array}\right\} \tag{17}$$

Here

$$\operatorname{dist}(\mathbf{A}, K) = \inf\{|\mathbf{A} - \mathbf{B}| : \mathbf{B} \in K\}$$

is the distance of the matrix $\mathbf{A}$ from the set K. It is appropriate to think of $f = \operatorname{dist}(\cdot, K)$ as of a prototype of an energy function that takes the minimum on K.

The definition of K^{qc} draws the attention to the fact that the set of *macroscopic* deformation gradients of minimum energy can be substantially larger than the set of minima of f. In Example 1.6 the set of minima of f as in (12) consists of the two matrices $\mathbf{B}, \mathbf{C}$ but if $\mathbf{B}, \mathbf{C}$ are compatible, then the lamination construction gives a sequence of deformations which satisfy (17) for any $\mathbf{A}$ from the line segment with endpoints $\mathbf{B}, \mathbf{C}$.

The structure of the set of deformation gradients of minimum energy is dictated by the symmetry group $\mathcal{G}$. We here describe two cases, a cubic crystal (with $\mathcal{G}$ the discrete cubic point group) and an isotropic nematic elastomer, where $\mathcal{G} = SO(3)$.

Shape memory alloys Martensitic transformations are temperature-induced, coherent solid-solid phase transformations involving a change in crystalline symmetry. If the binary alloy In-Tl is cooled through a critical temperature θ_c, it undergoes a martensitic transformation from cubic parent phase (austenite) to a tetragonal phase (martensite) which has three symmetry–related *variants*, differing in the orientation of the tetragonal axes with respect to the cubic axes, Figure 3(a).

Let $f = f(\mathbf{F}, \theta)$ be its free energy expressed as a function of the deformation gradient $\mathbf{F}$ and temperature θ; when the temperature is fixed, we write $f = f(\mathbf{F})$. Denote by $K = K_\theta \subset \mathbb{M}^{n\times n}$ the set of minima of $f(\cdot, \theta)$. Figure 2 shows the typical process of the

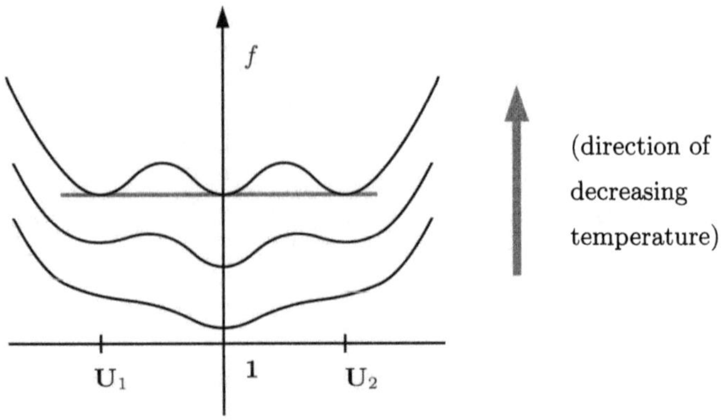

Figure 2. The temperature-evolution of the free energy

formation of new minima at the temperature of phase transformation. As a consequence of the symmetry requirements we have

$$\mathbf{Q}K = K, \quad \mathbf{Q} \in SO(n), \quad K\mathbf{H} = K, \quad \mathbf{H} \in \mathcal{G},$$

where $\mathcal{G}$ is the cubic symmetry group. For the cubic to tetragonal change

$$\left.\begin{array}{ll} \theta > \theta_c : & K = SO(3), \\ \theta = \theta_c : & K = SO(3) \cup SO(3)\mathbf{U}_1 \cup \cdots \cup SO(3)\mathbf{U}_3, \\ \theta < \theta_c : & K = SO(3)\mathbf{U}_1 \cup \cdots \cup SO(3)\mathbf{U}_3, \end{array}\right\} \tag{18}$$

where $SO(3)$, $SO(3)\mathbf{U}_i$ are the *energy wells,* corresponding to the austenite and the three variants of the martensite, respectively. The *transformation strains* $\mathbf{U}_i$ are tensors of the form

$$\mathbf{U}_1 = \operatorname{diag}(\mu, \eta, \eta), \quad \mathbf{U}_2 = \operatorname{diag}(\eta, \mu, \eta), \quad \mathbf{U}_3 = \operatorname{diag}(\eta, \eta, \mu),$$

with $\eta = \eta(\theta), \mu = \mu(\theta)$ experimentally determined quantities.

The specimen can exist separately in each of these phases. This does not automatically imply that it can also be in a state in which some of these phases coexist since the crystal lattices at the interface must match by Hadamard's lemma. In the above case, the martensite wells are mutually compatible (= contain rank 1 connections), but not compatible with the austenite well. The macroscopic deformations of minimum energy belong to $K^{qc} = K_\theta^{qc}$. The austenite well K in $(18)_1$ $(\theta > \theta_c)$ is the orthogonal group and so $K^{qc} = K$ by Example 1.20, the austenite does not form a microstructure. The relaxations of K from $(18)_{2,3}$ is an unsolved problem although many particular microstructures from K^{qc} are mathematically described. These include the famous austenite/martensite interface (Ball and James, 1987) and the relaxations of $SO(3)\mathbf{U}_i \cup SO(3)\mathbf{U}_j, i \neq j$, (Ball and James, 1992).

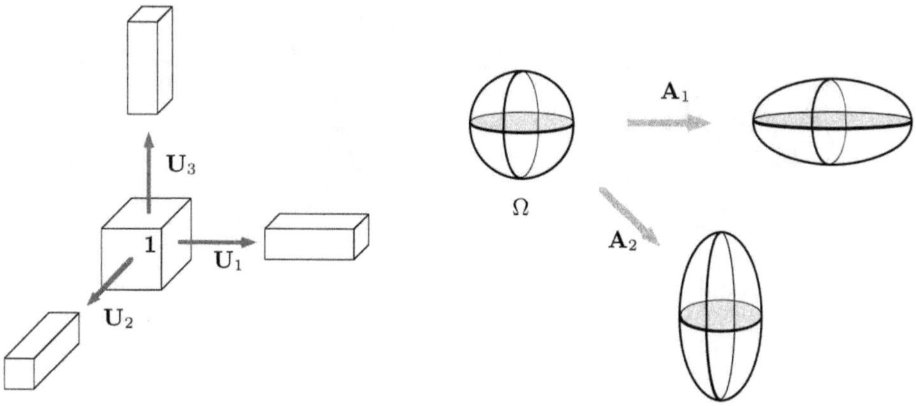

(a) Cubic to tetragonal (b) Isotropic to nematic

Figure 3. Changes in symmetry

Nematic elastomers Nematic elastomers are rubbery solids capable of a symmetry breaking phase transformation from a random isotropic phase to an aligned nematic phase. Their structure consists of cross-linked backbone polymer chains with attached nematic elements. The elements have random orientation in the high temperature isotropic phase while in the low temperature nematic phase the elements are aligned, see Figure 4.

The isotropic phase is described by the usual neo-Hookean constitutive equation

$$f(\mathbf{F}) = \tfrac{1}{2}\mu \operatorname{tr}(\mathbf{F}\mathbf{F}^T), \quad \det \mathbf{F} = 1,$$

while in the nematic phase the state variables are $\mathbf{F}$ and the director $\mathbf{d}$; the energy is of the form Bladon, Terentjev and Warner (1993)

$$g(\mathbf{F}, \mathbf{d}) = \frac{1}{2}\mu r^{1/3}\big(\operatorname{tr}(\mathbf{F}\mathbf{F}^T) - \frac{r-1}{r}|\mathbf{F}^T\mathbf{d}|^2\big), \quad \det \mathbf{F} = 1, |\mathbf{d}| = 1.$$

Here $r = r(\theta) > 1$ is a material constant, the step length anisotropy parameter. The director $\mathbf{d}$ is highly mobile and in equilibrium it attains the value as to minimize $f(\mathbf{F}, \mathbf{d})$ pointwise. It is easily found that the minimizer is an eigenvector of $\mathbf{F}\mathbf{F}^T$ corresponding to the largest eigenvalue. Thus if

$$f(\mathbf{F}) := \min\big\{g(\mathbf{F}, \mathbf{d}) : \mathbf{d} \in \mathbb{R}^3, |\mathbf{d}| = 1\big\}$$

we have

$$f(\mathbf{F}) = \frac{1}{2}\mu r^{1/3}\big(\alpha_1^2/r + \alpha_2^2 + \alpha_3^2\big), \tag{19}$$

where $\alpha_1 \geq \alpha_2 \geq \alpha_3$ are the singular values of $\mathbf{F}$ (see Section 2.1 for the definition). This is an isotropic energy function.

(a) Random (b) Aligned

Figure 4. Nematic units

It can be shown that an $\mathbf{F}$ attains a minimum of f on the shell $\det \mathbf{F} = 1$ if and only if its singular values are

$$\gamma := (r^{1/3}, r^{-1/6}, r^{-1/6}). \tag{20}$$

A local alignment of the mesogens thus causes a stretch of the network in a fixed direction and a contraction in the perpendicular directions. Figure 3(b) shows the unstable isotropic spherical reference configuration Ω and two deformations of minimum energy. They are isotropy–related to the diagonal matrix

$$\mathbf{L} = \operatorname{diag}(r^{1/3}, r^{-1/6}, r^{-1/6}), \tag{21}$$

the step length anisotropy tensor. Hence,

$$\left.\begin{array}{ll} \theta > \theta_c: & K = SO(3), \\ \theta = \theta_c: & K = SO(3) \cup SO(3)\mathbf{L}SO(3), \\ \theta < \theta_c: & K = SO(3)\mathbf{L}SO(3) \end{array}\right\} \tag{22}$$

with $\mathbf{L}$ given by (21). In the terminology of the preceding subsection, the set $K = K_\theta$ from $(22)_{2,3}$ contains infinitely many isotropy–related wells $W_{\mathbf{R}} := SO(3)\mathbf{L}\mathbf{R}$ where $\mathbf{R}$ ranges $SO(3)$. A description of K^{qc} from $(22)_{2,3}$ is available: a particular case of Example 3.14. Chapter 4 is devoted to a more detailed discussion of the nematic phase, including the relaxation of a class of energies which contains (19).

Minimization of the bulk elastic energy is a simplification, and mathematically results in infinitely fine microstructures. To penalize formations of too many interfaces, one can introduce an interfacial energy depending on the higher gradient terms, such as in the following Cahn–Hilliard type of energy:

$$I_\epsilon(\mathbf{u}) = \int_\Omega f(D\mathbf{u})\, d\mathbf{x} + \int_\Omega \epsilon |D^2\mathbf{u}|^2\, d\mathbf{x}.$$

This introduces the length scale in the model. Similarly, for the nematic elastomers one can consider

$$I_\epsilon(\mathbf{u}) = \int_\Omega f(D\mathbf{u})\, d\mathbf{x} + \int_\Omega \epsilon |\mathbf{F}^T D\mathbf{d}|^2\, d\mathbf{x}$$

depending on the gradient of the director in the Oseen–Zöcher–Frank way; see Anderson, Carlson and Fried (1999), Fried and Korchagin (2002) for more details.

1.3 Semiconvexity Properties

In the preceding sections we encountered the notions of quasiconvexity and, in a special case, polyconvexity. This section presents the semiconvexity notions systematically.

Consider the variational problem

$$\left.\begin{aligned} J &:= \inf\{I(\mathbf{u}) : \mathbf{u} \in \mathcal{A}\}, \\ \mathcal{A} &:= \left\{\mathbf{u} \in W^{1,p}(\Omega) : \mathbf{u} - \mathbf{v} \in W_0^{1,p}(\Omega)\right\} \end{aligned}\right\} \tag{23}$$

where $1 \le p < \infty, \mathbf{v} \in W^{1,p}(\Omega))$ and I is given by (4). The ideal program is to evaluate J and to find the minimizer, i.e., an element $\mathbf{u}_0 \in \mathcal{A}$ such that

$$I(\mathbf{u}_0) \le I(\mathbf{u}), \quad \mathbf{u} \in \mathcal{A}. \tag{24}$$

We say that (23) has a solution if a $\mathbf{u}_0$ as in (24) exists. The properties of the variational problem are related to the semiconvexity properties of the integrand f.

The semiconvexity notions are defined below at a given $\mathbf{A} \in \mathbb{M}^{m \times n}$, and it is implicitly understood as a part of the definitions that the property in question is defined to hold for f globally if it holds at every $\mathbf{A}$.

For a given matrix $\mathbf{A} \in \mathbb{M}^{m \times n}$, a minor is a determinant of any square submatrix (of arbitrary order) $\mathbf{B}$ of $\mathbf{A}$. The number of minors depend on the dimensions m, n of the matrix $\mathbf{A}$ and we denote it by $M(m,n)$. We order the minors in some fixed way and denote this collection by $\operatorname{Min}\mathbf{A}$. In dimension 2×2 there are five minors, the elements of the matrix $\mathbf{A}$ and $\det \mathbf{A}$, thus

$$\operatorname{Min}\mathbf{A} = (\mathbf{A}, \det\mathbf{A}) \in \mathbb{M}^{2\times 2} \times \mathbb{R}.$$

In dimension 3×3, there are 19 minors: nine elements of $\mathbf{A}$, nine elements of the cofactor matrix $\operatorname{cof}\mathbf{A}$ and $\det\mathbf{A}$. Thus

$$\operatorname{Min}\mathbf{A} = (\mathbf{A}, \operatorname{cof}\mathbf{A}, \det\mathbf{A}) \in \mathbb{M}^{3\times 3} \times \mathbb{M}^{3\times 3} \times \mathbb{R}.$$

Definitions 1.8.[4] The function $f : \mathbb{M}^{m\times n} \to \mathbb{R} \cup \{\infty\}$ is said to be
(i) convex at $\mathbf{A}$ if

$$f(\mathbf{A}) \le \sum_{i=1}^{p} t_i f(\mathbf{A}_i) \tag{25}$$

for all $p = 1, \ldots$, all $\mathbf{A}_i \in \mathbb{M}^{m\times n}, t_i \ge 0$, such that

$$\mathbf{A} = \sum_{i=1}^{p} t_i \mathbf{A}_i, \quad \sum_{i=1}^{p} t_i = 1;$$

[4]The convexity is due to Weierstrass and Jensen; rank 1 convexity due to Coral (1937) and Graves (1939); quasiconvexity and polyconvexity due to Morrey (1966) in the calculus of variations; for applications in nonlinear elasticity, see Ball (1977). The recognition of the importance of the semiconvexity notions for nonlinear elasticity can be regarded as a solution of *the main open problem of the theory of material behavior* (Truesdell and Noll, 1965, Section 20).

(ii) polyconvex at $\mathbf{A}$ if

$$f(\mathbf{A}) \le \sum_{i=1}^{p} t_i f(\mathbf{A}_i)$$

for all $p = 1, \dots,$ all $\mathbf{A}_i \in \mathbb{M}^{m\times n}$ and $t_i \ge 0$, such that

$$\operatorname{Min} \mathbf{A} = \sum_{i=1}^{p} t_i \operatorname{Min} \mathbf{A}_i, \quad \sum_{i=1}^{p} t_i = 1; \tag{26}$$

(iii) quasiconvex at $\mathbf{A}$ if

$$|E| f(\mathbf{A}) \le \int_E f(\mathbf{A} + D\mathbf{v}(\mathbf{x}))\, d\mathbf{x} \tag{27}$$

for each open $E \subset \mathbb{R}^n$ with $|\partial E| = 0$ and each $\mathbf{v} \in W_0^{1,\infty}(E)$, for which the integral makes sense as a Lebesgue integral;

(iv) rank 1 convex at $\mathbf{A}$ if

$$f(\mathbf{A}) \le t f(\mathbf{A} + (1-t)\mathbf{a} \otimes \mathbf{n}) + (1-t) f(\mathbf{A} - t\mathbf{a} \otimes \mathbf{n}), \tag{28}$$

for any $\mathbf{a} \in \mathbb{R}^m, \mathbf{n} \in \mathbb{R}^n$, $0 \le t \le 1$.

Remarks 1.9.

- The quasiconvexity says that the homogeneous deformation attains a minimum energy among all deformations with the same boundary values. The rank 1 convexity says that the homogeneous deformation has a minimum energy among all simple laminates with the same mean deformation. (See the proof of Proposition 1.11).
- For the theory of phase transformations, the pointwise versions of the semiconvexity properties, as opposed to the global versions, are important. For example, the van der Waals isotherm is rank 1 convex at every point outside the Maxwell interval, and not convex inside this interval.
- For $m = n = 2$ the conditions (26) read

$$\mathbf{A} = \sum_{i=1}^{p} t_i \mathbf{A}_i, \quad \det \mathbf{A} = \sum_{i=1}^{p} t_i \det \mathbf{A}_i, \quad \sum_{i=1}^{p} t_i = 1,$$

and f is polyconvex if and only if $f(\mathbf{A}) = g(\mathbf{A}, \det \mathbf{A})$ where $g : \mathbb{M}^{2\times 2} \times \mathbb{R} \to \mathbb{R} \cup \{\infty\}$ is a convex function (of 5 scalar variables).
- For $m = n = 3$ the conditions (26) read

$$\mathbf{A} = \sum_{i=1}^{p} t_i \mathbf{A}_i, \quad \operatorname{cof} \mathbf{A} = \sum_{i=1}^{p} t_i \operatorname{cof} \mathbf{A}_i, \quad \det \mathbf{A} = \sum_{i=1}^{p} t_i \det \mathbf{A}_i, \quad \sum_{i=1}^{p} t_i = 1$$

and f is polyconvex if and only if $f(\mathbf{A}) = g(\mathbf{A}, \operatorname{cof} \mathbf{A}, \det \mathbf{A})$ where $g : \mathbb{M}^{3\times 3} \times \mathbb{M}^{3\times 3} \times \mathbb{R} \to \mathbb{R} \cup \{\infty\}$ is a convex function (of 19 scalar variables).

Before we proceed to discussing the significance of the above concepts for the calculus of variations, let us clarify the relationships among them.

Theorem 1.10. *If* $\mathbf{u} \in W^{1,\infty}(\Omega)$ *satisfies* $\mathbf{u} = \mathbf{Fx}$ *on* $\partial\Omega$ *then*

$$\int_{\Omega} \operatorname{Min}(D\mathbf{u})\, d\mathbf{x} = |\Omega| \operatorname{Min}(\mathbf{F}). \tag{29}$$

The proof is omitted, but see Exercise 1.4.

Proposition 1.11. *Let* $f : \mathbb{M}^{m\times n} \to \mathbb{R}$ *and* $\mathbf{A} \in \mathbb{M}^{m\times n}$. *Then*

$$\text{convexity} \Rightarrow \text{polyconvexity} \Rightarrow \text{quasiconvexity} \Rightarrow \text{rank 1 convexity}, \tag{30}$$

where these conditions are interpreted as conditions at the point $\mathbf{A}$ *and the implication quasiconvexity* $\Rightarrow$ *rank 1 convexity holds only if* f *is finite-valued in some neighborhood of* $\mathbf{A}$. *If* $m = 1$ *or* $n = 1$ *we have actually equivalences.*

Šverák (1992) showed that for $m \geq 3, n \geq 2$ there exists a rank 1 convex function on $\mathbb{M}^{m\times n}$ that is not quasiconvex. The rank 1 convexity and polyconvexity thus provide technically simpler necessary and sufficient conditions, respectively, for quasiconvexity.

Proof The first implication is obvious. The second follows from Theorem 1.10. Namely, assume that f is a polyconvex function so that f is of the form

$$f(\mathbf{A}) = g(\operatorname{Min} \mathbf{A})$$

with g a convex function. If $\mathbf{A}, \Omega, \mathbf{u}$ are as in the definition of quasiconvexity then by (29) and Jensen's inequality

$$|\Omega| g(\operatorname{Min} \mathbf{A}) \leq \int_{\Omega} g(\operatorname{Min}(\mathbf{A} + D\mathbf{u}))\, d\mathbf{x}$$

which reduces to (27). To prove the last implication, let $\mathbf{a}, \mathbf{n}$ and t be as in the definition of rank 1 convexity. Let Ω be a unit cube with one edge in the direction $\mathbf{n}$, let $h : \mathbb{R} \to \mathbb{R}$ be defined by (14), and define the sequence of simple laminates $\mathbf{v}_j : \Omega \to \mathbb{R}^m$ with interface normal $\mathbf{n}$ by

$$\mathbf{v}_j(\mathbf{x}) = -\frac{1}{j}\mathbf{a} h(\mathbf{x} \cdot \mathbf{n}).$$

One finds that $D\mathbf{v}_j$ attains only the values $(1-t)\mathbf{a}\otimes\mathbf{n}$ and $-t\mathbf{a}\otimes\mathbf{n}$ on regions of weights $t, (1-t)$ respectively and thus

$$\int_{\Omega} f(\mathbf{A} + D\mathbf{v}_j)\, d\mathbf{x} = t f(\mathbf{A} + (1-t)\mathbf{a}\otimes\mathbf{n}) + (1-t) f(\mathbf{A} - t\mathbf{a}\otimes\mathbf{n}).$$

However, the functions $\mathbf{v}_j$ do not satisfy the boundary condition $\mathbf{v}_j = \mathbf{0}$ on $\partial\Omega$. A modification similar to that in Example 1.6 can be invoked to find a sequence $\mathbf{u}_j$ of modifications of $\mathbf{v}_j$ such that $\mathbf{u}_j = \mathbf{0}$ on $\partial\Omega$. If f is continuous, we have

$$\int_{\Omega} f(\mathbf{A} + D\mathbf{u}_j)\, d\mathbf{x} \to t f(\mathbf{A} + (1-t)\mathbf{a}\otimes\mathbf{n}) + (1-t) f(\mathbf{A} - t\mathbf{a}\otimes\mathbf{n}).$$

Thus (27) implies (28). A refinement of this construction (see Fonseca (1988)) shows that the conclusion holds also for f not necessarily continuous but finite–valued. □

1.4 A Motivation for Quasiconvexity: Minimizers

Two results below show the relevance of the quasiconvexity for the existence of minimizers. A necessary condition, the localization theorem, shows that the integrand must be quasiconvex along any (sufficiently smooth) minimizer. A sufficient condition shows that the (global) quasiconvexity, when augmented with suitable coercivity conditions, leads to the existence of minimizers.

Proposition 1.12.[5] (Localization theorem) *Let f be continuous and let $\mathbf{u}_0 \in W^{1,\infty}(\Omega)$ be a strong local minimizer of I, i.e.,*

$$I(\mathbf{u}_0) \leq I(\mathbf{u})$$

for all $\mathbf{u} \in W^{1,\infty}(\Omega)$ with $\mathbf{u} - \mathbf{u}_0 \in W_0^{1,\infty}(\Omega)$ and $\|\mathbf{u} - \mathbf{u}_0\|_{L^\infty(\Omega)} < \delta$ where $\delta > 0$ is arbitrarily small. Then for every $\mathbf{x} \in \Omega$ such that $\mathbf{u}_0$ is of class C^1 in a neighborhood of $\mathbf{x}$, f is quasiconvex at $D\mathbf{u}_0(\mathbf{x})$.

The quasiconvexity along the minimizer implies the rank 1 convexity and the following is a simple application of the rank 1 convexity.

Proposition 1.13. *Let f be of class C^1, let $\mathbf{u}_0$ be a continuous and piecewise continuously differentiable strong local minimizer of I (i.e., one satisfying hypotheses of Proposition 1.12) with a singular surface $\mathcal{S}$. Then*
(i) (Maxwell's relation) *the limiting values $\mathbf{F}^\pm$ of $D\mathbf{u}_0$ along $\mathcal{S}$ satisfy*

$$f(\mathbf{F}^+) = f(\mathbf{F}^-) + \hat{\mathbf{S}}(\mathbf{F}^+) \cdot (\mathbf{F}^+ - \mathbf{F}^-) \tag{31}$$

where $\hat{\mathbf{S}} := Df$ is the stress and $\mathbf{n}$ the normal to $\mathcal{S}$;
(ii) (Eshelby's conservation law) *for every part $P \subset \Omega$,*

$$\int_{\partial P} \mathbf{P}\mathbf{n}\, dA = \mathbf{0} \tag{32}$$

where $\mathbf{P} := f\mathbf{1} - \mathbf{F}^T\hat{\mathbf{S}}$ is the Eshelby tensor.

Remarks 1.14.

- In one dimension, Maxwell's relation reduces to Erdman's corner condition; in fluids to the equality of chemical potentials of the phases.
- The local forms of (32) read

$$\operatorname{Div} \mathbf{P} = \mathbf{0}, \quad [\mathbf{P}]\mathbf{n} = \mathbf{0}, \tag{33}$$

outside and on the singular surface, respectively. Note that the bulk equation $(33)_1$ is a consequence of the balance of forces $\operatorname{Div} \mathbf{S} = \mathbf{0}$, but $(33)_2$ is not a consequence of $[\mathbf{S}]\mathbf{n} = \mathbf{0}$.

[5] Ball (1977).

- The *conditions of thermodynamic equilibrium* on the phase interface read:
  – geometrical compatibility

$$[\mathbf{F}] = \mathbf{a} \otimes \mathbf{n} \text{ for some } \mathbf{a} \in \mathbb{R}^m;$$

  – force balance

$$[\mathbf{S}]\mathbf{n} = \mathbf{0};$$

  – Maxwell's relation

$$[\mathbf{P}]\mathbf{n} = \mathbf{0}.$$

Exercise 1.15.
(i) Prove that if $f : \mathbb{M}^{m\times n} \to \mathbb{R}$ is of class C^1 then f is rank 1 convex at $\mathbf{A} \in \mathbb{M}^{m\times n}$ if and only if

$$f(\mathbf{B}) \geq f(\mathbf{A}) + \hat{\mathbf{S}}(\mathbf{A}) \cdot (\mathbf{B} - \mathbf{A}) \tag{34}$$

for each $\mathbf{B}$ compatible with $\mathbf{A}$ where $\hat{\mathbf{S}} = Df$.
(ii) Prove (31).
(iii) Prove (32).

Hints: (i): Denote by $\varphi(t)$ the right–hand side of (28) and note that $\varphi(t) \geq f(\mathbf{A})$ for all $t \in [0,1]$ with the equality at $t = 0$. Thus $d\varphi(0)/dt \geq 0$. (ii): The equality of forces says

$$\mathbf{s} := \hat{\mathbf{S}}(\mathbf{F}^+)\mathbf{n} = \hat{\mathbf{S}}(\mathbf{F}^-)\mathbf{n}. \tag{35}$$

We have $\mathbf{F}^+ = \mathbf{F}^- + \mathbf{a} \otimes \mathbf{n}$ by Hadamard's lemma. Applying (34) once with $\mathbf{A} = \mathbf{F}^-$ and once with $\mathbf{A} = \mathbf{F}^+$ and using (35), we obtain

$$f(\mathbf{F}^+) \geq f(\mathbf{F}^-) + \mathbf{s} \cdot \mathbf{a} \geq f(\mathbf{F}^+) - \mathbf{s} \cdot \mathbf{a} + \mathbf{s} \cdot \mathbf{a} = f(\mathbf{F}^+).$$

Thus we have the equality signs throughout. (iii): Rewrite (31) as $(33)_2$ and use the divergence theorem.

Theorem 1.16.[6] (Existence of minimizers) *Let $f : \mathbb{M}^{m\times n} \to \mathbb{R}$ be quasiconvex and satisfy*

$$c|\mathbf{F}|^p + d \leq f(\mathbf{F}) \leq C(|\mathbf{F}|^p + 1),$$

where $1 < p < \infty, c, C > 0, d \in \mathbb{R}$. Then for any $\mathbf{v} \in W^{1,p}(\Omega, \mathbb{R}^m)$ the problem (23) has a solution.

The proof constitutes the *direct method of the calculus of variations.* One seeks to prove that each minimizing sequence $\mathbf{u}_j$ converges weakly to an element $\mathbf{u}$ and that $\liminf I(\mathbf{u}_j) \geq I(\mathbf{u})$. The last property, the *sequential weak lower semicontinuity* (swlsc) of I, is crucial to conclude that $I(\mathbf{u})$ is a minimizer. Morrey's original motivation for quasiconvexity was exactly swlsc, which is essentially equivalent to it. Ball's existence theorems Ball (1977) in elasticity include the local invertibility constraint $\det \mathbf{F} > 0$, $f(\mathbf{F}) \to \infty$ as $\det \mathbf{F} \to 0$. Because of this, a stronger condition of polyconvexity is necessary.

[6] I refer to Morrey (1966), Dacorogna (1989), Pedregal (1997), Pedregal (2000), Roubíček (1997) for various versions of this theorem.

1.5 Relaxation of Functions and Sets

We here complete the definitions of relaxations of functions and sets from the preceding sections with definitions using the semiconvexity notions.

The basis for the considerations in this section is that the pointwise supremum of any family of functions having one semiconvexity property has the same semiconvexity property. Thus if $f : \mathbb{M}^{m\times n} \to \mathbb{R} \cup \{\infty\}$, there exists a largest convex (polyconvex, quasiconvex, rank 1 convex) function f^c (f^{pc}, f^{qc}, f^{rc}) not exceeding f. The chain (30) implies

$$f^c \le f^{pc} \le f^{qc} \le f^{rc} \tag{36}$$

with the inequality $f^{qc} \le f^{rc}$ only provided f is finite–valued. The experience shows that it often happens that f^{pc} and f^{rc} can be calculated and shown to be equal, thus leading also to f^{qc}.

The following result describes the general construction of the semiconvex hulls.

Proposition 1.17. *If* $f : \mathbb{M}^{m\times n} \to \mathbb{R} \cup \{\infty\}$ *and* $\mathbf{A} \in \mathbb{M}^{m\times n}$ *then*
(i)

$$f^c(\mathbf{A}) = \inf\left\{\sum_{i=1}^{p} t_i f(\mathbf{A}_i)\right\},$$

where the infimum is taken over all $p = 1, \ldots,$ *all* $\mathbf{A}_i \in \mathbb{M}^{m\times n}$ *and* $t_i \ge 0$ *such that*

$$\mathbf{A} = \sum_{i=1}^{p} t_i \mathbf{A}_i, \quad \sum_{i=1}^{p} t_i = 1;$$

(ii)[7]

$$f^{pc}(\mathbf{A}) = \inf\left\{\sum_{i=1}^{p} t_i f(\mathbf{A}_i)\right\},$$

where the infimum is taken over all $p = 1, \ldots,$ *all* $\mathbf{A}_i \in \mathbb{M}^{m\times n}$ *and* $t_i \ge 0$ *such that*

$$\operatorname{Min} \mathbf{A} = \sum_{i=1}^{p} t_i \operatorname{Min} \mathbf{A}_i, \quad \sum_{i=1}^{p} t_i = 1;$$

(iii)[7] *if additionally* f *is finite-valued then*

$$f^{qc}(\mathbf{A}) = \inf\left\{|E|^{-1}\int_E f(\mathbf{A} + D\mathbf{v})\,d\mathbf{x} : \mathbf{v} \in W_0^{1,\infty}(E)\right\},$$

where E *is any fixed bounded open subset of* $\mathbb{R}^n$ *with* $|\partial E| = 0$;
(iv)[8]

$$f^{rc}(\mathbf{A}) = \lim_{k\to\infty} f_k(\mathbf{A}),$$

[7] Dacorogna (1989).
[8] Kohn and Strang (1986).

where $f_0 = f$, $f_{k+1} = f_k^{ks}$, *and for each* $g : \mathbb{M}^{m\times n} \to \mathbb{R} \cup \{\infty\}$,

$$g^{ks}(\mathbf{A}) := \inf \{tg(\mathbf{A} + (1-t)\mathbf{a}\otimes\mathbf{n}) + (1-t)g(\mathbf{A} - t\mathbf{a}\otimes\mathbf{n})\},$$

where the infimum is taken over all $t \in [0,1]$, $\mathbf{a} \in \mathbb{R}^m, \mathbf{n} \in \mathbb{R}^n$.

Item (iv) is the procedure of sequential lamination. The occurrence of the limit in the calculation of f^{rc}, which has no counterpart in the cases of f^c, f^{pc}, and f^{qc}, is due to the fact that the rank 1 convexity inequality does not cover the whole state space $\mathbb{M}^{m\times n}$ (there are tensors inaccessible by rank 1 perturbations from a given $\mathbf{A}$). As a consequence, if f is convex (polyconvex, quasiconvex) at $\mathbf{A} \in \mathbb{M}^{m\times n}$ then $f^c(\mathbf{A}) = f(\mathbf{A})$ ($f^{pc}(\mathbf{A}) = f(\mathbf{A})$, $f^{qc}(\mathbf{A}) = f(\mathbf{A})$) but it often happens that f is rank 1 convex at $\mathbf{A}$ and yet $f^{rc}(\mathbf{A}) < f(\mathbf{A})$.

Let us return to the minimum problem (23). As far as the infimum J is concerned, one can always pass to a relaxed variational problem with a quasiconvex integrand.

Theorem 1.18.[9] (Relaxation) *Let* f *satisfy*

$$c|\mathbf{F}|^p \le f(\mathbf{F}) \le C(|\mathbf{F}|^p + 1),$$

where $1 < p < \infty, c > 0$. *Then for every* $\mathbf{v} \in W^{1,p}(\Omega)$,

$$J = \min \{\bar{I}(\mathbf{u}) : \mathbf{u} \in \mathcal{A}\}$$

where

$$\bar{I}(\mathbf{u}) = \int_\Omega f^{qc}(D\mathbf{u})\,d\mathbf{x};$$

here J *and* $\mathcal{A}$ *are as in* (23).

Recall that in Section 1.2 we defined the quasiconvex hull K^{qc} of a compact set K of matrices via (17) and gave it an interpretation of the set of minima of f^{qc} if the set of minima of the original function f was K. We here utilize this interpretation to give a formally different, but equivalent, definition of K^{qc}; at the same time we also introduce the hulls corresponding to the other semiconvexity notions.

The lamination convex, rank 1 convex, quasiconvex, polyconvex, and closed convex hulls $M^{lc}, M^{rc}, M^{qc}, M^{pc}, M^c$, respectively, of $M \subset \mathbb{M}^{m\times n}$, are defined by

$$M^s = \{\mathbf{A} \in \mathbb{M}^{m\times n} : f(\mathbf{A}) \le 0 \text{ for each } f \in F^s, f|M = 0\},$$

$s = lc, rc, qc, pc, c$, where for $s = lc$, F^s is the set of all rank 1 convex functions $f : \mathbb{M}^{m\times n} \to \mathbb{R} \cup \{\infty\}$, while for $s = rc, qc, pc, c$, F^s is the set of all *finite-valued* $f : \mathbb{M}^{m\times n} \to \mathbb{R}$ with the corresponding semiconvexity property. One has

$$M^{lc} \subset M^{rc} \subset M^{qc} \subset M^{pc} \subset M^c$$

[9] Dacorogna (1982), Acerbi and Fusco (1984).

and $M^{rc}, M^{qc}, M^{pc}, M^c$ are closed. Here a remark similar to that made in connection with hulls of functions applies: It often happens that M^{rc}, M^{pc} are calculable and coincide, thus giving also M^{qc}.

Alternatively

$$M^s = \{\mathbf{A} \in \mathbb{M}^{m\times n} : f(\mathbf{A}) = 0 \text{ for each } f \in F^s, f|M = 0, f \geq 0\},$$

and

$$M^s = \{\mathbf{A} \in \mathbb{M}^{m\times n} : f(\mathbf{A}) \leq 0 \text{ for each } f \in F^s, f|M \leq 0\}.$$

A set M is called lamination convex, rank 1 convex, quasiconvex, polyconvex, and closed convex if it coincides with the corresponding hull. The lamination convex, rank 1 convex, quasiconvex, polyconvex, and closed convex hull of M is then the smallest lamination convex, rank 1 convex, quasiconvex, polyconvex, and closed convex sets, respectively, containing M.

Proposition 1.19. *Let $M \subset \mathbb{M}^{m\times n}$. Then*

(i) *M is lamination convex if and only if the conditions $\mathbf{A}, \mathbf{B} \in M$ and $\operatorname{rank}(\mathbf{B} - \mathbf{A}) = 1, 0 \leq t \leq 1$ imply $t\mathbf{A} + (1-t)\mathbf{B} \in M$; we have*

$$M^{lc} = \bigcup_{i=0}^{\infty} M^i,$$

where $M^0 = M, M^{i+1} = [M^i]^{ks}$ and for any $K \subset \mathbb{M}^{m\times n}$,

$$K^{ks} := \{t\mathbf{A} + (1-t)\mathbf{B} : \mathbf{A}, \mathbf{B} \in K, \operatorname{rank}(\mathbf{B} - \mathbf{A}) \leq 1, 0 \leq t \leq 1\};$$

(ii) *M is polyconvex if and only if it is closed and each $\mathbf{A} \in \mathbb{M}^{m\times n}$ satisfying*

$$\operatorname{Min} \mathbf{A} = \sum_{i=1}^{p} t_i \operatorname{Min} \mathbf{A}_i, \quad \sum_{i=1}^{p} t_i = 1 \tag{37}$$

for some $\mathbf{A}_i \in M$ and $t_i \geq 0$ is in M; we have

$$M^{pc} = \operatorname{cl}\{\mathbf{A} \in \mathbb{M}^{m\times n} : \mathbf{A} \text{ satisfies (37) for some } \mathbf{A}_i \in M, t_i \geq 0\};$$

equivalently,

$$M^{pc} = \left\{\mathbf{A} : \operatorname{Min} \mathbf{A} \in (\hat{M})^c\right\} \quad \text{where} \quad \hat{M} = \{M(\mathbf{A}) : \mathbf{A} \in M\}.$$

Example 1.20.

$$[SO(n)]^{qc} = SO(n).$$

Indeed,

$$SO(n) = \{\mathbf{A} \in \mathbb{M}^{n\times n} : f(\mathbf{A}) = 0, g(\mathbf{A}) = 0\} \tag{38}$$

where

$$f(\mathbf{A}) = |\mathbf{A}|^n - c_n \det \mathbf{A}, \quad c_n := n^{n/2}, \quad g(\mathbf{A}) = (\det \mathbf{A} - 1)^2 \tag{39}$$

are nonnegative polyconvex functions; recall that $|\mathbf{A}| = \sqrt{\operatorname{tr}(\mathbf{A}\mathbf{A}^T)}$. For $n \geq 2$, $SO(n)$ is not convex.

Exercise 1.21.
(i) Prove (38). Hints: use the arithmetic–geometric mean inequality to show that f is nonnegative and $f(\mathbf{A}) = 0$ if and only if $\mathbf{A} = \lambda \mathbf{Q}$ where $\lambda \geq 0, \mathbf{Q} \in SO(n)$.
(ii) Prove that the functions f, g in (39) are polyconvex.

Exercise 1.22. The set $\mathbb{R}_+ SO(n), \mathbb{R}_+ := [0, \infty)$, is polyconvex:

$$\mathbb{R}_+ SO(n) = \{\mathbf{A} \in \mathbb{M}^{n \times n} : f(\mathbf{A}) = 0\}$$

where f is as in (38). For $n = 1, 2$, $\mathbb{R}_+ SO(n)$ is convex. Indeed the distance from $\mathbb{R}_+ SO(2)$ is convex (see Exercise 2.5).

2 Semiconvexity of Isotropic Functions

Isotropic (invariant) functions of a matrix argument are defined by the requirement that they remain unchanged under multiplication from the left and right by (proper) orthogonal tensors, and isotropic sets are defined similarly. The stored energies of isotropic solids and sets of deformation gradients of minimum energy are examples. Invariant functions can be represented as functions $\tilde{f} = \tilde{f}(\alpha)$ of the signed singular values of the tensor $\mathbf{A}$. The goal of this chapter is to translate the semiconvexity properties of f into the corresponding properties of $\tilde{f}$. Full answers (equivalences) are available for the convexity and rank 1 convexity, for the polyconvexity there is an efficient sufficient condition in arbitrary dimension and an equivalent condition in dimension 2 and 3. No general result is available for the quasiconvexity. It is known that within the class of invariant functions the convexity, polyconvexity and quasiconvexity are different. The relationship between quasiconvexity and rank 1 convexity is an open problem.

2.1 The Representation Theorem

From this point on we consider exclusively the square case: $m = n, n \geq 2$. A set $M \subset \mathbb{M}^{n \times n}$ is said to be invariant if $\mathbf{QAR} \in M$ for every $\mathbf{A} \in M$ and $\mathbf{Q}, \mathbf{R} \in SO(n)$. A function $f : M \to \mathbb{R} \cup \{\infty\}$ is said to be invariant if M is invariant and $f(\mathbf{QAR}) = f(\mathbf{A})$ for each $\mathbf{A} \in M$ and $\mathbf{R}, \mathbf{Q} \in SO(n)$. If the above assertions hold for every $\mathbf{Q}, \mathbf{R} \in O(n)$, M or f is called fully invariant. Here $SO(n)$ is the set of all orthogonal matrices with $\det = 1$ and $O(n)$ is the set of all orthogonal matrices.

The singular values, also called principal stretches, of a tensor $\mathbf{F} \in \mathbb{M}^{n \times n}$ are the eigenvalues of $\sqrt{\mathbf{F}^T \mathbf{F}}$ ordered in a nonincreasing way. The signed singular values $(\tau_1, \ldots, \tau_n)$ of $\mathbf{F}$ are defined as an n-tuple such that $\tau_1, \ldots, \tau_{n-1}, |\tau_n|$ are the singular values of $\mathbf{F}$ and $\operatorname{sgn} \tau_n = \operatorname{sgn} \det \mathbf{F}$. We write $\tau(\mathbf{F}) = (\tau_1(\mathbf{F}), \ldots, \tau_n(\mathbf{F}))$ for the n-tuple of the signed singular values of $\mathbf{F}$. The signed singular values remember the sign of the determinant, thus $\det \mathbf{A} = \tau_1(\mathbf{F}) \cdots \tau_n(\mathbf{F})$, while the product of the ordinary singular values gives only $|\det \mathbf{F}|$. Let

$$\mathbb{G}^n = \{\alpha \in \mathbb{R}^n : \alpha_1 \geq \ldots \alpha_{n-1} \geq |\alpha_n|\}$$

and note that $\tau(\mathbf{F}) \in \mathbb{G}^n$. Let further

$$\mathbb{G}^n_+ = \{\alpha \in \mathbb{G}^n : \alpha_n \geq 0\}, \quad \mathbb{G}^n_{++} = \{\alpha \in \mathbb{G}^n : \alpha_n > 0\}.$$

The functions $\tau : \mathbb{M}^{n\times n} \to \mathbb{G}^n$ are invariant. Moreover, τ are Lipschitz continuous on $\mathbb{M}^{n\times n}$ and of class C^∞ in the region where the components of τ are mutually distinct. For $n = 2$ the representation space $\mathbb{G}^2$ is the wedge delimited by the lines $\alpha_2 = \pm\alpha_1$.

By the singular value decomposition theorem (= the combination of the polar and spectral decomposition theorems), each matrix $\mathbf{A} \in \mathbb{M}^{n\times n}$ can be written as

$$\mathbf{A} = \mathbf{Q}\,\mathrm{diag}(\tau(\mathbf{A}))\mathbf{R}$$

where $\mathbf{Q}, \mathbf{R} \in SO(n)$. If M is an invariant set and f an invariant function we define the representation $\tilde{M} \subset \mathbb{R}^n$ of M and the representation $\tilde{f} : \mathbb{R}^n \to \mathbb{R} \cup \{\infty\}$ of f by

$$\tilde{M} = \{\alpha \in \mathbb{R}^n : \mathrm{diag}(\alpha) \in M\}, \quad \tilde{f}(\alpha) = f(\mathrm{diag}(\alpha)), \quad \alpha \in \mathbb{R}^n.$$

The representations determine the corresponding invariant objects; in fact

$$M = \left\{\mathbf{A} \in \mathbb{M}^{n\times n} : \tau(\mathbf{A}) \in \tilde{M}\right\}, \quad f(\mathbf{A}) = \tilde{f}(\tau(\mathbf{A})), \quad \mathbf{A} \in \mathbb{M}^{n\times n}.$$

However, $\tilde{M}$ and $\tilde{f}$ still have some residual invariance properties. To state them, we say that a subset $\tilde{M}$ of $\mathbb{R}^n$ or a function $\tilde{f} : \mathbb{R}^n \to \mathbb{R} \cup \{\infty\}$ is symmetric if $P\alpha \in \tilde{M}$ or $f(P\beta) = f(\beta)$ for any $\alpha \in \tilde{M}, \beta \in \mathbb{R}^n$ and every n by n permutation matrix P. The set $\tilde{M}$ or the function $\tilde{f}$ is said to be even if $\epsilon\alpha \in \tilde{M}$ or $f(\epsilon\beta) = f(\beta)$ for every $\alpha \in \tilde{M}, \beta \in \mathbb{R}^n$ and $\epsilon \in \{-1, 1\}^n$ with $\epsilon_1 \cdots \epsilon_n = 1$, where we abbreviate $\epsilon\alpha := \mathrm{diag}(\epsilon)\alpha$. If the same holds for every $\epsilon \in \{-1, 1\}^n$ without the restriction $\epsilon_1 \cdots \epsilon_n = 1$, we say that $\tilde{M}$ or $\tilde{f}$ is fully even.

The representations $\tilde{M}, \tilde{f}$ of an invariant set M and an invariant function f are symmetric and even. They are fully even if and only if M and f are fully invariant. The correspondences $\tilde{} : M \mapsto \tilde{M}, f \mapsto \tilde{f}$ are one-to-one. The symmetry and even nature of the representations $\tilde{M}, \tilde{f}$, imply that $\tilde{M}, \tilde{f}$ are fully determined by $\tilde{M} \cap \mathbb{G}^n$ and by the restriction $\tilde{f}|\mathbb{G}^n$ of $\tilde{f}$ to $\mathbb{G}^n$. If the matter is clear from the context, we use the term representation for these restricted objects.

Exercise 2.1. Prove that for $n = 2$

(i) we have the expressions

$$\tau_{1,2}(\mathbf{A}) = \frac{1}{2\sqrt{2}}\big(|\mathbf{A} + \mathrm{cof}\,\mathbf{A}| \pm |\mathbf{A} - \mathrm{cof}\,\mathbf{A}|\big) = \frac{1}{2}\big(\sqrt{|\mathbf{A}|^2 + 2\det\mathbf{A}} \pm \sqrt{|\mathbf{A}|^2 - 2\det\mathbf{A}}\big);$$

(ii) the functions $\tau_1, \tau_1 + \tau_2, \tau_1 - \tau_2$ are convex on $\mathbb{M}^{2\times 2}$;

(iii) if f is invariant then there exists a function $h : [0, \infty) \times \mathbb{R} \to \mathbb{R}$ such that

$$f(\mathbf{A}) = h(|\mathbf{A}|, \det\mathbf{A})$$

for every $\mathbf{A} \in \mathbb{M}^{2\times 2}$.

Hints: (i): In view of the invariance of the above expressions, it suffices to verify them on matrices of the form $\mathbf{A} = \mathrm{diag}(\alpha_1, \alpha_2)$ where $\alpha \in \mathbb{G}^2$. (ii): The norm is convex and the cofactor is linear if $n = 2$. (iii): The values $|\mathbf{A}|, \det\mathbf{A}$ determine $\tau(\mathbf{A})$ uniquely.

Exercise 2.2. Show that the function $\mathbf{A} \mapsto |\mathbf{A}|$ is represented by $\alpha \mapsto |\alpha|$.

If f is invariant and has one of the semiconvexity properties at an $\mathbf{A} \in \mathbb{M}^{n\times n}$, it has the same semiconvexity property at all $\mathbf{B} \in \mathbb{M}^{n\times n}$ with the same signed singular values. Based on this, we say that an invariant function has a semiconvexity property at an $\alpha \in \mathbb{G}^n$ if it has that semiconvexity property at some $\mathbf{A} \in \mathbb{M}^{n\times n}$ with the signed singular values α.

Invariant semiconvex functions have been examined by different methods; an exhaustive account is outside the scope of this chapter. However, to motivate the choice of the material that follows, we here briefly review some of the results.

One line of research examines the differential conditions for the rank 1 convexity, and in particular the Legendre–Hadamard condition

$$D^2 f(\mathbf{A})(\mathbf{a}\otimes\mathbf{b},\mathbf{a}\otimes\mathbf{b}) \geq 0$$

for all $\mathbf{A} \in \mathbb{M}^{n\times n}$, $\mathbf{a},\mathbf{b} \in \mathbb{R}^n$, which is equivalent to the rank 1 convexity in the class C^2 functions of any symmetry. In dimension $n = 2$ this condition leads to the inequalities of Knowles and Sternberg (1977). For $n \geq 3$ this leads to more complicated conditions, for $n = 3$ see Simpson and Spector (1983, Theorem 4.1), Ogden (1984, Subsection 6.2.7), and for n arbitrary see Šilhavý (1999, Proposition 6.4) and Dacorogna (2001). The differential conditions, however, are of limited use in calculating the hulls and this line is not pursued here.

Another line, and the one that is adopted here, seeks to establish global conditions for the semiconvexity properties, involving finite differences of arguments. Such are the sufficient conditions for the convexity and polyconvexity of invariant functions in Ball (1977), based on the von Neumann inequality (see Section 2.2). Further results of this type are Propositions 2.6 and 2.7, below. Of full use are only conditions in the forms of equivalences for the semiconvexity notion in question, because only then one can use them to adapt the procedures in Proposition 1.17 to invariant functions. Such equivalences are available only for the convexity and rank 1 convexity. For polyconvexity, equivalent conditions are available only in dimensions 2 and 3 (see Section 2.3 and the references). Finally, no general result is available for quasiconvexity. For a review of *fully* invariant functions, with references, see Chapter 3 of Dacorogna and Marcellini (1999).

An interesting question is the relationship among the semiconvexity notions for invariant and fully invariant functions. The first result is an example by Aubert (1987) which shows that the rank 1 convexity and polyconvexity are different in the class of invariant functions in dimension 2 (see also Dacorogna and Marcellini (1987)). The difference between the quasiconvexity and polyconvexity in the class of invariant functions is established in Alibert and Dacorogna (1992). The cited examples are fully invariant, and leave open the possibility that the notions coincide in the narrower class of fully invariant functions. The result Cardaliaguet and Tahraoui (2002) shows that the polyconvexity and rank 1 convexity coincide in the class of compact fully invariant sets, and under a simple condition also in the class of invariant sets. The papers Šilhavý (2002a), Šilhavý (2002c) establish the difference of the rank 1 convexity and polyconvexity and of the quasiconvexity and polyconvexity in the class of fully invariant functions in any dimension $n \geq 2$:

Example 2.3. *Let $1 \leq p < \infty, n \geq 2$, and let f be a fully invariant function on $\mathbb{M}^{n\times n}$ represented by*

$$\tilde{f}(\alpha) = \begin{cases} \alpha_1^p + |\alpha_2|^p - 1 & \text{if} \quad \alpha_1 \geq 1, \\ |\alpha_1\alpha_2|^p & \text{if} \quad \alpha_1 \leq 1, \end{cases}$$

for each $\alpha \in \mathbb{G}^n$. Then

(i) *f is rank 1 convex;*
(ii) *if $p \geq 2$, f is polyconvex; if $1 \leq p < 2$, f is not polyconvex;*
(iii) *if $1 \leq p < 2$ then f^{qc} is not polyconvex.*

A general open problem is the relationship between the rank 1 convexity and quasiconvexity for invariant functions.

2.2 Convexity

The treatment of convex and polyconvex invariant functions is based on the following

Proposition 2.4.[10] *For each $\mathbf{A}, \mathbf{B} \in \mathbb{M}^{n\times n}$,*

$$\mathbf{A} \cdot \mathbf{B} \leq \tau(\mathbf{A}) \cdot \tau(\mathbf{B}). \tag{40}$$

The classical von Neumann inequality[11] is (40) with the signed singular values τ replaced by the ordinary singular values, and clearly follows from (40). The equality in (40) occurs if and only if $\mathbf{A} = \operatorname{diag}(\tau(\mathbf{A})), \mathbf{B} = \operatorname{diag}(\tau(\mathbf{B}))$ in the same orthonormal basis.

Exercise 2.5.
(i) The distance $\operatorname{dist}(\cdot, M)$ from an invariant set $M \subset \mathbb{M}^{n\times n}$ is represented by the distance $\operatorname{dist}(\cdot, \tilde{M})$ from the (symmetric even) representation $\tilde{M} \subset \mathbb{M}^n$. On $\mathbb{G}^n$, $\operatorname{dist}(\cdot, \tilde{M}) = \operatorname{dist}(\cdot, \tilde{M} \cap \mathbb{G}^n)$.
(ii) The distance from the set $M = \{\mathbf{B} \in \mathbb{M}^{n\times n} : \tau(\mathbf{B}) = \gamma\}$, where $\gamma \in \mathbb{G}^n$, is represented by $\tilde{f}(\alpha) = |\alpha - \gamma|, \alpha \in \mathbb{G}^n$.
(iii) The distance from the set $M = \mathbb{R}_+ SO(n)$, where $\mathbb{R}_+ = [0, \infty)$, is represented by $\tilde{f}$ where

$$\tilde{f}^2(\alpha) = \Big(\alpha_1 - \frac{\alpha_1 + \cdots + \alpha_n}{n}\Big)^2 + \cdots + \Big(\alpha_n - \frac{\alpha_1 + \cdots + \alpha_n}{n}\Big)^2,$$

$\alpha \in \mathbb{G}^n$. For $n = 2$ we obtain $\tilde{f}^2(\alpha) = \frac{1}{2}(\alpha_1 - \alpha_2)^2$ which is a convex function. Hence $\mathbb{R}_+ SO(2)$ is convex.

Proposition 2.6.[12] *Let f be a finite–valued invariant function on $\mathbb{M}^{n\times n}$ represented by $\tilde{f}$ and let $\mathbf{A} \in \mathbb{M}^{n\times n}$ have the signed singular values α. The following conditions are equivalent:*
(i) *f is convex at $\mathbf{A}$, i.e., there exists a subgradient $\mathbf{S} \in \mathbb{M}^{n\times n}$ such that*

$$f(\mathbf{B}) \geq f(\mathbf{A}) + \mathbf{S} \cdot (\mathbf{B} - \mathbf{A}); \tag{41}$$

(ii) *the restriction of $\tilde{f}$ to $\mathbb{G}^n$ has a subgradient s that is in $\mathbb{G}^n$; i.e., there exists an $s \in \mathbb{G}^n$ such that*

$$\tilde{f}(\beta) \geq \tilde{f}(\alpha) + s \cdot (\beta - \alpha) \tag{42}$$

for each $\beta \in \mathbb{G}^n$. If f is fully invariant and $\alpha \in \mathbb{G}^n_+$ then s can be chosen as to satisfy even $s \in \mathbb{G}^n_+$.

[10] Šilhavý (1997, Proposition 18.3.2(ii)), Rosakis (1997).
[11] See, e.g., Marshall and Olkin (1979).
[12] Cf. Šilhavý (1997, Proposition 18.3.5).

Thus f is convex at $\mathbf{A}$ if and only if the restriction to $\mathbb{G}^n$ of its representation is convex at α *and* the subgradient s satisfies an *extra* restriction $s \in \mathbb{G}^n$. The last can be restated equivalently as

$$s_i - s_j \geq 0, \quad s_i + s_j \geq 0, \quad 1 \leq i < j \leq n.$$

Note that if f is an energy function of an isotropic body, then s_i are called the principal forces (as opposed to principal stresses, the eigenvalues of the Cauchy stress). The inequalities $s_i \geq s_j, i < j$, are called the ordered–forces (O–F) inequalities: the principal forces are ordered in the same way as the principal stretches α_i. It is known that O–F inequalities are violated by nearly incompressible materials, such as conventional rubbers: their stored energies are not convex functions of the deformation gradient. On the other hand, the inequalities $s_i + s_j \geq 0, i \neq j$, are violated in states under compression where $s_i < 0$ for all i. However, stored energies of well-compressible real materials often have a large supply of points of convexity in the region of deformation gradients corresponding to tension.

To avoid these undesirable consequences, Coleman and Noll (1959) and Hill (1968) postulated a convexity restricted in one or another way. These and related works greatly contributed to the understanding of the nonlinear elastic response; important tools have been developed on the way. However, none of these *apriori* inequalities was found to be directly related to some qualitative property of the behavior of the material, such as existence, uniqueness, etc.

Proof We can assume that $\mathbf{A} = \mathrm{diag}(\alpha)$. (i) $\Rightarrow$ (ii): If we write (41) for $\mathbf{B} = \mathrm{diag}(\beta)$ where $\beta \in \mathbb{R}^n$, we obtain

$$\tilde{f}(\beta) \geq \tilde{f}(\alpha) + s \cdot (\beta - \alpha), \tag{43}$$

where s are the diagonal elements of $\mathbf{S}$. Let i, j be any pair of distinct indices and apply (43) to the β which differs from α by the interchange of the i- and j- components of α. Then $\tilde{f}(\beta) = \tilde{f}(\alpha)$ and (43) leads to

$$(s_i - s_j)(\alpha_i - \alpha_j) \geq 0. \tag{44}$$

Applying (43) to β which differs from α by the interchange and a simultaneous multiplication by -1 of the i- and j- components of α, we obtain

$$(s_i + s_j)(\alpha_i + \alpha_j) \geq 0. \tag{45}$$

Since $\alpha \in \mathbb{G}^n$, we have

$$\alpha_i + \alpha_j \geq 0, \quad \alpha_i - \alpha_j \geq 0, \quad 1 \leq i < j \leq n. \tag{46}$$

If we assume the strict version of these inequalities throughout, then (44) and (45) imply

$$s_i - s_j \geq 0, \quad s_i + s_j \geq 0, \quad 1 \leq i < j \leq n, \tag{47}$$

i.e., $s \in \mathbb{G}^n$. Thus (42) holds in this case. If one or several of the inequalities in (46) hold with the equality sign, then the above argument fails, but one can still pass to an s that satisfies $s \in \mathbb{G}^n$ as an easy omitted argument shows. (ii) $\Rightarrow$ (i): Prove that (41) holds with $\mathbf{S} = \operatorname{diag}(s)$. If $\mathbf{B} \in \mathbb{M}^{n\times n}$ and $\beta = \tau(\mathbf{B})$ then

$$f(\mathbf{B}) - f(\mathbf{A}) = \tilde{f}(\beta) - \tilde{f}(\alpha) \geq s \cdot (\beta - \alpha) \geq \mathbf{S} \cdot (\mathbf{B} - \mathbf{A})$$

where the last inequality follows from (40). □

Proposition 2.7.[13] *A finite-valued invariant function f is convex on $\mathbb{M}^{n\times n}$ if and only if the the restriction $\tilde{f}_0$ of its representation $\tilde{f}$ to $\mathbb{G}^n$ is convex and for each $\alpha \in \mathbb{G}^n$, $\tilde{f}_0$ has a subgradient at α that is in $\mathbb{G}^n$.*

This follows from Proposition 2.6.

Exercise 2.8. (Convexity and fluids)
(i) Show that if f is of the form

$$f(\mathbf{A}) = \varphi(\det \mathbf{A}), \quad \mathbf{A} \in \mathbb{M}^{n\times n}, \quad n \geq 2, \tag{48}$$

where $\varphi : \mathbb{R} \to \mathbb{R} \cup \{\infty\}$ is some function, then f is convex if and only if f is constant.
(ii) If $\tilde{f}$ is differentiable then the subgradient s as in (42), if it exists, is $s = D\tilde{f}(\alpha)$. Show that if f is given by (48) with φ differentiable and $s := D\tilde{f}(\alpha) \in \mathbb{G}^n$ for every $\alpha \in \mathbb{G}^n$, then f is constant. Thus for φ differentiable, the conclusion (i) above follows from $s \in \mathbb{G}^n$ alone.

Hint: Assume first $n = 2$ and for every $a \in \mathbb{R}^2, \lambda \in \mathbb{R}$, let $\mathbf{A} = \operatorname{diag}(a)$,

$$\mathbf{A}_\pm = \begin{bmatrix} a_1 & \pm\lambda \\ \mp\lambda & a_2 \end{bmatrix}, \quad \mathbf{B}_\pm = \begin{bmatrix} a_1 & \pm\lambda \\ \pm\lambda & a_2 \end{bmatrix}.$$

Then

$$\begin{gathered} \mathbf{A} = \tfrac{1}{2}(\mathbf{A}_+ + \mathbf{A}_-), \quad \mathbf{A} = \tfrac{1}{2}(\mathbf{B}_+ + \mathbf{B}_-), \\ \det \mathbf{A}_\pm = \det \mathbf{A} + \lambda^2, \quad \det \mathbf{B}_\pm = \det \mathbf{A} - \lambda^2. \end{gathered} \tag{49}$$

Hence the convexity inequality (25) for the convex combinations (49) implies

$$\varphi(a_1 a_2) \leq \varphi(a_1 a_2 \pm \lambda^2).$$

Thus φ is both nondecreasing and nonincreasing. If $n > 2$, it suffices to embed the above two-dimensional matrices into n-dimensional matrices using the block-diagonal form.

Exercise 2.9. The functions $\tau_1 + \cdots + \tau_n, \tau_1 + \cdots + \tau_{n-1} - \tau_n$ are convex on $\mathbb{M}^{n\times n}$.

2.3 Polyconvexity

For polyconvexity the available results provide a necessary and sufficient condition only in dimension $n = 2, 3$; in higher dimensions, only sufficient conditions are known, and the proofs reduce the question to invariant convex functions in one or another way. The following proposition gives a necessary and sufficient condition for an invariant function to be polyconvex at a single point in dimension 2.

[13] Cf. Šilhavý (1997, Proposition 18.3.4), Rosakis (1997). See also Dacorogna and Koshigoe (1993) for $n = 2$.

Proposition 2.10.[14] *Let f be a finite–valued invariant function on $\mathbb{M}^{2\times 2}$ represented by $\tilde{f}$. Then the following conditions are equivalent:*
(i) *f is polyconvex at $\mathbf{A} = \operatorname{diag}(\alpha), \alpha \in \mathbb{G}^2$, i.e., there exist $\mathbf{C} \in \mathbb{M}^{2\times 2}, c \in \mathbb{R}$ such that*

$$f(\mathbf{B}) \geq f(\mathbf{A}) + \mathbf{C} \cdot (\mathbf{B} - \mathbf{A}) + c \det(\mathbf{B} - \mathbf{A}) \tag{50}$$

for every $\mathbf{B} \in \mathbb{M}^{2\times 2}$;
(ii) *there exist $s \in \mathbb{R}^2, c \in \mathbb{R}$ satisfying*

$$s_1 + s_2 \geq c(\alpha_1 + \alpha_2), \quad c(\alpha_1 - \alpha_2) \geq s_2 - s_1 \tag{51}$$

such that

$$\tilde{f}(\beta) - \tilde{f}(\alpha) \geq s \cdot (\beta - \alpha) + c(\beta_1 - \alpha_1)(\beta_2 - \alpha_2) \tag{52}$$

for every $\beta \in \mathbb{G}^2$; equivalently, there exist $d \in \mathbb{G}^2, c \in \mathbb{R}$ such that

$$\tilde{f}(\beta) - \tilde{f}(\alpha) \geq d \cdot (\beta - \alpha) + c(\beta_1\beta_2 - \alpha_1\alpha_2). \tag{53}$$

The condition will be used numerously in the subsequent text. If f is the stored energy of an isotropic material, then s_1, s_2 are the principal forces (eigenvalues of the Piola–Kirchhoff stress at $\mathbf{A} = \operatorname{diag}(\alpha)$). Inequalities (51) imply the Baker–Ericksen inequality

$$\alpha_1 s_1 - \alpha_2 s_2 \geq 0; \tag{54}$$

see Exercise 2.11. In contrast to the O–F inequality $s_1 \geq s_2$, (54) is a realistic requirement that can be satisfied also by nearly incompressible materials and in particular by rubbers. In dimension 2, the principal stresses (eigenvalues of the Cauchy stress) are $t_i = \alpha_i s_i / \alpha_1 \alpha_2$ and thus (54) reads $t_1 \geq t_2$ if $\alpha_1 \geq \alpha_2$. For an n-dimensional version of the Baker–Ericksen inequalities, see (59).

Proof Let $\mathbf{A} := \operatorname{diag}(\alpha)$. Note first that if f is polyconvex at $\mathbf{A}$ then $f^{pc}(\mathbf{A}) = f(\mathbf{A})$ as a consequence of the definition of the polyconvexity at $\mathbf{A}$. The function $h := f^{pc}$ thus can be represented as $h(\mathbf{A}) = g(\mathbf{A}, \det \mathbf{A})$ where $g : \mathbb{M}^{2\times 2} \times \mathbb{R} \to \mathbb{R}$ is a convex function. Hence g has a subgradient at $(\mathbf{A}, \det \mathbf{A})$, i.e.,

$$g(\mathbf{B}, \delta) \geq g(\mathbf{A}, \det \mathbf{A}) + \mathbf{D} \cdot (\mathbf{B} - \mathbf{A}) + c(\delta - \det \mathbf{A})$$

for all $(\mathbf{B}, \delta) \in \mathbb{M}^{2\times 2} \times \mathbb{R}$ where $\mathbf{D} \in \mathbb{M}^{2\times 2}$ and $c \in \mathbb{R}$ are constants. Consequently,

$$f(\mathbf{B}) \geq h(\mathbf{B}) \geq f(\mathbf{A}) + \mathbf{D} \cdot (\mathbf{B} - \mathbf{A}) + c(\det \mathbf{B} - \det \mathbf{A}) \tag{55}$$

for all $\mathbf{B} \in \mathbb{M}^{2\times 2}$. It now suffices to use $\det \mathbf{B} - \det \mathbf{A} = \det(\mathbf{B} - \mathbf{A}) + \operatorname{cof} \mathbf{A} \cdot (\mathbf{B} - \mathbf{A})$ to find that (50) holds with $\mathbf{C} = \mathbf{D} + c \operatorname{cof} \mathbf{A}$. Conversely, (50) implies (55) and hence the polyconvexity of f at $\mathbf{A}$. (i) $\Rightarrow$ (ii): If we set $\mathbf{B} = \operatorname{diag}(\beta)$ in (50) where $\beta \in \mathbb{R}^2$, we obtain (52) with s_1, s_2 the diagonal elements of $\mathbf{C}$. Next prove that s satisfies (51). If one inserts $\beta = \pm(\alpha_2, \alpha_1)$ into (52) and uses $\tilde{f}(\beta) = \tilde{f}(\alpha)$, one obtains, respectively,

$$c(\alpha_1 - \alpha_2)^2 \geq (s_2 - s_1)(\alpha_1 - \alpha_2),$$

[14] Šilhavý (1999).

$$(s_2 + s_1)(\alpha_1 + \alpha_2) \geq c(\alpha_1 + \alpha_2)^2$$

and hence (51). To prove (53), define $d \in \mathbb{R}^2$ by $d_{1,2} = s_{1,2} - c\alpha_{2,1}$, note that (53) is a rearrangement of (52) and that (51) implies $d \in \mathbb{G}^2$. (ii) $\Rightarrow$ (i): Thus assume that (53) holds for every $\beta \in \mathbb{G}^2$ and some $d \in \mathbb{G}^2, c \in \mathbb{R}$. We shall prove that

$$f(\mathbf{B}) \geq f(\mathbf{A}) + \mathbf{D} \cdot (\mathbf{B} - \mathbf{A}) + c(\det \mathbf{B} - \det \mathbf{A}) \tag{56}$$

for every $\mathbf{B} \in \mathbb{M}^{2\times 2}$ where $\mathbf{D} := \operatorname{diag}(d), \mathbf{A} = \operatorname{diag}(\alpha)$. Let $\mathbf{B} \in \mathbb{M}^{2\times 2}$ be arbitrary and denote by $\beta \in \mathbb{G}^2$ the signed singular values of $\mathbf{B}$. Using $d \in \mathbb{G}^2$ one finds that the signed singular values of $\mathbf{D}$ are d. Hence (40) implies $d \cdot \beta \geq \mathbf{D} \cdot \mathbf{B}$; also $d \cdot \alpha = \mathbf{D} \cdot \mathbf{A}$. Thus (53) implies (56). □

Exercise 2.11. Let $\alpha \in \mathbb{G}^2$.
(i) If $s \in \mathbb{R}^2, c \in \mathbb{R}$ satisfy (51) then s satisfies the Baker–Ericksen inequality (54);
(ii) if $\tilde{f} : \mathbb{R}^2 \to \mathbb{R}$ is of class C^1 in a neighborhood of α and (52) holds for some $s \in \mathbb{R}^2, c \in \mathbb{R}$ and all $\beta \in \mathbb{R}^2$ then $s_i = \tilde{f}_i(\alpha)$;
(iii) if $s \in \mathbb{R}^2$ satisfies the Baker–Ericksen inequality (54) with the equality sign then (52) reduces to

$$\tilde{f}(\beta) \geq \tilde{f}(\alpha) + c(\beta_1\beta_2 - \alpha_1\alpha_2); \tag{57}$$

(iv) if $\tilde{f}(\beta) = \varphi(\beta_1\beta_2)$ in a neighborhood of α where φ is a differentiable function, then (52) holds in the form (57) where $c = \varphi'(\alpha_1\alpha_2)$.

Proposition 2.10 leads to the following necessary and sufficient condition for global polyconvexity in dimension 2.

Proposition 2.12.[15] *Let f be a finite–valued function on $\mathbb{M}^{2\times 2}$ represented by $\tilde{f}$. Then the following conditions are equivalent:*
(i) *f is polyconvex;*
(ii) *there exists a convex function $\tilde{g} : \mathbb{R}^3 \to \mathbb{R}$ such that $\tilde{g}(\cdot, \delta)$ is symmetric and even for each $\delta \in \mathbb{R}$ and*

$$\tilde{f}(\alpha) = \tilde{g}(\alpha_1, \alpha_2, \alpha_1\alpha_2)$$

for every $\alpha \in \mathbb{R}^2$;
(iii) *there exists a convex function $\tilde{h} : \mathbb{R}^2_+ \times \mathbb{R} \to \mathbb{R}$ such that $\tilde{h}(\cdot, \delta)$ is nondecreasing for each $\delta \in \mathbb{R}$ and*

$$\tilde{f}(\alpha) = \tilde{h}(\alpha_1 + \alpha_2, \alpha_1 - \alpha_2, \alpha_1\alpha_2)$$

for every $\alpha \in \mathbb{G}^2$.

A recent result of Mielke (2002) gives a necessary and sufficient condition for polyconvexity in dimension 3.

[15]Cf. Dacorogna and Koshigoe (1993), Rosakis (1997), Šilhavý (1999).

2.4 Rank 1 Convexity

To proceed to the discussion of the rank 1 convexity, we first address the question of possible signed singular values of rank 1 perturbation of a given matrix. The pair $\alpha, \beta \in \mathbb{G}^n$ is said to be compatible if[16]

$$|\beta_1| \geq |\alpha_2|, \quad |\alpha_1| \geq |\beta_2| \geq |\alpha_3|, \quad \ldots, \quad |\alpha_{n-1}| \geq |\beta_n|.$$

We also say that α, β satisfy the *interlacing inequalities.*

Proposition 2.13.[17] *If* $\mathbf{A} = \operatorname{diag}(\alpha), \alpha \in \mathbb{G}^n$, *then* $\beta \in \mathbb{G}^n$ *are the signed singular values of some rank 1 perturbation of* $\mathbf{A}$ *if and only if* α, β *are compatible.*

For a β compatible with α it is possible to construct explicitly a rank 1 perturbation of $\operatorname{diag}(\alpha)$ with the signed singular values β.

The condition for the rank 1 convexity of an invariant function in terms of $\tilde{f}$ consists of two parts, one being the Baker–Ericksen inequalities and the other a convexity on certain curves in $\mathbb{G}^n$. We first discuss the former condition.

The Baker–Ericksen inequalities (see (59), below) are restrictions on the *first* derivatives of $\tilde{f}$ and are standard in this context; however, the "integral" form in terms of ordering is recent. If $\alpha, \beta \in \mathbb{G}^n$ satisfy

$$\prod_{i=1}^{k} \alpha_i \leq \prod_{i=1}^{k} \beta_i, \quad k = 1, \ldots, n-1, \qquad \text{and} \qquad \prod_{i=1}^{n} \alpha_i = \prod_{i=1}^{n} \beta_i, \tag{58}$$

we write $\alpha \leq_{be} \beta$; if instead of the last equation in (58) we have

$$\prod_{i=1}^{n} |\alpha_i| \leq \prod_{i=1}^{n} |\beta_i|,$$

we write $\alpha \leq_{|be|} \beta$. We say that a function $\tilde{f} : \mathbb{G}^n \to \mathbb{R} \cup \{\infty\}$ is BE–monotonic if $\tilde{f}(\alpha) \leq \tilde{f}(\beta)$ whenever $\alpha \leq_{be} \beta$; if the same holds whenever $\alpha \leq_{|be|} \beta$, we say that $\tilde{f}$ fully BE–monotonic. The ordering $\leq_{be}$ is shown in Figure 5: The sequence of states preserves volume, increases the largest edge, and increases the largest face.

Exercise 2.14. If $\tilde{f} : \mathbb{G}^n \to \mathbb{R} \cup \{\infty\}$ is a fully BE–monotonic function then $\alpha \mapsto \tilde{f}(\alpha)$ is non-decreasing on $\mathbb{G}^n_+$ in each variable from the list $\alpha_1, \ldots, \alpha_n$.

Exercise 2.15.[18] Let $\tilde{f} : \mathbb{G}^n \to \mathbb{R}$ be of class C^1. Show that
(i) $\tilde{f}$ satisfies the Baker–Ericksen inequalities if and only if it is non–decreasing along every smooth curve $\eta : [0, 1] \to \mathbb{G}^n$ such that for some $1 \leq i < j \leq n$,

$$\eta_i \eta_j = \text{const}, \quad \eta_k = \text{const} \qquad \text{if} \qquad i \neq k \neq j, \quad \dot{\eta}_i > 0$$

[16] The absolute values are effective only in the case of α_n, β_n; otherwise α_i, β_i are nonnegative.
[17] $n = 2$: Aubert and Tahraoui (1980), Aubert and Tahraoui (1987), $n = 3$: Rosakis (1990), n arbitrary: Šilhavý (1999).
[18] Šilhavý (2002b).

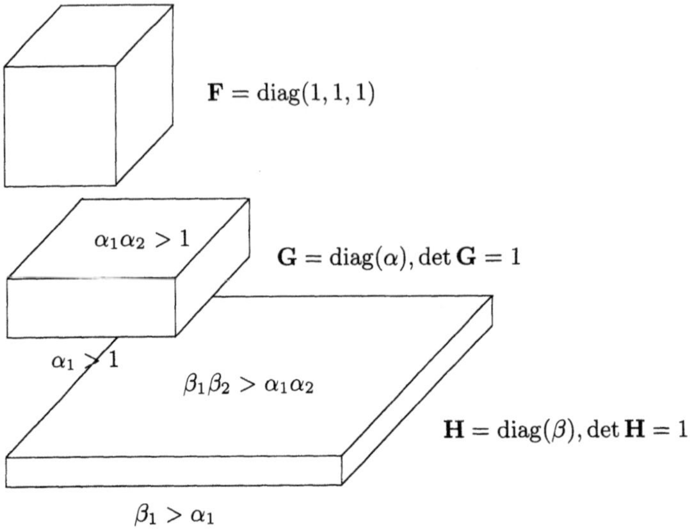

Figure 5. The sequence $\mathbf{F} \to \mathbf{G} \to \mathbf{H}$ increases energy

for all $t \in [0,1]$;
(ii) $\tilde{f}$ is BE–monotonic if and only if it satisfies the Baker–Ericksen inequalities

$$(s_i\alpha_i - s_j\alpha_j)(\alpha_i - \alpha_j) \geq 0, \quad i \neq j \tag{59}$$

where $s_i = \tilde{f}_i(\alpha)$;
(iii) if $\tilde{f} : \mathbb{G}^n \to \mathbb{R}$ is of class C^1 and satisfies the Baker–Ericksen inequalities (59) with the equality sign for all $i \neq j$ and all $\alpha \in \mathbb{G}^n$, then $\tilde{f}(\alpha) = \varphi(\alpha_1 \cdots \alpha_n)$ for some $\varphi : \mathbb{R} \to \mathbb{R}$ and all $\alpha \in \mathbb{G}^n$ (in other words, the material is a "fluid").

Hint: (i) is immediate. If $\tilde{f}$ is BE–monotonic and η a curve as in (i), then $\eta(s) \leq_{be} \eta(t)$ for each $s \leq t$ and thus the Baker–Ericksen inequalities follow from (i). Conversely, let $\tilde{f}$ satisfy the Baker–Ericksen inequalities, let $\alpha, \beta \in \mathbb{G}^n$ satisfy (58) and prove $\tilde{f}(\alpha) \leq \tilde{f}(\beta)$. Assume temporarily that the common value π of the product in $(58)_2$ positive, which is equivalent to $\alpha_n > 0, \beta_n > 0$. Introduce the change of variables $\gamma \to x$ where $\gamma \in \mathbb{G}^n_{++}$, by $x_i = \ln \gamma_i$, which maps $\mathbb{G}^n_{++}$ onto $\mathbb{D}^n := \{x \in \mathbb{R}^n : x_1 \geq \cdots \geq x_n\}$. The transformed function $g(x) = \tilde{f}(\gamma)$ then satisfies

$$g_1 \geq g_2 \geq \cdots \geq g_n \tag{60}$$

throughout $\mathbb{D}^n$ as a consequence of the Baker–Ericksen inequalities. Let ξ be a line segment joining the images of α, β under the transformation, $\xi(t) = (1-t)z + ty = z + tu$, where $z_i = \ln \alpha_i$, $y_i = \ln \beta_i$, $u = y - z$. Introducing $\psi(t) = g(z + tu), t \in [0,1]$, one obtains a partial summation formula

$$\dot{\psi}(t) = \sum_{i=1}^{n-1} (g_i - g_{i+1}) w_i$$

where $w_i := u_1 + \cdots + u_i \geq 0, i = 1, \ldots, n-1$, by (58). Thus (60) lead to $\dot{\psi}(t) \geq 0$ which gives $\tilde{f}(\alpha) \leq \tilde{f}(\beta)$. (iii) Note that (60) holds with the equality signs throughout and follow the rest of the proof in (ii).

Let $S = (S^1, \dots, S^n)$ be the $n$tuple of elementary symmetric functions of n variables,

$$S^0(\alpha_1, \dots, \alpha_n) = 1,$$

$$S^k(\alpha_1, \dots, \alpha_n) = \sum_{1 \le i_1 < \dots < i_k \le n} \alpha_{i_1} \dots \alpha_{i_n}, \quad \text{if} \quad k = 1, \dots, n.$$

If $\epsilon = (\epsilon_1, \dots, \epsilon_n) \in \{-1, 1\}^n$ and $\alpha \in \mathbb{R}^n$ we abbreviate $\epsilon\alpha := \operatorname{diag}(\epsilon)\alpha$. For an $x \in \mathbb{R}^n$ we write $x \ge 0$ if the components of x are nonnegative.

Theorem 2.16.[19] *Let $f : \mathbb{M}^{n\times n} \to \mathbb{R}$ be an invariant function on $\mathbb{M}^{n\times n}$ with the representation $\tilde{f}$. Then f is rank 1 convex if and only if the following two conditions hold:*
(i) *$\tilde{f}$ is BE–monotonic;*
(ii) *we have*

$$\tilde{f}(\gamma) \le (1-t)\tilde{f}(\alpha) + t\tilde{f}(\beta)$$

for each compatible α, β, and each $\gamma \in \mathbb{G}^n, \epsilon \in \{-1, 1\}^n$ and $t \in [0, 1]$ such that $\epsilon(\beta - \alpha) \ge 0$ and

$$S(\epsilon\gamma) = (1-t)S(\epsilon\alpha) + tS(\epsilon\beta). \tag{61}$$

Theorem 2.16, stated here without proof, is a basis for the relaxation procedure in Section 3.1. Of its two condition, the BE–monotonicity is easier to verify. It will be frequently used in this text. Roughly, Condition (ii) says that under the passage from $\mathbb{M}^{n\times n}$ to the representation space $\mathbb{G}^n$, the line segments $\mathbf{C} = (1-t)\mathbf{A} + t\mathbf{B}, 0 \le t \le 1$, with rank 1 connected endpoints $\mathbf{A}, \mathbf{B}$, change to *curves* consisting of points γ satisfying (61). For other conditions in dimension 2 see Aubert (1995), Šilhavý (2001). For fully invariant functions the BE–monotonicity can be replaced by full monotonicity:

Theorem 2.17.[20] *If $f : \mathbb{M}^{n\times n} \to \mathbb{R} \cup \{\infty\}$ is an invariant rank 1 convex function then $\tilde{f}$ is fully BE monotonic.*

Exercise 2.18. (Rank 1 convexity and fluids)
(i) If $f : \mathbb{M}^{m\times n} \to \mathbb{R} \cup \{\infty\}$ is a (generally noninvariant) rank 1 convex function, then $f(\mathbf{A})$, considered as a function of the $m \times n$ matrix elements of $\mathbf{A} \in \mathbb{M}^{m\times n}$, is convex in each of these elements holding the remaining elements fixed ("separate convexity");
(ii) if f is an invariant rank 1 function on $\mathbb{M}^{n\times n}$ then its representation $\tilde{f}$ is separately convex on $\mathbb{R}^n$;
(iii) if f is a function on $\mathbb{M}^{n\times n}$ of the form

$$f(\mathbf{A}) = \varphi(\det \mathbf{A}), \quad \mathbf{A} \in \mathbb{M}^{n\times n},$$

where $\varphi : \mathbb{R} \to \mathbb{R} \cup \{\infty\}$, then f is rank 1 convex $\Leftrightarrow$ f is quasiconvex $\Leftrightarrow$ f polyconvex $\Leftrightarrow$ φ is convex.

Hint: To prove (iii), use (ii) to show that φ is convex.

[19] Šilhavý (2003b).
[20] Dacorogna and Marcellini (1999).

Exercise 2.19.[21] Let f be an invariant function on $\mathbb{M}^{n\times n}$ with the representation of the form $\tilde{f}(\alpha) = g(\alpha_2, \ldots, |\alpha_n|), \alpha \in \mathbb{G}^n$, where $g : \mathbb{G}_+^{n-1} \to \mathbb{R} \cup \{\infty\}$. Then f is rank 1 convex if and only if it is constant. More precisely, f is fully BE-monotonic if and only if it is constant.

Hint: Prove that $g(\xi) = g(\eta)$ for any $\xi, \eta \in \mathbb{G}_+^{n-1}$. Let $s \geq \xi_1, t \geq \eta_1$ and set $\alpha = (s, \xi), \beta = (t, \eta)$ so that $\alpha, \beta \in \mathbb{G}_+^n$. If t is sufficiently large, then $\alpha \leq_{|be|} \beta$ so that $\tilde{f}(\alpha) \leq \tilde{f}(\beta)$ gives $g(\xi) \leq g(\eta)$. If s is sufficiently large then we have the opposite ordering.

3 Relaxation of Invariant Functions and Sets

The conditions for semiconvexity of invariant functions of the preceding chapter lead to simplified procedures for finding semiconvex hulls in some cases. A simplification is available in the case of convexification and rank 1 convexification in arbitrary dimension and in the case of polyconvexification in dimension 2.[22] No reduction is available for the quasiconvexification and polyconvexification in arbitrary dimension. A common method for evaluation the quasiconvex hull is to find the rank 1 convex hull and then to prove that it is polyconvex. This underlines the importance of the rank 1 convex hull. This chapter, therefore, presents an adaptation of the iterative procedure of Kohn & Strang (Item (iv) of Proposition 1.17) for invariant functions. This procedure reflects the equivalent conditions for the rank 1 convexity of the preceding chapter. Hence each step of this procedure splits into the BE–monotonization and the convexification on the curve segments determined by the elementary symmetric functions. Of these two, the Baker–Ericksen monotonization is much easier to perform in concrete cases, and several examples and one general condition will be presented for which the Baker–Ericksen monotonization already provides a polyconvex function, thereby terminating the iterative procedure. The examples presented here are mostly in dimension 2. However, in Chapter 4 the monotonization will be shown to provide the relaxation of a class of invariant functions related to the nematic elastomers.

3.1 Rank 1 Convex Hull of Invariant Functions

One of the first nontrivial examples of a relaxation was given by Kohn & Strang in the context of optimal design theory. I present here only the restriction to the 2×2 case:

Example 3.1.[23] *Let $f : \mathbb{M}^{2\times 2} \to \mathbb{R}$ be given by*

$$f(\mathbf{A}) = \begin{cases} 0 & \text{if} \quad \mathbf{A} = \mathbf{0}, \\ 1 + |\mathbf{A}|^2 & \text{if} \quad \mathbf{A} \neq \mathbf{0}, \end{cases}$$

$\mathbf{A} \in \mathbb{M}^{2\times 2}$. *Then $f^{rc} = f^{qc} = f^{pc}$ where*

$$f^{rc}(\mathbf{A}) = \begin{cases} 2\sqrt{\sigma(\mathbf{A})} - 2|\det \mathbf{A}| & \text{if} \quad \sigma(\mathbf{A}) \leq 1, \\ 1 + |\mathbf{A}|^2 & \text{if} \quad \sigma(\mathbf{A}) \geq 1, \end{cases}$$

[21] Buttazzo, Dacorogna and Gangbo (1994).

[22] See Buttazzo, Dacorogna and Gangbo (1994) for convexification of fully invariant functions in arbitrary dimension and polyconvexification in dimension 2.

[23] Kohn and Strang (1986).

and $\sigma(\mathbf{A}) := |\mathbf{A}|^2 + 2|\det \mathbf{A}|$.

This was calculated by using Proposition 1.17(iv).

To proceed to the general discussion of hulls of invariant functions, note first that the relaxation preserves the invariance: If $f : \mathbb{M}^{n\times n} \to \mathbb{R} \cup \{\infty\}$ or $M \subset \mathbb{M}^{n\times n}$ is (fully) invariant then so also are $f^s, s = rc, qc, pc, c,$ or $M^s, s = lc, rc, qc, pc, c.$[24] We denote by $\tilde{f}^s, \tilde{M}^s$ the corresponding representations. For one special case of invariant functions there is a simple way to determine the hulls:

Exercise 3.2. (Thermostatics of fluids) Consider an invariant function of the form

$$f(\mathbf{A}) = \varphi(\det \mathbf{A}), \quad \mathbf{A} \in \mathbb{M}^{n\times n}, \tag{62}$$

where $\varphi : \mathbb{R} \to \mathbb{R} \cup \{\infty\}$. Prove:
(i) any hull of f is again of the form (62);
(ii) $g := f^{pc} = f^{qc} = f^{rc}$ where

$$g(\mathbf{A}) = \varphi^c(\det \mathbf{A}),$$

where φ^c is the convex hull of φ;
(iii) $f^c = c = \text{const}$ where $c = \inf\{\varphi(\delta) : \delta \in \mathbb{R}\}$.

Remark. Item (ii) shows that the relaxation f^{qc} reduces to the Maxwell–Gibbs common tangent construction on the graph of energy versus specific volume.

Hint: (i): we have $f(\mathbf{QAH}) = f(\mathbf{A})$ for every $\mathbf{Q} \in SO(n), \mathbf{A} \in \mathbb{M}^{n\times n}, \mathbf{H} \in SL(n)$. Let g be the semiconvex hull of f. If $\mathbf{Q} \in SO(n), \mathbf{H} \in SL(n)$ are fixed then the function $h(\mathbf{A}) = g(\mathbf{QAH})$ has the same semiconvexity property and $h \le f$. Since g is the maximal semiconvex function below f, we have $h \le g$, i.e., $g(\mathbf{QAH}) \le g(\mathbf{A})$. The arbitrariness implies the equality $g(\mathbf{QAH}) = g(\mathbf{A})$. Hence g has the representation (62). (ii): Use (i) and Exercise 2.18. (iii): Use (i) and Exercise 2.8.

Exercise 3.3. (i) Use the function h as in the proof of Exercise 3.2(i) to show that any hull of an invariant function is invariant. (ii) Use (i) and the definition of hulls of sets in Section 1.5 to show that any hull of an invariant set is invariant.

We now proceed to the adaptation of the Kohn–Strang procedure Proposition 1.17(iv) for invariant functions as suggested by Theorem 2.16. The procedure reflects the two conditions (i), (ii) in 2.16. Thus each step splits into two substeps of which the first is the "Baker–Ericksen monotonization" $\tilde{f}^{be}$ and the second a convexification described as $\tilde{f}^e$ below. For each function $\tilde{f} : \mathbb{G}^n \to \mathbb{R} \cup \{\infty\}$ denote by $\tilde{f}^{be} : \mathbb{R}^n \to \mathbb{R} \cup \{\infty\}$ the function defined by

$$\tilde{f}^{be}(\alpha) = \inf\left\{\tilde{f}(\beta) : \beta \in \mathbb{G}^n, \alpha \le_{be} \beta\right\}, \quad \alpha \in \mathbb{G}^n,$$

and by $\tilde{f}^{|be|}$ the function on $\mathbb{G}^n$ defined by by the same formula with $\le_{be}$ replaced by $\le_{|be|}$. The functions $\tilde{f}^{be}$ and $\tilde{f}^{|be|}$ are, respectively, the largest BE–monotonic (fully BE–monotonic) function below $\tilde{f}$. For each $\tilde{f} : \mathbb{G}^n \to \mathbb{R} \cup \{\infty\}$ define a symmetric even function $\tilde{f}^e : \mathbb{R}^n \to \mathbb{R} \cup \{\infty\}$ by

$$\tilde{f}^e(\alpha) = \inf\{(1-t)f(\beta) + tf(\gamma) : (t, \beta, \gamma) \in \mathcal{A}(\alpha)\}$$

[24] See Exercise 3.3.

where the infimum is taken over the set $\mathcal{A}(\alpha)$ of all (t,β,γ) such that β,γ satisfy the interlacing inequalities, $0\le t\le 1$, and $S(\epsilon\alpha)=(1-t)S(\epsilon\beta)+tS(\epsilon\gamma)$ where

$$\epsilon_i := \begin{cases} +1 & \text{if} \quad \beta_i \ge \gamma_i, \\ -1 & \text{if} \quad \beta_i < \gamma_i. \end{cases}$$

We finally set $\tilde f^u = [\tilde f^{be}]^e$ and $\tilde f^{|u|} = [\tilde f^{|be|}]^e$.

Proposition 3.4.[25] *Let $f : \mathbb{M}^{n\times n} \to \mathbb{R}\cup\{\infty\}$ be an invariant function, bounded from below by a rank 1 convex function and define a sequence $\tilde f_k$, $k = 0,\ldots,$ by $\tilde f_0 = \tilde f$, $\tilde f_{k+1} = \tilde f_k^u$. Suppose furthermore that either $n = 2$ or $n > 2$ and f is finite-valued. Then*

$$\tilde f^{rc} = \lim_{k\to\infty} \tilde f_k.$$

If f is fully invariant, one can replace the above definition of $\tilde f_k$ by $\tilde f_{k+1} = \tilde f_k^{|u|}$.

Since $\tilde f^{|u|} \le \tilde f^u$, the sequence based on $\tilde f^{|u|}$ gives generally a faster convergence. Also $\tilde f^{|be|}$ is easier to calculate. The procedure $\tilde f_k$ is usually useful only if it terminates: if after a few steps, one obtains a stable value of $\tilde f_k$ which then coincides with $\tilde f^{rc}$. Such will be the case in all examples considered in this text.

Exercise 3.5. Prove Proposition 3.4.

Let us demonstrate the termination of the procedure in 3.4 on the following example:

Example 3.6.[26] (Saint Venant–Kirchhoff material) *Let $f : \mathbb{M}^{2\times 2} \to \mathbb{R}$ be represented by*

$$\tilde f(\alpha) = (\alpha_1^2 - 1)^2 + (\alpha_2^2 - 1)^2 + \frac{\nu}{1-2\nu}(\alpha_1^2 + \alpha_2^2 - 2)^2,$$

$\alpha \in \mathbb{G}^2$, where $\nu \in [0, \frac{1}{2})$. Then $f^{|be|} = f^{rc} = f^{qc} = f^{pc} = f^c$ where

$$\tilde f^{rc}(\alpha) = \begin{cases} 0 & \text{if} \quad \alpha_1 \le 1, \\ \dfrac{1}{1-\nu}(\alpha_1^2 - 1)^2 & \text{if} \quad \alpha_1 \ge 1,\ \nu\alpha_1^2 + (1-\nu)\alpha_2^2 \le 1, \\ \tilde f(\alpha) & \text{otherwise}. \end{cases} \tag{63}$$

f is a fully invariant function; hence $\tilde f^{rc} \le \tilde f^{|be|}$. In fact in this case $\tilde f^{|be|}$ already represents a convex function on $\mathbb{M}^{2\times 2}$ and the procedure terminates.

Exercise 3.7. Prove that $\tilde f^{|be|}$ is given by the right–hand side of (63) and represents a convex function on $\mathbb{M}^{2\times 2}$. Hint: Since $\tilde f$ represents a fully invariant function, it suffices to consider only the restriction of $\tilde f^{|be|}$ on $\mathbb{G}^2_+$. By Example 2.14 one has $\tilde f^{rc} \le g$ on $\mathbb{G}^2_+$ where $g : \mathbb{G}^2 \to \mathbb{R}$ is a monotonization of $\tilde f$ with respect to the α_1, α_2 variables given by

$$g(\alpha) = \inf\Big\{\tilde f(\beta) : \beta \in \mathbb{G}^2_+, \alpha_1 \le \beta_1, \alpha_2 \le \beta_2\Big\}. \tag{64}$$

[25] Šilhavý (2003a).

[26] Le Dret and Raoult (1994); see also Le Dret and Raoult (1995) for $n = 3$.

To evaluate the function g, one first notices that if $\alpha_1 \leq 1$, then the set of β occurring in (64) contains the point $\beta_1 = \beta_2 = 1$ where $\tilde{f}(\beta) = 0$. Thus $g(\alpha) = 0$ if $\alpha_1 \leq 1$. By examining the derivative $\tilde{f}_2(\alpha)$ one finds that if $\alpha_1 \geq 1, \nu\alpha_1^2 + (1-\nu)\alpha_2^2 \leq 1$, then then the set of β occurring in (64) contains the point β with $\beta_1 = \alpha_1$ such that $\nu\beta_1^2 + (1-\nu)\beta_2^2 = 1$. Thus

$$g(\alpha) \leq \tilde{f}(\beta) = \frac{1}{1-\nu}(\alpha_1^2 - 1)^2.$$

Finally, in the last regime in (63) we have $\tilde{f}_1 \geq 0, \tilde{f}_2 \geq 0$ and thus $g(\alpha) = \tilde{f}(\alpha)$. This proves that g is given by the right–hand side of (63). That functions is convex in α. Since also the derivatives of g satisfy the condition on the subgradient stated in Proposition 2.7, g represents a convex function on $\mathbb{M}^{2\times 2}$.

3.2 Relaxation of Functions Violating the Baker–Ericksen Inequalities

Recall that the Baker–Ericksen inequalities are a necessary condition for the rank 1 convexity of invariant functions. Thus the violation of them is a particular mode of loss of stability in a material. In this and the next sections we consider relaxation of materials in dimension 2 which lose their stability in this way. In dimension 2 there is just one Baker–Ericksen inequality,

$$\alpha_1\tilde{f}_1 - \alpha_2\tilde{f}_2 \geq 0 \tag{65}$$

and on the border separating the region $\mathcal{E}$ where (65) holds from the region where (65) is violated we have

$$\alpha_1\tilde{f}_1 - \alpha_2\tilde{f}_2 = 0. \tag{66}$$

The stress tensor $\mathbf{S} := Df(\mathbf{A})$ at the deformation gradient $\mathbf{A}$ for which (66) holds then reduces to the hydrostatic pressure $\mathbf{S} = -p \operatorname{cof} \mathbf{A}$ where $p = -\tilde{f}_1/\alpha_2 = -\tilde{f}_2/\alpha_1$. This form of the stress tensor characterizes fluids, and we call such points liquefaction points. Theorem 3.8, below, presents a relaxation of a class of materials for which the boundary $\partial\mathcal{T}$ of the phase transition zone $\mathcal{T}$, defined in (7), is determined by (66) and in the interior of the phase transition zone the relaxation f^{qc} displays a fluid–like behavior, i.e.,

$$f^{qc}(\mathbf{A}) = \varphi(\det \mathbf{A})$$

where φ is determined by the compatibility with the original function on $\partial\mathcal{T}$. The relaxation reduces to the BE–monotonization $\tilde{f}^{be}$. Moreover, for this class of functions it turns out that the quasiconvexification of an arbitrary power $f^q, q \geq 1$, (assuming $f \geq 0$) is the corresponding power of the quasiconvexification,

$$(f^q)^{qc} = (f^{qc})^q; \tag{67}$$

actually (67) holds with the power replaced by any increasing convex function. The theorem will be applied in the next section to determine the relaxations of some functions having minima on the isotropic set $W = \{\mathbf{A} \in \mathbb{M}^{2\times 2} : \tau(\mathbf{A}) = \gamma\}$, where $\gamma \in \mathbb{G}^2$ is fixed. Recall from Section 1.2 that sets of this type are minima of the energy functions of nematic elastomers.

Let $f : \mathbb{M}^{2\times 2} \to \mathbb{R}$ be an invariant function represented by $\tilde{f} : \mathbb{G}^2 \to \mathbb{R}$. A point $\rho \in \mathbb{G}^2$ is said to be a polyconvex liquefaction point if there exists a constant $c \in \mathbb{R}$ such that

$$\tilde{f}(\beta) \geq \tilde{f}(\rho) + c(\beta_1\beta_2 - \rho_1\rho_2) \tag{68}$$

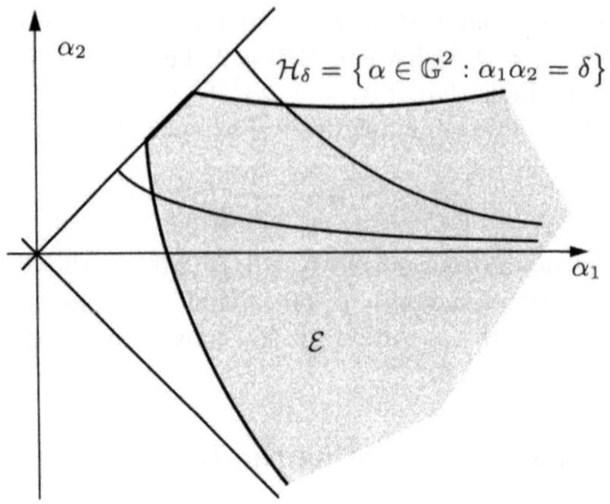

Figure 6. An absorbing set $\mathcal{E}$

for all $\beta \in \mathbb{G}^2$. Each polyconvex liquefaction point is a point of polyconvexity, cf. Exercise 2.11. A closed set $\mathcal{E} \subset \mathbb{G}^2$ is said to be absorbing if (i) for every $\delta \in \mathbb{R}$ there exists an $\alpha \in \mathcal{E}$ with $\delta = \alpha_1\alpha_2$ and (ii) for every $\alpha \in \mathcal{E}$ and every $\beta \geq_{be} \alpha$ one has $\beta \in \mathcal{E}$. Figure 6 shows an absorbing set $\mathcal{E}$ (shaded). For every $\delta \in \mathbb{R}$, the hyperbola $\mathcal{H}_\delta$ intersects $\mathcal{E}$ at least once; starting from the 'first' point of intersection, $\mathcal{H}_\delta$ stays in $\mathcal{E}$ forever. (Recall that for $n = 2$, $\beta \geq_{be} \alpha$ means that $\beta_1 \geq \alpha_1, \beta_1\beta_2 = \alpha_1\alpha_2$.)

Theorem 3.8. *Let $f : \mathbb{M}^{2\times 2} \to \mathbb{R}$ be an invariant function represented by $\tilde{f}$. Suppose that there exists an absorbing set $\mathcal{E} \subset \mathbb{G}^2$ such that*
(i) *each point $\alpha \in \mathcal{E}$ is a point of polyconvexity of f;*
(ii) *each point of $\partial\mathcal{E} \setminus \partial\mathbb{G}^2$ is a polyconvex liquefaction point.*
Then

$$f^{rc} = f^{qc} = f^{pc} = g \tag{69}$$

where g is an invariant function represented by

$$\tilde{g}(\alpha) = \begin{cases} \tilde{f}(\alpha) & \text{if } \alpha \in \mathcal{E}, \\ \tilde{h}(\alpha_1\alpha_2) & \text{if } \alpha \in \mathbb{G}^2 \setminus \mathcal{E}, \end{cases} \tag{70}$$

and $\tilde{h} : \mathbb{R} \to \mathbb{R}$ is a convex function such that

$$\tilde{h}(\alpha_1\alpha_2) = \tilde{f}(\alpha) \quad \text{if} \quad \alpha \in \partial\mathcal{E} \setminus \partial\mathbb{G}^2. \tag{71}$$

Moreover, if $f \geq 0$, then for any $q \geq 1$,

$$(f^q)^{rc} = (f^q)^{qc} = (f^q)^{pc} = g^q. \tag{72}$$

Thus in the complement of $\mathcal{E}$, the macroscopic response $\tilde{f}^{qc}$ is liquid–like. The corresponding density $\tilde{h}$ is uniquely determined on the set of $\delta \in \mathbb{R}$ for which there exists and

$\alpha \in \partial\mathcal{E} \setminus \partial\mathbb{G}^2$ such that $\delta = \alpha_1\alpha_2$, and (71) says how to determine the corresponding energy density.

Proof Since $\mathcal{E}$ is an absorbing set, for each $\delta \in \mathbb{R}$ there exists a unique point $\rho = \bar{\rho}(\delta) \in \partial\mathcal{E}$ with $\rho_1\rho_2 = \delta$ such that $\beta \in \mathcal{E}$ for all $\beta \in \mathbb{G}^2$ with $\beta \geq_{be} \rho$ and $\beta \notin \mathcal{E}$ for all $\beta \in \mathbb{G}^2$ with $\beta \leq_{be} \rho, \beta \neq \rho$. Let $\tilde{h}_1 : \mathbb{R} \to \mathbb{R}$ be defined by

$$\tilde{h}_1(\delta) = \tilde{f}(\bar{\rho}(\delta))$$

for each $\delta \in \mathbb{R}$, let

$$I := \{\delta \in \mathbb{R} : \bar{\rho}(\delta) \notin \partial\mathbb{G}^2\},$$

and note that each element of I is a point of convexity of $\tilde{h}_1$. Indeed, if $\delta \in I$ then $\rho := \bar{\rho}(\delta)$ is a polyconvex liquefaction point; taking arbitrary $\omega \in \mathbb{R}$, setting $\beta = \bar{\rho}(\omega)$ in (53) and using the definition of $\tilde{h}_1$ we obtain

$$\tilde{h}_1(\omega) \geq \tilde{h}_1(\delta) + c(\omega - \delta).$$

Let $\tilde{h}$ be the convex hull of $\tilde{h}_1$. From the above we see that $\tilde{h} = \tilde{h}_1$ on I and that $\tilde{h}$ satisfies (71). Let g be defined by (70). Note first that $g \leq f$. On $\mathcal{E}$ we have $\tilde{f} = \tilde{g}$ and thus let $\alpha \in \mathbb{G}^2 \setminus \mathcal{E}$ and put $\delta = \alpha_1\alpha_2$. Then $\bar{\rho}(\delta)$ is a polyconvex liquefaction point and (68) gives $\tilde{f}(\alpha) \geq \tilde{f}(\bar{\rho}(\delta)) = \tilde{h}(\delta) = \tilde{g}(\alpha)$. Thus $g \leq f$. Next we wish to show that $f^{rc} \leq g$. This inequality is obvious on $\mathcal{E}$, for there we have $g = f$ and $f^{rc} \leq f$ always. To prove $f^{rc} \leq g$ on the complement of $\mathcal{E}$, we invoke the BE–monotonicity of invariant rank 1 convex functions. Thus let $\alpha \in \mathbb{G}^2 \setminus \mathcal{E}$ and let $\delta := \alpha_1\alpha_2$. Then $\alpha \leq_{be} \bar{\rho}(\delta)$; hence

$$\tilde{f}^{rc}(\alpha) \leq \tilde{f}^{rc}(\bar{\rho}(\delta)) \leq \tilde{f}(\bar{\rho}(\delta)) = \tilde{h}_1(\delta) = \tilde{h}(\delta) = \tilde{g}(\alpha),$$

where the first inequality is the BE–monotonicity and the equalities are the definitions of $\tilde{h}_1, \tilde{h}$ and $\tilde{g}$. Thus $f^{rc} \leq g$. To prove (69), it suffices to prove that g is a polyconvex function, for then the inequality $g \leq f$ gives $g \leq f^{pc}$; combining with $f^{rc} \leq g$ and with the relevant inequalities in (36) completes the proof. We shall prove that each point $\mathbf{A} \in \mathbb{M}^{2\times 2}$ is a point of polyconvexity of g. If the signed singular values α of $\mathbf{A}$ are in $\mathcal{E}$, then $\mathbf{A}$ is a point of polyconvexity of f. Therefore, let d, c be as in (53) and prove that

$$\tilde{g}(\beta) \geq \tilde{g}(\alpha) + d \cdot (\beta - \alpha) + c(\beta_1\beta_2 - \alpha_1\alpha_2) \tag{73}$$

for each $\beta \in \mathbb{G}^2$. This follows immediately from (53) if $\beta \in \mathcal{E}$. Let $\beta \in \mathbb{G}^2 \setminus \mathcal{E}$, set $\delta = \beta_1\beta_2$ and $\rho = \bar{\rho}(\delta)$. Then $\tilde{g}(\beta) = \tilde{h}(\delta) = \tilde{h}_1(\delta) = \tilde{f}(\rho)$ and thus

$$\tilde{g}(\beta) \geq \tilde{f}(\alpha) + d \cdot (\rho - \alpha) + c(\delta - \alpha_1\alpha_2). \tag{74}$$

It now suffices to note that $d \in \mathbb{G}^2$ and $\beta \leq_{be} \rho$ imply $d \cdot \beta \leq d \cdot \rho$; taking into account that $\tilde{f}(\alpha) = \tilde{g}(\alpha)$ we then see that (74) implies (73). Thus every $\mathbf{A}$ whose signed singular values are in $\mathcal{E}$ is a point of polyconvexity of g. Next prove that every $\mathbf{A}$ with the signed singular values α in $\mathbb{G}^2 \setminus \mathcal{E}$ is a point of polyconvexity of g. Set $\delta = \alpha_1\alpha_2$ and recall that $\tilde{h}$ is convex, so that there exists a $c \in \mathbb{R}$ such that

$$\tilde{h}(\omega) \geq \tilde{h}(\delta) + c(\omega - \delta) \tag{75}$$

for all $\omega \in \mathbb{R}$. Prove that

$$\tilde{g}(\beta) \geq \tilde{g}(\alpha) + c(\beta_1\beta_2 - \alpha_1\alpha_2) \tag{76}$$

for every $\beta \in \mathbb{G}^2$. Noting that $\tilde{g}(\alpha) = \tilde{h}(\delta)$ and setting $\omega = \beta_1\beta_2$, we see that (76) is immediate if $\beta \in \mathbb{G}^2 \setminus \mathcal{E}$, for then $\tilde{g}(\beta) = \tilde{h}(\omega)$ and (75), (76) become identical. Thus let $\beta \in \mathcal{E}$ and set $\rho = \bar{\rho}(\omega)$; hence $\rho \leq_{be} \beta$. On $\mathcal{E}$, $\tilde{f}$ is polyconvex, hence rank 1 convex, and hence BE monotonic. Consequently, $\tilde{f}(\rho) \leq \tilde{f}(\beta)$ and (75) implies (76). A referencc to Theorem 2.10 completes the proof of the polyconvexity of g and hence also of (69). To prove (72), it suffices to note that if f satisfies the hypotheses of the preceding part of the Theorem then so also does f^q and the function $\tilde{h}$ corresponding to $\tilde{f}$ can be chosen to the q-th power of the function $\tilde{h}$ corresponding to f. □

3.3 Relaxation of Distances from Isotropic Wells in Dimension 2

This sections uses Theorem 3.8 to calculate relaxations of some model energy functions that attain minima on the set

$$W := \{\mathbf{A} \in \mathbb{M}^{2\times 2} : \tau(\mathbf{A}) = \gamma\} \tag{77}$$

where $\gamma \in \mathbb{G}^2$ is fixed. Recall that the distance from W is represented by $\tilde{f}(\alpha) = |\alpha - \gamma|$, see Exercise 2.5(ii).

Example 3.9. *Let $f : \mathbb{M}^{2\times 2} \to \mathbb{R}$ be the distance from the set W defined by (77) where $\gamma \in \mathbb{G}^2$. Then for any $p \geq 2$, $(f^p)^{rc} = (f^p)^{qc} = (f^p)^{pc} = g^p$ where $g : \mathbb{M}^{2\times 2} \to \mathbb{R}$ is an invariant function represented by*

$$\tilde{g}(\alpha) = \begin{cases} |\alpha - \gamma| & \text{if} \quad \alpha \in \mathcal{E} \\ \tilde{h}(\alpha_1\alpha_2) & \text{if} \quad \alpha \in \mathbb{G}^2 \setminus \mathcal{E}, \end{cases}$$

where

$$\mathcal{E} = \{\alpha \in \mathbb{G}^2 : \alpha_1^2 - \alpha_2^2 - \gamma_1\alpha_1 + \gamma_2\alpha_2 > 0\}, \tag{78}$$

and $\tilde{h} : \mathbb{R} \to \mathbb{R}$ is a convex function determined by

$$\tilde{h}(\alpha_1\alpha_2) = |\alpha - \gamma| \quad \text{if} \quad \alpha \in \partial\mathcal{E}. \tag{79}$$

The region $\mathcal{E}$ is shown in Figure 7; if $\gamma_1 > \gamma_2$, its boundary $\partial\mathcal{E}$ is a hyperbola in $\mathbb{G}^2$ with asymptotes $\alpha_1 + \alpha_2 = \frac{1}{2}(\gamma_1 + \gamma_2), \alpha_1 - \alpha_2 = \frac{1}{2}(\gamma_1 - \gamma_2)$. The asymptotes intersect at $\frac{1}{2}\gamma$. To see it, note that introducing the variables $\alpha_\pm := \alpha_1 \pm \alpha_2$, the condition $\alpha \in \mathbb{G}^2$ reads $\alpha_+ \geq 0, \alpha_- \geq 0$ and

$$\mathcal{E} := \{\alpha \in \mathbb{G}^2 : (\alpha_+ - \tfrac{1}{2}\gamma_+)(\alpha_- - \tfrac{1}{2}\gamma_-) \geq \tfrac{1}{4}\gamma_+\gamma_-\}. \tag{80}$$

For $\gamma_1 = \gamma_2$ the hyperbola collapses to its asymptotes. $\partial\mathcal{E}$ passes through γ and intersects the α_1 axis at γ_1.

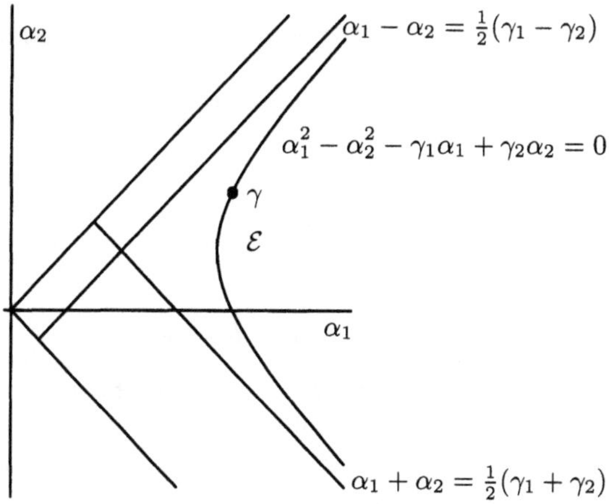

Figure 7. The region $\mathcal{E}$ in (78)

Remarks 3.10.

• The function $\tilde{h}$ is determined only implicitly by (79). To evaluate $\tilde{h}(\delta)$ for a given $\delta \in \mathbb{R}$, one has to find the point α of intersection of $\partial\mathcal{E}$ with the hyperbola $\mathcal{H}_\delta = \{\alpha \in \mathbb{G}^2 : \alpha_1\alpha_2 = \delta\}$; this provides a quartic equation

$$\alpha_1^4 - \gamma_1\alpha_1^3 + \gamma_2\delta\alpha_1 - \delta^2 = 0.$$

This equation can be solved explicitly; however, the formulas are too complicated to be recorded here. If $\gamma_1 = \gamma_2$ then $\tilde{h}$ is explicit, see Exercise 3.13.

• Theorem 3.8 does not enable one to calculate $(f^p)^{rc}$ and $(f^p)^{qc}$ if $1 \le p < 2$; however, it can be shown that

$$(f^p)^{qc} \ne (f^p)^{pc} = (f^p)^c$$

where

$$(\tilde{f^p})^c(\alpha) = \frac{1}{2^{p/2}}\left([\alpha_+ - \gamma_+]_+^2 + [\alpha_- - \gamma_-]_+^2\right)^{p/2}, \tag{81}$$

$\alpha \in \mathbb{G}^2$, where $[x]_+ = \max\{x, 0\}$ for each $x \in \mathbb{R}$. In particular, $(f^p)^{qc}$ is not polyconvex. Note also that (81) gives the convex hull of f^p for all $p \ge 1$.

Proof of Example 3.9 Let $F : \mathbb{M}^{2\times 2} \to \mathbb{R}$ be the square of distance from W, represented by $\tilde{F}(\alpha) = |\alpha - \gamma|^2$. Then $f^p = F^{p/2} = F^q$ and since $p = 2q \ge 2$, it suffices to prove that F satisfies the hypothesis of Theorem 3.8 with $\mathcal{E}$ given by (78) and with $\tilde{h} = \tilde{H}$ given by

$$\tilde{H}(\alpha_1\alpha_2) = |\alpha - \gamma|^2 \quad \text{if} \quad \alpha \in \partial\mathcal{E}. \tag{82}$$

It is found that $\mathcal{E}$ is an absorbing set. Prove that if $\mathbf{A} \in \mathbb{M}^{2\times 2}$ has the signed singular values α in $\mathcal{E}$ then $\mathbf{A}$ is a point of polyconvexity of F. Applying Theorem 2.10, we seek to prove the existence of $d \in \mathbb{G}^2, c \in \mathbb{R}$ such that

$$\tilde{F}(\beta) \ge \tilde{F}(\alpha) + d \cdot (\beta - \alpha) + c(\beta_1\beta_2 - \alpha_1\alpha_2) \tag{83}$$

for all $\beta \in \mathbb{G}^2$. Denoting by $s := D\tilde{F}(\alpha) = 2(\alpha - \gamma)$ the partial derivatives of $\tilde{F}$, using

$$\tilde{F}(\beta) - \tilde{F}(\alpha) = s \cdot (\beta - \alpha) + (\beta_1 - \alpha_1)^2 + (\beta_2 - \alpha_2)^2$$

and noting that for any $c \in \mathbb{R}$ with $|c| \leq 2$,

$$(\beta_1 - \alpha_1)^2 + (\beta_2 - \alpha_2)^2 \geq c(\beta_1 - \alpha_1)(\beta_2 - \alpha_2),$$

we obtain (83) with

$$d_1 = s_1 - c\alpha_2, \quad d_2 = s_2 - c\alpha_1. \tag{84}$$

Thus it remains to prove that for every $\alpha \in \mathcal{E}$ there exists a c with $|c| \leq 2$ such that the d from (84) satisfies $d \in \mathbb{G}^2$. We note that the condition $\alpha \in \mathbb{G}^2$ is equivalent to $\alpha_1 + \alpha_2 \geq 0, \alpha_1 - \alpha_2 \geq 0$. For the rest of the proof assume that $\alpha_1 + \alpha_2 > 0, \alpha_1 - \alpha_2 > 0$; the boundary cases are similar. One finds that $d \in \mathbb{G}^2$ if and only if

$$c_+ := \frac{s_1 + s_2}{\alpha_1 + \alpha_2} \geq c \geq -\frac{s_1 - s_2}{\alpha_1 - \alpha_2} := c_-. \tag{85}$$

A necessary and sufficient condition for the existence of a $c \in [-2, 2]$ satisfying (85) is

$$c_+ \geq c_-, \quad c_+ \geq -2, \quad c_- \leq 2. \tag{86}$$

One finds that $c_\pm = \pm 2(\alpha_\pm - \gamma_\pm)/\alpha_\pm$ where $\alpha_\pm = \alpha_1 \pm \alpha_2, \gamma_\pm = \gamma_1 \pm \gamma_2$. Inequality $(86)_1$ is then found to be equivalent to (80), i.e., to $\alpha \in \mathcal{E}$, which we assume, and $(86)_{2,3}$ are equivalent to $\alpha_\pm \geq \frac{1}{2}\gamma_\pm$ which follow from $\alpha \in \mathcal{E}$. In summary, if $\alpha \in \mathcal{E}$ then there exist $d \in \mathbb{G}^2$ and $c \in \mathbb{R}$ such that (83) holds for all $\beta \in \mathbb{G}^2$. Thus $\mathbf{A}$ is a point of polyconvexity of F. Finally prove that each $\alpha \in \partial\mathcal{E} \setminus \partial\mathbb{G}^2$ is a polyconvex liquefaction point. For this it suffices to note that $d = 0$ whenever $\alpha \in \partial\mathcal{E}$. Thus the hypotheses of Theorem 3.8 are satisfied; moreover (71) implies (82). □

The function f of the following example is similar to that of Example 3.9 in that it vanishes exactly on W. Also the relaxation is similar, but the liquid–like response can be calculated explicitly. We assume that $\gamma_1\gamma_2 = 1$; the general case is obtained by rescaling.

Example 3.11. *Let $p \geq 2$ and let $f : \mathbb{M}^{2\times 2} \to \mathbb{R}$ be given by*

$$\tilde{f}(\alpha) = \sqrt{(\bar{\alpha}_1 - 1)^2 + (\bar{\alpha}_2 - 1)^2}, \quad \alpha \in \mathbb{G}^2$$

where $\gamma \in \mathbb{G}^2_{++}$ satisfies $\gamma_1\gamma_2 = 1$, and $\bar{\alpha} := (\alpha_1/\gamma_1, \alpha_2/\gamma_2)$. Then $(f^p)^{rc} = (f^p)^{qc} = (f^p)^{pc} = g^p$ where g is an invariant function represented by

$$\tilde{g}(\alpha) = \begin{cases} \tilde{f}(\alpha) & \text{if } \ \bar{\alpha}_1 \geq \bar{\alpha}_2,\, \bar{\alpha}_1 + \bar{\alpha}_2 \geq 1, \\ \tilde{h}(\alpha_1\alpha_2) & \text{otherwise,} \end{cases}$$

$\alpha \in \mathbb{G}^2$, where

$$\tilde{h}(\delta) = \begin{cases} \sqrt{1 - 2\delta} & \text{if } \ \delta \leq 1/4, \\ \sqrt{2}|\sqrt{\delta} - 1| & \text{if } \ \delta > 1/4. \end{cases} \tag{87}$$

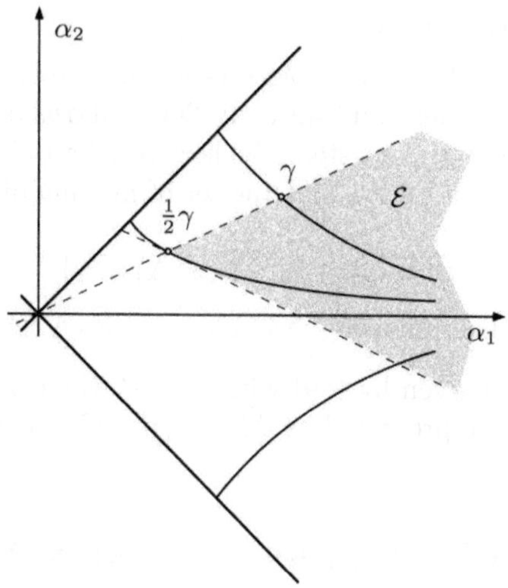

Figure 8. The the region $\mathcal{E}$ in (88)

If

$$\mathcal{E} := \{\alpha \in \mathbb{G}^2 : \bar{\alpha}_1 \geq \bar{\alpha}_2, \bar{\alpha}_1 + \bar{\alpha}_2 \geq 1\}, \tag{88}$$

then the effective response coincides with the original one on $\mathcal{E}$ while outside $\mathcal{E}$ the effective response is fluid-like. See Figure 8.

Exercise 3.12. Prove 3.11.

Exercise 3.13.[27] Prove: *Let $f : \mathbb{M}^{2\times 2} \to \mathbb{R}$ be the distance from the orthogonal group, represented by $\tilde{f}(\alpha) = \sqrt{(\alpha_1 - 1)^2 + (\alpha_2 - 1)^2}$, $\alpha \in \mathbb{G}^2$. If $p \geq 2$, then $(f^p)^{rc} = (f^p)^{qc} = (f^p)^{pc} = \tilde{g}^p$ where*

$$\tilde{g}(\alpha) = \begin{cases} (1 - 2\alpha_1\alpha_2)^{1/2} & \text{if} \quad \alpha_1 + \alpha_2 \leq 1, \\ \tilde{f}(\alpha) & \text{otherwise.} \end{cases}$$

This is a special case of Example 3.9 or 3.11 with $\gamma_1 = \gamma_2 = 1$. The response is fluid-like in the complement of

$$\mathcal{E} = \{\alpha \in \mathbb{G}^2 : \alpha_1 + \alpha_2 \geq 1\}.$$

The convexification of f^p calculated in (81) reduces to

$$(\tilde{f}^p)^c(\alpha) = \begin{cases} \dfrac{1}{2^{p/2}}(\alpha_1 - \alpha_2)^p & \text{if} \quad \alpha_1 + \alpha_2 \leq 2, \\ \tilde{f}^p(\alpha) & \text{otherwise,} \end{cases}$$

for any $1 \leq p < \infty, \alpha \in \mathbb{G}^2$.

[27]See Šilhavý (2001) for $p = 2$.

3.4 Relaxation of Isotropic Sets

No general results on hulls of isotropic sets are available in dimension $n > 2$; for $n = 2$ the result by Cardaliaguet and Tahraoui (2002) allows one to determine the hulls $M^{rc} = M^{qc} = M^{pc}$ in essentially one step. We here restrict only to some well–motivated special cases. Recall from Section 1.2 that the set of minima of nematic elastomers is of the form

$$M = \{\mathbf{A} \in \mathbb{M}^{n\times n} : \tau(\mathbf{A}) = \gamma\} \tag{89}$$

where γ is given by (20). The following example provides a relaxation of this set.

Example 3.14.[28] If M is given by (89) with $\gamma \in \mathbb{G}^n$ then $M^{lc} = M^{rc} = M^{qc} = M^{pc}$ and these identical sets are represented by $\tilde{M}^{lc} = \{\alpha \in \mathbb{G}^n : \alpha \le_{be} \gamma\}$.

Example 3.15.[29] If

$$M = \{\mathbf{A} \in \mathbb{M}^{n\times n} : (\tau_1(\mathbf{A}), \ldots, \tau_{n-1}(\mathbf{A}), |\tau_n(\mathbf{A})|) = \gamma\}$$

where $\gamma \in \mathbb{G}^n_+$ then $M^{lc} = M^{rc} = M^{qc} = M^{pc}$ and these identical sets are represented by $\tilde{M}^{lc} = \{\alpha \in \mathbb{G}^n : \alpha \le_{|be|} \gamma\}$. In particular, $O(n)^{rc} = O(n)^c = \{\mathbf{A} \in \mathbb{M}^{n\times n} : \tau_1(\mathbf{A}) \le 1\}$. The set M is fully invariant.

4 Nematic Elastomers

Nematic elastomers are polymeric elastic solids with strong orientational properties. Their energy functions, derived from statistical considerations, is nonelliptic. As it is also isotropic, nematic elastomers provide an ideal topic to illustrate the above general theory. This chapter first briefly reviews the molecular composition, phenomenology, and constitutive equations of nematic elastomers. Some of the interesting instabilities, such as stripped domain patterns and a soft liquid behavior are described, emphasizing their roots in the nonellipticity. Then a relaxation of a large class of energy functions is presented which includes those of nematic elastomers. Finally a phase diagram coming from the relaxation is described and its relation to the behavior of nematic elastomers is explained.

4.1 The Energy Function

Nematic elastomers are rubber-like materials consisting of cross-linked liquid crystal polymeric molecules. They are solids, as opposed to liquid crystal fluids. The first synthesis was performed by Finkelmann, Kock and Rehage (1981) by the cross-linking of a polymeric fluid that includes liquid crystalline molecules as elements of its main-chain or as pendant side-groups. The reader is referred to the review paper Warner and Terentjev (1996) for more details.

Liquid crystal fluids can be imagined as a continuous distribution of inextensible rods characterized by the director field **d**. Since nematic elastomers are solids, their description requires the deformation gradient **F** in addition to the director **d**. The constitutive

[28] Šilhavý (2002b).

[29] Dacorogna and Tanteri (1996).

equation of Bladon, Terentjev and Warner (1993) reads

$$g(\mathbf{F},\mathbf{d}) = \frac{1}{2}\mu r^{1/3}\big(\mathrm{tr}(\mathbf{F}\mathbf{F}^T) - \frac{r-1}{r}|\mathbf{F}^T\mathbf{d}|^2\big), \quad \det\mathbf{F} = 1, |\mathbf{d}| = 1, \tag{90}$$

where $\mu > 0$ is the shear modulus and r the step length anisotropy parameter, describing the average shape of polymeric molecular coils. We assume $r > 1$, although the case $r < 1$ is also encountered and admits a parallel treatment.[30] As explained in Section 1.2, upon a minimization of g with respect to $\mathbf{d}$ at a fixed $\mathbf{F}$ one obtains an isotropic energy function

$$f(\mathbf{F}) = \frac{1}{2}\mu r^{1/3}(\alpha_1^2/r + \alpha_2^2 + \alpha_3^2), \tag{91}$$

where $\alpha_1 \geq \alpha_2 \geq \alpha_3, \alpha_1\alpha_2\alpha_3 = 1$, are the singular values of $\mathbf{F}$. To incorporate the incompressibility constraint into the energy function, we put the energy equal to ∞ outside the shell $\alpha_1\alpha_2\alpha_3 = 1$: in this way we are led to the isotropic energy function on $\mathbb{M}^{3\times3}$ represented by

$$\tilde{f}(\alpha) = \begin{cases} r^{1/3}(\alpha_1^2/r + \alpha_2^2 + \alpha_3^2) & \text{if} \quad \alpha_1\alpha_2\alpha_3 = 1, \\ \infty & \text{otherwise}, \end{cases} \tag{92}$$

$\alpha \in \mathbb{G}^3$, where we have also removed unessential factors.

Exercise 4.1.
(i) Show that if $r \neq 1$, then the invariant function f on $\mathbb{M}^{3\times3}$ represented by the right–hand side of (91) for all $\alpha \in \mathbb{G}^3$, is not differentiable. Hint: If f is differentiable, then $\tilde{f} : \mathbb{R}^n \to \mathbb{R}$ is differentiable and symmetric, since $\tilde{f}(\alpha) = f(\mathrm{diag}(\alpha))$. However, the right–hand side of (91) cannot be extended to a differentiable *symmetric* function of α.
(ii) Show that the invariant function f on $\mathbb{M}^{3\times3}$ represented by (92) is not BE–monotonic and hence not rank 1 convex. Hint: show that if $\gamma := (r^{1/3}, r^{-1/6}, r^{-1/6})$ then $\tilde{f}(\alpha) \geq \tilde{f}(\gamma)$ for all $\alpha \in \mathbb{G}^3$. On the other hand, the BE–monotonicity requires $\tilde{f}(\alpha) \leq \tilde{f}(\gamma)$ for all $\alpha \leq_{be} \gamma$ and in particular for $\alpha = (1, 1, 1)$.

4.2 Soft Deformation Paths

Associated with the nonellipticity of the energy (91) or (92) is the formation of striped domain patterns observed by Kundler and Finkelmann (1995) and explained by Warner and Terentjev (1996). This implies the ability of the nematic phase to behave like a liquid in a special class of deformations.

Consider a thin sheet parallel to the x_1, x_2 plane which is initially in the zero energy configuration described by the deformation gradient $\mathbf{F}^i$ with the maximal eigenvalue along the x_3 direction. This is also the orientation of the director. Suppose that the sheet is stretched along the x_1 direction so that the macroscopic deformation is

$$\mathbf{F}_t = \begin{bmatrix} \eta_1 & 0 & 0 \\ 0 & \eta_2 & 0 \\ 0 & 0 & \eta_3 \end{bmatrix}. \tag{93}$$

[30] The general theory of constitutive equations nematic elastomers was developed by Anderson, Carlson and Fried (1999), in parallel with the theory of nematic fluids by Ericksen and Leslie.

See Figure 9, where the actual shape of the sheet is replaced by blocks reflecting the eigenvalues of $\mathbf{F}_t$. Here η depends on t in such a way that η_2 is constant and equal to the smallest singular value $\eta_2 = r^{-1/6}$, while the above initial condition (that the maximum eigenvalue is in the x_3 direction) requires $\eta^i = (r^{-1/6}, r^{-1/6}, r^{1/3})$; assume that the process terminates as soon as $\mathbf{F}_t$ reaches the form (93) with $\eta = \eta^f = (r^{1/3}, r^{-1/6}, r^{-1/6})$, i.e., the maximum eigenvalue of $\mathbf{F}$ is in the x_1–direction. The uniform deformation $\mathbf{F}_t$ does not have the minimum energy, recall that the singular values of the minimizing deformation gradient are

$$\gamma := (r^{1/3}, r^{-1/6}, r^{-1/6}). \tag{94}$$

However, it has been observed that that the sheet does not undergo this homogeneous deformation path. Rather, the specimen creates a simple laminate with the normal direction $\mathbf{n} = \mathbf{e}_3$. This laminate is described by the deformation gradients

$$\mathbf{F}_t^\pm = \begin{bmatrix} \eta_1 & 0 & \pm\delta \\ 0 & \eta_2 & 0 \\ 0 & 0 & \eta_3 \end{bmatrix}$$

where $\delta \geq 0$ is determined from the condition that the singular values of $\mathbf{F}_t^\pm$ are $(r^{1/3}, r^{-1/6}, r^{-1/6})$. This gives

$$\delta^2 = r^{2/3} + r^{-1/3} - \eta_1^2 - \eta_3^2 = \frac{(\eta_1^2 - r^{-1/3})(r^{2/3} - \eta_1^2)}{\eta_1^2}.$$

The minimizing director $\mathbf{d}^\pm$ (the eigenvector of $\mathbf{F}_t^\pm \mathbf{F}_t^{\pm T}$ corresponding to the largest eigenvalue) is

$$\mathbf{d}^\pm = (\mp\eta_3\delta, 0, r^{-1/3} - \eta_3^2).$$

One finds that

$$\mathbf{F}_t^+ - \mathbf{F}_t^- = 2\delta \mathbf{e}_1 \otimes \mathbf{e}_3, \quad \mathbf{F}_t = \frac{1}{2}(\mathbf{F}_t^+ + \mathbf{F}_t^-),$$

and thus the simple laminate with relative weights $\frac{1}{2}$ and deformation gradients $\mathbf{F}_t^\pm$ provides the average deformation $\mathbf{F}_t$. By creating this laminate, the specimen reduces its energy to the (pointwise) absolute minimum value. Since the force required to produce the deformation is the derivative of the current constant minimum energy with respect to the elongation, the process requires negligible force, a feature shared with a similar process performed with an incompressible fluid.

Remark 4.2. If $\mathbf{G}_t^\pm := \mathbf{F}_t^{\pm T}$, then also

$$\mathbf{G}_t^+ - \mathbf{G}_t^- = 2\delta \mathbf{e}_3 \otimes \mathbf{e}_1, \quad \mathbf{F}_t = \frac{1}{2}(\mathbf{G}_t^+ + \mathbf{G}_t^-)$$

and $f(\mathbf{G}_t^\pm) = \tilde{f}(\gamma)$. Thus the laminate $\mathbf{G}_t^\pm$ provides another energy–minimizing state compatible with the macroscopic deformation of the specimen, now with the stripes normal to $\mathbf{e}_1$. These laminates are not observed, presumably since they are incompatible with the details of the arrangement of the experiment (clamps).

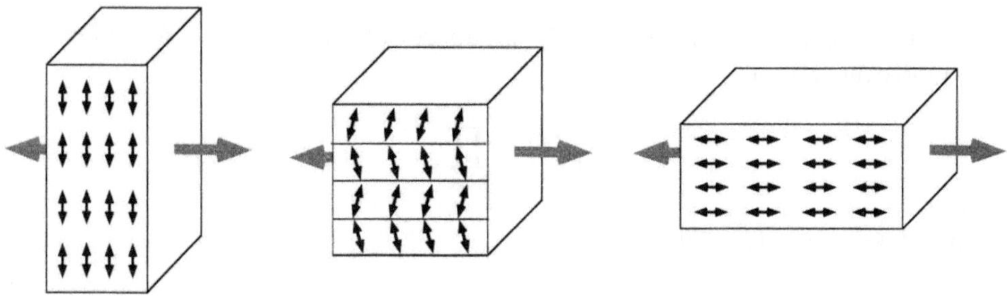

Figure 9. The experiment of Kundler & Finkelmann (1995)

Exercise 4.3.
(i) If $\omega(t) = \tau(\mathbf{F}_t)$ are the signed singular values of $\mathbf{F}_t$ during the experiment, show that $\omega(t) \leq_{be} \gamma$.
(ii) Combine (i) with Proposition 3.4 to show that

$$\tilde{f}^{rc}(\omega(t)) = \tilde{f}(\gamma). \tag{95}$$

Hint: $\tilde{f}^{rc}(\omega(t)) \leq \tilde{f}^{be}(\omega(t)) \leq \tilde{f}(\gamma)$. On the other hand, $\tilde{f}(\gamma)$ is the absolute minimum value of $\tilde{f}$. Thus by (95) there exists a sequence of microstructures compatible with the macroscopic deformation $\mathbf{F}_t$ which gives the minimum energy $\tilde{f}(\gamma)$. The above constructed $\mathbf{F}_t^{\pm}$ determines the form of this microstructure. Notice that the proof of BE–monotonicity of rank 1 convex functions in Section 2.4 constructs exactly the same sequence in the general context.

For a bifurcation analysis of the experiment within the framework of the director–dependent energy (90), see Fried and Korchagin (2002).

4.3 Relaxation: Arbitrary Dimension

This section describes the relaxation of a class of energy functions that includes (92) as a special case. Namely, let $n \geq 2$ be arbitrary and consider an energy function f on $\mathbb{M}_+^{n\times n}$ with a representation of the form

$$\tilde{f}(\alpha) = U\Big(\frac{\alpha_1^p}{\gamma_1^p}, \dots, \frac{\alpha_n^p}{\gamma_n^p}\Big) + h(\alpha_1 \cdots \alpha_n), \quad \alpha \in \mathbb{G}_{++}^n, \tag{96}$$

where $h : \mathbb{R}_{++} \to \mathbb{R} \cup \{\infty\}$ is convex, $U : \mathbb{R}_{++}^n \to \mathbb{R}$ is a strictly increasing symmetric convex function of class C^1, $\gamma \in \mathbb{G}_{++}^n, \gamma_1 \cdots \gamma_n = 1$, and $p \geq n$ for compressible materials and $p \geq n-1$ for incompressible materials. The energy (92) is obtained by the following identifications: γ is given by (94), $p = 2$, $U(\xi) = \xi_1 + \xi_2 + \xi_3$, and

$$h(\delta) = \begin{cases} 0 & \text{if } \delta = 1, \\ \infty & \text{otherwise.} \end{cases} \tag{97}$$

Assume that γ is fixed. For any $\alpha \in \mathbb{R}^n$ we abbreviate $\bar{\alpha} := (\bar{\alpha}_1, \ldots, \bar{\alpha}_n)$ where $\bar{\alpha}_i := \alpha_i/\gamma_i$; moreover, we set

$$\pi_i(\alpha) = \prod_{k=1}^{i} \alpha_k$$

for any $\alpha \in \mathbb{G}^n_{++}, i = 1, \ldots, n$.

Exercise 4.4.
(i) Prove that for every $\delta > 0$, the restriction of the energy function (96) to the set $\mathcal{D}_\delta := \{\alpha \in \mathbb{G}^n_{++} : \alpha_1 \ldots \alpha_n = \delta\}$ has a minimum at $\alpha = \delta^{1/n}\gamma$;
(ii) Prove that if h is of class C^1 then $\tilde{f}$ satisfies the Baker–Ericksen inequalities at $\alpha \in \mathbb{G}^n_{++}$ if and only if $\alpha \in \mathcal{E} := \{\alpha \in \mathbb{G}^n_{++} : \bar{\alpha}_i \geq \bar{\alpha}_{i+1}, \quad i = 1, \ldots, n-1\}$.

Hint: the function U satisfies the ordered-forces inequalities (cf.(47))

$$(U_i(\xi) - U_j(\xi))(\xi_i - \xi_j) \geq 0, \quad 1 \leq i < j \leq n.$$

Proposition 4.5. *For each $\alpha \in \mathbb{G}^n_{++}$ there exists a unique $\beta := \alpha^* \in \mathcal{E}$ such that $\alpha \leq_{be} \beta$, and*

$$(\pi_i(\alpha) - \pi_i(\beta))(\bar{\beta}_i - \bar{\beta}_{i+1}) = 0, \quad i = 1, \ldots, n-1.$$

If $\alpha \in \mathcal{E}$ then $\alpha^* = \alpha$.

The point α^* is to be interpreted as a nonlinear projection of α onto the region $\mathcal{E}$, where the energy satisfies the Baker–Ericksen inequalities.

Theorem 4.6.[31] *Let f be an invariant function on $\mathbb{M}^{n\times n}_+$ with the representation of the form* (96) *where $h : \mathbb{R}_{++} \to \mathbb{R} \cup \{\infty\}$ is convex and $U : \mathbb{R}^n_{++} \to \mathbb{R}$ is a symmetric convex function of class C^1 which is strictly increasing in the sense that*

$$U_i > 0 \quad \text{on} \quad \mathbb{R}^n_{++}, \quad i = 1, \ldots, n.$$

If $p \geq n$ then

$$\tilde{f}^{be}(\alpha) = \tilde{f}^{rc}(\alpha) = \tilde{f}^{qc}(\alpha) = \tilde{f}^{pc}(\alpha) = \tilde{f}(\alpha^*) \tag{98}$$

for each $\alpha \in \mathbb{G}^n_{++}$. In the incompressible case, i.e., if h is of the form (97)*, the conclusion holds under the weaker condition $p \geq n - 1$.*

Thus the relaxation in this class of energies reduces to the BE–monotonization and the value of the monotonization is the value of the original energy at the projected point $\alpha^* \in \mathcal{E}$.

[31] Šilhavý (2001).

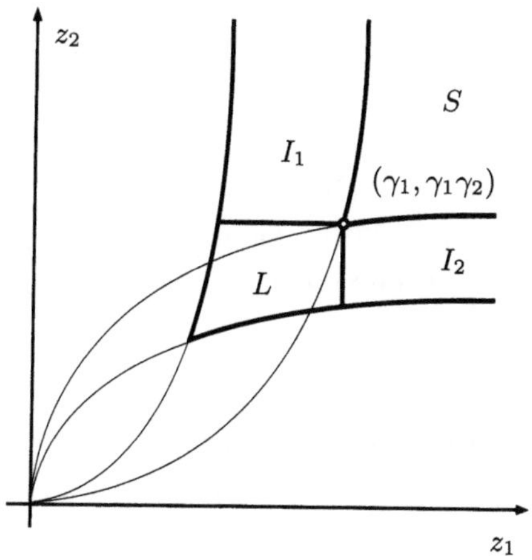

Figure 10. The phase diagram of (99)

4.4 Relaxation in Dimension 3 and the Phase Diagram

By determining the projection $\alpha^* \in \mathcal{E}$ as in (98) we obtain explicit formulas of the relaxation of f in dimension $n = 3$. Only the incompressible case is considered, with the minimum possible value of the exponent $p = 2$. Thus the energy is of the form

$$\tilde{f}(\alpha) = \begin{cases} S(\bar{\alpha}_1^2, \bar{\alpha}_2^2, \bar{\alpha}_3^2) & \text{if} \quad \alpha_1\alpha_2\alpha_3 = 1, \\ \infty & \text{otherwise,} \end{cases}$$

with some convex, symmetric and strictly increasing function $S : \mathbb{R}^3_{++} \to \mathbb{R}$. Then for α with $\alpha_1\alpha_2\alpha_3 = 1$ we have

$$\tilde{f}^{be}(\alpha) = \begin{cases} \tilde{f}(\gamma) & \text{if} \quad \bar{\alpha}_1 \le 1, \quad \bar{\alpha}_1\bar{\alpha}_2 \le 1, \\ S(1/\bar{\alpha}_3, 1/\bar{\alpha}_3, \bar{\alpha}_3^2) & \text{if} \quad \bar{\alpha}_1\bar{\alpha}_2 \ge 1, \ \bar{\alpha}_2 \ge \bar{\alpha}_1, \\ S(\bar{\alpha}_1^2, 1/\bar{\alpha}_1, 1/\bar{\alpha}_1) & \text{if} \quad \bar{\alpha}_1 \ge 1, \quad \bar{\alpha}_2 \le \bar{\alpha}_3, \\ \tilde{f}(\alpha) & \text{if} \quad \bar{\alpha}_1 \ge \bar{\alpha}_2 \ge \bar{\alpha}_3, \end{cases} \tag{99}$$

and

$$\tilde{f}^{be}(\alpha) = \infty \qquad \text{otherwise.}$$

The relaxed energy (99) displays a full fluid–like effective behavior in the regime $(99)_1$ and two different partial fluid–like responses in $(99)_{2,3}$.

To make the result more explicit, we use the phase diagram Figure 10,[32] which shows the three regimes in (99). The diagram depicts states $\alpha \in \mathbb{G}^3_{++}$ with $\alpha_1\alpha_2\alpha_3 = 1$ under the change of variables

$$z_1 = \alpha_1, \quad z_2 = \alpha_1\alpha_2$$

[32] This diagram was established in DeSimone and Dolzmann (2002) for a special class of energy functions.

with the inverses

$$\alpha_1 = z_1, \quad \alpha_2 = z_2/z_1, \quad \alpha_3 = 1/z_2.$$

The condition $\alpha \in \mathbb{G}^3_{++}$ reads

$$\sqrt{z_1} \le z_2 \le z_1^2, \tag{100}$$

see Figure 10, where the region (100) is delimited by the outer bold parabolas. The boundary of the region $\mathcal{E}$ is determined by equations $\bar{\alpha}_1 = \bar{\alpha}_2$ or $\bar{\alpha}_2 = \bar{\alpha}_3$. The equation $\bar{\alpha}_1 = \bar{\alpha}_2$ reads

$$z_2 = \frac{\gamma_2}{\gamma_1} z_1^2,$$

which is the right convex parabola, and the equation $\bar{\alpha}_2 = \bar{\alpha}_3$ reads

$$z_2 = \sqrt{\gamma_2/\gamma_3}\sqrt{z_1},$$

which is the upper concave parabola. The two parabolas intersect at the point $(\gamma_1, \gamma_1\gamma_2)$ which is the image of γ. The regions L, I_1, I_2, S correspond to the regimes $(99)_{1,2,3,4}$, respectively. In in the liquid region L, the relaxed energy has the value of the energy at γ. The region L is the image of the relaxation of the set

$$\{\mathbf{A} \in \mathbb{M}^{3\times 3} : \tau(\mathbf{A}) = \gamma\},$$

cf. Example 3.14. The state of minimum energy $\tilde{f}(\gamma)$ can be approximately achieved also at all macroscopic deformation gradients from L by creating (sequential) laminates with deformation gradients of singular values γ. In particular, $\mathbf{A} = \mathbf{1}$ belongs to L and thus the unstable homogeneous deformation Ω from Figure 3(b) is in reality replaced by a deformation that is homogeneous only macroscopically but actually has the described microstructure. For γ of the form (94) the region I_2 vanishes. The soft deformation path of Section 4.2 is realized in the boundary between I_1 and L : in that path, the smallest singular value of $\mathbf{F}_t$ remains constant and thus the effective energy is constant (and of minimum value for this particular path) by $(99)_2$.

Bibliography

E. Acerbi and N. Fusco. Semicontinuity problems in the calculus of variations. *Arch. Rational Mech. Anal.,* 83:125–145, 1984.

J.-J. Alibert and B. Dacorogna. An example of a quasiconvex function that is not polyconvex in two dimensions. *Arch. Rational Mech. Anal.,* 117:155–166, 1992.

D. R. Anderson, D. E. Carlson and E. Fried. A continuum mechanical theory for nematic elastomers. *J. Elasticity,* 56:33–58, 1999.

G. Aubert. On a counterexample of a rank 1 convex function which is not polyconvex in the case $N = 2$. *Proc. R. Soc. Edinburgh,* 106A:237–240, 1987.

G. Aubert. Necessary and sufficient conditions for isotropic rank-one convex functions in dimension 2. *J. Elasticity,* 39:31–46, 1995.

G. Aubert and R. Tahraoui. Sur la faible fermeture de certains ensembles de contrainte en élasticité non-linéaire plane. *C. R. Acad. Sci. Paris,* 290:537–540, 1980.

G. Aubert and R. Tahraoui. Sur la faible fermeture de certains ensembles de contrainte en élasticité non-lineaire plane. *Arch. Rational Mech. Anal.,* 97:33–58, 1987.

J. M. Ball. Convexity conditions and existence theorems in nonlinear elasticity. *Arch. Rational Mech. Anal.,* 63:337–403, 1977.

J. M. Ball and R. D. James. Fine phase mixtures as minimizers of energy. *Arch. Rational Mech. Anal.,* 100:13–52, 1987.

J. M. Ball and R. D. James. Proposed experimental tests of a theory of fine microstructure and the two-well problem. *Phil. Trans. R. Soc. Lond.,* 338:389–450, 1992.

P. Bladon, E. M. Terentjev and M. Warner. Transitions and instabilities in liquid–crystal elastomers. *Phys. Rev.,* E 47:R3838–R3840, 1993.

G. Buttazzo, B. Dacorogna and W. Gangbo. On the envelopes of functions depending on singular values of matrices. *Bolletino U. M. I.,* 7:17–35, 1994.

P. Cardaliaguet and R. Tahraoui. Equivalence between rank-one convexity and polyconvexity for isotropic sets of $\mathbb{R}^{2\times 2}$ (Parts I, II). *Nonlinear. Anal.,* 50:1179–1200, 1201–1239, 2002.

B. D. Coleman and W. Noll. On the thermostatics of continuous media. *Arch. Rational Mech. Anal.,* 4:97–128, 1959.

B. Dacorogna. Quasiconvexity and relaxation of nonconvex problems in the calculus of variations. *J. Functional. Anal.,* 46:102–118, 1982.

B. Dacorogna. *Direct methods in the calculus of variations.* Springer, Berlin 1989.

B. Dacorogna. Necessary and sufficient conditions for strong ellipticity of isotropic functions in any dimension. *Discrete Contin. Dyn. Syst.,* B1:257–263, 2001.

B. Dacorogna and H. Koshigoe. On the different notions of convexity for rotationally invariant functions. *Ann. Fac. Sci. Toulouse,* II:163–184, 1993.

B. Dacorogna and P. Marcellini. A counterexample in the vectorial calculus of variations. In J. M. Ball, editor, *Material Instabilities in Continuum Mechanics,* pages 77–83. Clarendon Press, Oxford 1987.

B. Dacorogna and P. Marcellini. *Implicit partial differential equations.* Birkhäuser, Basel 1999.

B. Dacorogna and C. Tanteri. On the different convex hulls of sets involving singular values. *Proc. R. Soc. Edinburgh,* 128A:1261–1280, 1996.

A. DeSimone and G. Dolzmann. Macroscopic response of nematic elastomers via relaxation of a class of $SO(3)$–invariant energies. *Arch. Rational Mech. Anal.,* 161:181–204, 2002.

H. Finkelmann, H. J. Kock and G. Rehage. Liquid crystalline elastomers—a new type of liquid crystalline material. *Makromolekulare Chemie, Rapid Communications,* 2:317–322, 1981.

I. Fonseca. The lower quasiconvex envelope of the stored energy functions for an elastic crystal. *J. Math. pures et appl.,* 67:175–195, 1988.

A. B. Freidin, E. N. Vilchevskaya and L. L. Sharipova. Two–phase deformations within the framework of phase transition zones. *Theor. Appl. Mech.,* 28–29:145–167, 2002.

E. Fried and V. Korchagin. Striping of nematic elastomers. *Int. J. Solids Structures,* 2002. In the press.

R. Hill. Constitutive inequalities for simple materials. *J. Mech. Phys. Solids,* 16:229–242, 1968.

J. K. Knowles and E. Sternberg. On the failure of ellipticity of the equations for finite elastostatic plane strain. *Arch. Rational Mech. Anal.,* 63:321–326, 1977.

R. V. Kohn. The relaxation of a double-well energy. *Continuum Mech. Thermodyn.,* 3:193–236, 1991.

R. V. Kohn and G. Strang. Optimal design and relaxation of variational problems, I, II, III. *Comm. Pure Appl. Math.,* 39:113–137, 139–182, 353–377, 1986.

I. Kundler and H. Finkelmann. Strain-induced director reorientation in nematic liquid single crystal elastomers. *Makromolekulare Chemie, Rapid Communications,* 16:853, 1995.

H. Le Dret and A. Raoult. Enveloppe quasi-convexe de la densité d'énergie de Saint Venant–Kirchhoff. *C. R. Acad. Sci. Paris,* 318:93–98, 1994.

H. Le Dret and A. Raoult. The quasiconvex envelope of the Saint Venant–Kirchhoff energy function. *Proc. R. Soc. Edinburgh,* 125A:1179–1192, 1995.

K. A. Lurie and A. V. Cherkaev. On a certain variational problem of phase equilibrium. In J. M. Ball, editor, *Material Instabilities in Continuum Mechanics,* pages 257–268. Clarendon Press, Oxford 1987.

A. W. Marshall and I. Olkin. *Inequalities: theory of majorization and its applications.* Academic Press, New York 1979.

A. Mielke. Necessary and sufficient conditions for polyconvexity of isotropic functions. 2002. Preprint, University of Stuttgart.

C. B. Morrey, Jr. *Multiple integrals in the calculus of variations.* Springer, New York 1966.

R. W. Ogden. *Non-linear elastic deformations.* Ellis Horwood Ltd., Chichester 1984.

P. Pedregal. *Parametrized Measures and Variational Principles.* Birkhäuser, Basel Boston Berlin 1997.

P. Pedregal. *Variational Methods in Nonlinear Elasticity.* SIAM, Philadelphia 2000.

A. C. Pipkin. Elastic materials with two preferred states. *Quart. J. Mech. Appl. Math.,* 44:1–15, 1991.

P. Rosakis. Ellipticity and deformations with discontinuous gradients in finite elastostatics. *Arch. Rational Mech. Anal.,* 109:1–37, 1990.

P. Rosakis. Characterization of convex isotropic functions. *J. Elasticity,* 49:257–267, 1997.

T. Roubíček. *Relaxation in optimization theory and variational calculus.* W. de Gruyter, Berlin 1997.

M. Šilhavý. *The mechanics and thermodynamics of continuous media.* Springer, Berlin 1997.

M. Šilhavý. Convexity conditions for rotationally invariant functions in two dimensions. In A. Sequeira, H. Beirao da Veiga & J. Videman, editor, *Applied nonlinear functional analysis,* pages 513–530. Kluwer Press, New York 1999.

M. Šilhavý. On isotropic rank 1 convex functions. *Proc. R. Soc. Edinburgh,* 129A:1081–1105, 1999.

M. Šilhavý. Rank 1 Convex hulls of isotropic functions in dimension 2 by 2. *Mathematica Bohemica,* 126:521–529, 2001.

M. Šilhavý. Relaxation of a class of $SO(n)$ invariant energies related to nematic elastomers. 2001. Preprint, 2001.

M. Šilhavý. An O(n) invariant rank 1 convex function that is not polyconvex. *Theor. Appl. Mech.,* 28–29:325–336, 2002.

M. Šilhavý. Monotonicity of rotationally invariant convex and rank 1 convex functions. *Proc. R. Soc. Edinburgh,* 132A:419–435, 2002.

M. Šilhavý. An O(n) invariant quasiconvex function that is not polyconvex. 2002. Preprint.

M. Šilhavý. Rank 1 convex hulls of rotationally invariant functions. *Solid Mechanics and Its Applications,* Kluwer, 2002. In the press.

M. Šilhavý. On SO(n)-Invariant Rank 1 Convex Functions. *J. Elasticity,* 2003. In the press.

H. C. Simpson and S. Spector. On copositive matrices and strong ellipticity for isotropic elastic materials. *Arch. Rational Mech. Anal.,* 84:55–68, 1983.

V. Šverák. Rank-one convexity does not imply quasiconvexity. *Proc. R. Soc. Edinburgh,* 120:185–189, 1992.

C., Truesdell and W. Noll. The non-linear field theories of mechanics. In S. Fluegge, editor, *Handbuch der Physik III/3,* Springer, Berlin 1965.

M. Warner and E. M. Terentjev. Nematic elastomers—a new state of matter? *Prog. Polym. Sci.,* 21:853–891, 1996.

M. Warner and E. M. Terentjev. The physics of nematic elastomers. *Prog. Polym. Sci.,* 21:853, 1996.

Variational problems of crack equilibrium and crack propagation

Khanh Chau Le

Lehrstuhl für Allgemeine Mechanik, Ruhr-Universität Bochum
Universitätsstr. 150, D-44780 Bochum, Germany

1 Safe equilibrium

Let us start with the well-known principle of minimum energy in linear elastostatics. For simplicity we restrict ourselves to the 2-D plane strain problems by considering a homogeneous elastic body of cylindrical shape, whose generator is directed along the x_3-axis. Let the cross section of the body occupy a 2-D regular open region $\mathcal{B}$ of the (x_1, x_2)-plane, and assume that all quantities we are looking for depend only on x_1 and x_2. The boundary of $\mathcal{B}$, $\partial\mathcal{B}$, is decomposed into two curves ∂_w and ∂_τ such that

$$\partial\mathcal{B} = \overline{\partial_w \cup \partial_\tau}, \quad \partial_w \cap \partial_\tau = \emptyset.$$

The part ∂_w of the boundary is clamped, so the displacement is zero there, while on the remaining part ∂_τ the external traction τ is prescribed (see Fig. 1).

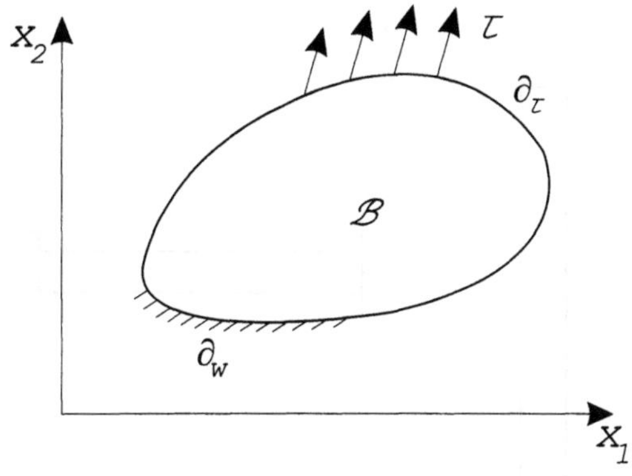

Figure 1. Cross section of a loaded body of cylindrical shape.

Under the action of external traction τ the elastic body will undergo the displacement $\check{w}(x)$ from the undeformed configuration to the equilibrium one. If cracks or cavities do not occur, we expect that $\check{w}(x)$ belongs to the Sobolev space $H(\mathcal{B})$ of vector-valued functions defined on $\mathcal{B}$ which are square integrable and have square integrable first order derivatives. The latter guarantee the finiteness of elastic energy of the body. The norm of an element w in this space is defined as

$$\|w\|^2 = \int_{\mathcal{B}} (w_\alpha w_\alpha + w_{\alpha,\beta} w_{\alpha,\beta})\, d^2x,$$

where w_α correspond to the cartesian components of the vector w. Greek indices run from 1 to 2, the comma preceding indices denotes partial differentiation, and summation over repeated indices is understood. We call the subset $\mathcal{C}$ of $H(\mathcal{B})$ such that

$$\mathcal{C} = \{w \in H(\mathcal{B}) |\, w|_{\partial_w} = 0\}$$

the set of kinematically admissible displacements.

The principle of minimum energy states that, among all kinematically admissible displacements $w \in \mathcal{C}$, the true displacement $\check{w}(x)$ minimizes the energy functional

$$I[w] = \int_{\mathcal{B}} W(e(w(x)))\, d^2x - \int_{\partial_\tau} \tau_\alpha w_\alpha\, ds.$$

Here e is the symmetric strain tensor, with components

$$e_{\alpha\beta}(w) = \frac{1}{2}(w_{\alpha,\beta} + w_{\beta,\alpha}),$$

and $W(e)$ is a positive definite quadratic form of e called stored energy density. In short we shall write this energy principle as

$$I[\check{w}] = \min_{w \in \mathcal{C}} I[w]. \tag{1}$$

Using Korn's inequality, one can prove that the minimization problem (1) has a unique solution.

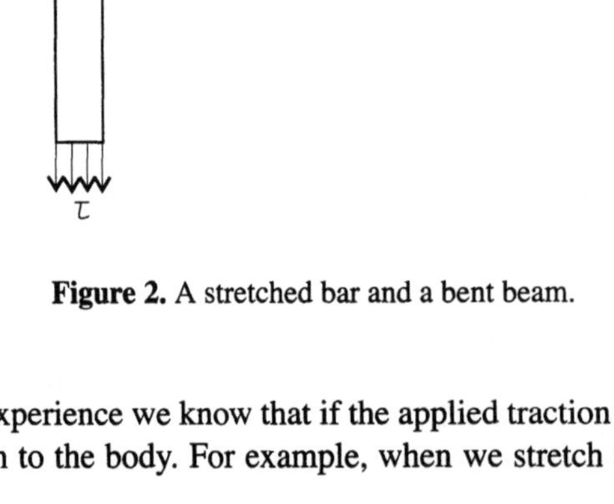

Figure 2. A stretched bar and a bent beam.

However, from our experience we know that if the applied traction is sufficiently large, some "bad" things will happen to the body. For example, when we stretch a bar or bend a beam, we know that for a sufficiently large traction cracks may occur due to the bond rupture (see Fig. 2). In the worst-case scenario the body, after a while, may be broken into pieces.

To be sure that no such things will happen to the body for a given loading we must enlarge the set of admissible displacements by taking into account the possibility of crack formation and compare the energy of the body without and with cracks. Let us introduce a crack as a curve $\mathcal{S}$ of broken bond between adjacent material points. Once the bond is broken, it is irreversible, so cracks cannot be healed. Because of the broken bond the displacement field $w(x)$ may have jump on $\mathcal{S}$, so let us call $\mathcal{S}$ discontinuity curve. We admit the crack kinking and branching, therefore $\mathcal{S}$ in general is assumed to be a piecewise smooth curve which may have more than one branch, as shown in Fig. 3. The penetration of crack faces into each other is not allowed, consequently

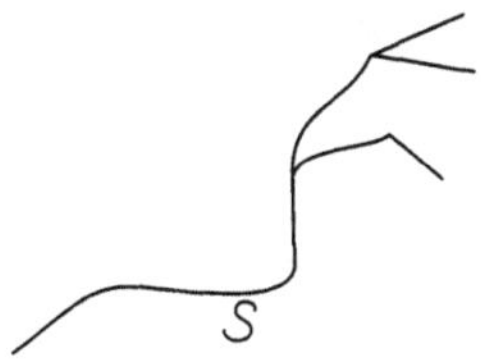

Figure 3. A crack path.

the constraint

$$[[w_\alpha]]n_\alpha = (w_\alpha^+ - w_\alpha^-)n_\alpha \geq 0$$

is expected to be fulfilled almost everywhere on $\mathcal{S}$. Here n is the normal vector which points in the direction $+$, $[[w]] = w^+ - w^-$ denotes the jump in displacement, with w^+ and w^- being the limiting values of w on two sides of the curve $\mathcal{S}$. Note that w belongs to the space of functions of special bounded variation $SBV(\mathcal{B})$. We shall not give here the definitions of norm and topology of this function space. For mathematically minded readers the paper by De Giorgi and Ambrosio (1988) could be recommended. We define the set of kinematically admissible displacements of the body with newly formed cracks as

$$\mathcal{D} = \{w \in SBV(\mathcal{B}) |\, [[w_\alpha]]n_\alpha \geq 0 \,\text{on}\, \mathcal{S}, w|_{\partial_w} = 0\},$$

with $\mathcal{S}$ being an arbitrary curve of discontinuity.

It is clear that energy is required to break bond and produce new cracks, therefore we define the energy of the cracked body by

$$I_c[w] = \int_{\mathcal{B}\backslash\mathcal{S}} W(e(w))\, d^2x + \int_{\mathcal{S}} \Theta([[w]], n)\, ds - \int_{\partial_\tau} \tau_\alpha w_\alpha \, ds, \tag{2}$$

in accordance with Griffith (1920), Griffith (1924). In comparison with $I[w]$ the only new term in (2) is the second one which corresponds to the surface energy, whose density $\Theta([[w]], n)$ may depend on the jump in displacement and on the normal vector n. We say that the body is in *safe* equilibrium if there exist a neighborhood $\mathcal{D}_\epsilon \subset \mathcal{D}$ of $\breve{w} \in \mathcal{C}$ such that the energy functional reaches a *local minimum* at $\breve{w}$ (see Le (1989), Le (1990))

$$I[\breve{w}] = \min_{w \in \mathcal{D}_\epsilon} I_c[w] \tag{3}$$

If this is not the case, we say that the crack nucleation begins.

Our variational formulation can also be easily adapted to the situation when the body contains a pre-existing crack $\mathcal{S}_0$. Let $\mathcal{C}(\mathcal{S}_0)$ be the subset of the Sobolev space which consists of all displacement fields $w \in H(\mathcal{B} \setminus \mathcal{S}_0)$ obeying the kinematical constraints

$$\mathcal{C}(\mathcal{S}_0) = \{w \in H(\mathcal{B} \setminus \mathcal{S}_0) | \, [[w_\alpha]]n_\alpha \geq 0 \,\text{on}\, \mathcal{S}_0, w = 0 \,\text{on}\, \partial_w\}.$$

This definition admits only displacement fields of the body with a pre-existing crack which have finite elastic energies. Since the region $\mathcal{B} \setminus \mathcal{S}_0$ is non-smooth, this does not rule out possible singularities of $w_{\alpha,\beta}$ at the crack tips $\partial\mathcal{S}_0$.

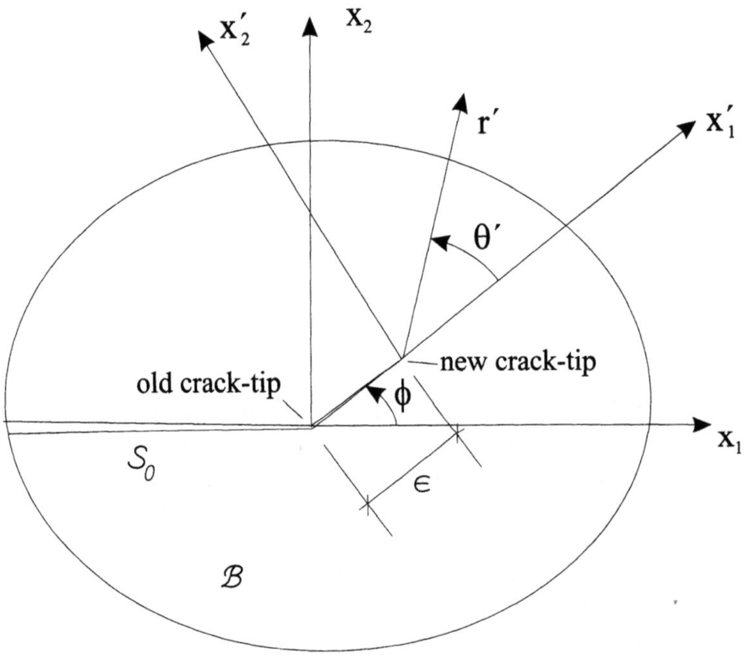

Figure 4. A virtual kink crack.

We call

$$\mathcal{D}(\mathcal{S}_0) = \{w \in SBV(\mathcal{B}) | \, \mathcal{S} \supseteq \mathcal{S}_0, [[w_\alpha]]n_\alpha \geq 0 \,\text{on}\, \mathcal{S}, w = 0 \,\text{on}\, \partial_w\}$$

the set of kinematically admissible displacements of the cracked body. The one-side constraint $\mathcal{S} \supseteq \mathcal{S}_0$ tells us that the pre-existing crack can only grow; there is no way to heal it. The energy of the cracked body is defined in accordance with (2). As before, we say that the body with the pre-existing crack $\mathcal{S}_0$ is in safe equilibrium if there exist a neighborhood $\mathcal{D}_\epsilon \subset \mathcal{D}(\mathcal{S}_0)$ of $\check{w} \in \mathcal{C}(\mathcal{S}_0)$ such that the energy functional reaches a *local minimum* at $\check{w}$

$$I_c[\check{w}] = \min_{w \in \mathcal{D}_\epsilon} I_c[w]. \tag{4}$$

If this is not the case, we say that the crack begins to grow.

Let us establish the necessary conditions for the displacement $\check{w}$ of a body with a pre-existing crack to be in safe equilibrium in accordance with the principle of minimum energy given above. For simplicity we consider a body with a pre-existing crack $\mathcal{S}_0$ located on the negative x_1-axis, and assume that $\Theta([[w]], n) = \Theta_0 = \text{const}$. Let us introduce a one parameter family of admissible displacements $\epsilon \mapsto w(., \epsilon) \in \mathcal{D}$, whose discontinuity curves describe a virtually growing crack $\mathcal{S}_\epsilon$. The crack kinking is admitted here, with ϕ being the kink angle which may be arbitrary (see Fig. 4). Since the crack can only grow, we require that

$$\mathcal{S}_{\epsilon'} \supseteq \mathcal{S}_\epsilon \supseteq \mathcal{S}_0 \quad \text{for } \epsilon' > \epsilon > 0, \quad \mathcal{S}_\epsilon = \mathcal{S}_0 \quad \text{when } \epsilon = 0,$$
$$w(x, 0) = \check{w}(x) \in \mathcal{C}.$$

After substituting this one parameter family of admissible displacements into the energy functional (2) it becomes a function of ϵ, $\epsilon \geq 0$. If $\check{w}$ is in safe equilibrium, this function has an end-point minimum at $\epsilon = 0$ for arbitrary family of admissible displacements in accordance with our definition. Therefore the following necessary condition for safe equilibrium must be fulfilled:

$$\delta I_c = \lim_{\epsilon \to 0} \frac{d}{d\epsilon} I_c[w(., \epsilon)] \geq 0 \quad \text{for arbitrary families of } w(., \epsilon) \in \mathcal{D}. \tag{5}$$

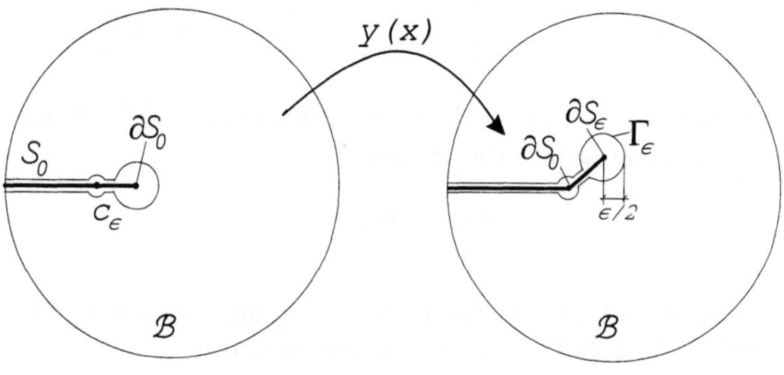

Figure 5. Parametrization and integration domains.

In order to derive consequences from (5), we must be able to calculate the derivative of $I_c[w(., \epsilon)]$ with respect to ϵ and then take the limit as $\epsilon \to 0$. The difficulties of this calculation are due to the changeable region $\mathcal{B} \setminus \mathcal{S}_\epsilon$ and curve $\mathcal{S}_\epsilon$. In order to overcome these difficulties we introduce one-parameter families of one-to-one mappings of $\mathcal{B}$ onto itself

$$\epsilon \mapsto y(., \epsilon)$$

so that

$$\mathcal{B} \setminus \mathcal{S}_0 \overset{y}{\mapsto} \mathcal{B} \setminus \mathcal{S}_\epsilon, \quad \mathcal{S}_0 \overset{y}{\mapsto} \mathcal{S}_\epsilon,$$
$$y(x, \epsilon) = x, \quad \text{when } \epsilon = 0 \text{ or } x \in \partial\mathcal{B}.$$

Since the crack kinking is admitted, functions $y(.,\epsilon)$ are assumed to be smooth everywhere except at the point c_ϵ lying on the x_1-axis which are mapped to the old crack tip $\partial\mathcal{S}_0$ (see Fig. 5). We call $y(.,\epsilon)$ parametrizations of medium. These mappings will be used as changes of variables for the 2-D and 1-D integrals in (2) which become then integrals over the fixed region $\mathcal{B}\setminus\mathcal{S}_0$ and curve $\mathcal{S}_0$. According to the transformation rule we have

$$\int_{\mathcal{B}\setminus\mathcal{S}_\epsilon} W(e_{\alpha\beta}(y))\,d^2y = \int_{\mathcal{B}\setminus\mathcal{S}_0} W(e_{\alpha\beta}(y(x,\epsilon)))J\,d^2x,$$

where $J = \det y_{\alpha,\beta}$ is the Jacobian. Since the region of integration, after this change of variables, does not depend on ϵ, the order of differentiation and integration can be interchanged so that

$$\lim_{\epsilon\to 0}\frac{d}{d\epsilon}\int_{\mathcal{B}\setminus\mathcal{S}_\epsilon} W(e_{\alpha\beta})\,d^2x = \lim_{\epsilon\to 0}\int_{\mathcal{B}\setminus\mathcal{S}_0}(J\delta W + W\delta J)\,d^2x.$$

Symbol δ under integral signs, called for short variation, is used to denote the partial derivative with respect to ϵ at fixed x. The variation of J reads

$$\delta J = \frac{\partial J}{\partial y_{\alpha,\beta}}\delta y_{\alpha,\beta} = (\mathrm{Cof})_{\alpha\beta}\delta y_{\alpha,\beta} = J\frac{\partial x_\beta}{\partial y_\alpha}\delta y_{\alpha,\beta} = J\frac{\partial \delta y_\alpha}{\partial y_\alpha},$$

where $(\mathrm{Cof})_{\alpha\beta}$ is the cofactor of $y_{\alpha,\beta}$. Consider now the variation of W

$$\delta W = \frac{\partial W}{\partial e_{\alpha\beta}}\,\delta e_{\alpha\beta} = \frac{1}{2}\sigma_{\alpha\beta}\,\delta\Big(\frac{\partial w_\alpha}{\partial y_\beta} + \frac{\partial w_\beta}{\partial y_\alpha}\Big) = \sigma_{\alpha\beta}\,\delta\frac{\partial w_\alpha}{\partial y_\beta}.$$

Here σ is the symmetric stress tensor field with components $\sigma_{\alpha\beta} = \frac{\partial W}{\partial e_{\alpha\beta}}$. In order to calculate the variation $\delta\frac{\partial w_\alpha}{\partial y_\beta}$ we recall the following identity

$$\frac{\partial w_\alpha}{\partial x_\gamma} = \frac{\partial w_\alpha}{\partial y_\beta}\frac{\partial y_\beta}{\partial x_\gamma}.$$

Applying the product rule of differentiation to this identity and remembering that the variation and partial derivative with respect to x are commutative we obtain

$$\frac{\partial\delta w_\alpha}{\partial x_\gamma} = \delta\frac{\partial w_\alpha}{\partial y_\beta}\frac{\partial y_\beta}{\partial x_\gamma} + \frac{\partial w_\alpha}{\partial y_\beta}\frac{\partial \delta y_\beta}{\partial x_\gamma}.$$

Multiplying this equation with $\partial x_\gamma/\partial y_\kappa$ and rearranging indices and terms we get

$$\delta\frac{\partial w_\alpha}{\partial y_\beta} = \frac{\partial\delta w_\alpha}{\partial y_\beta} - \frac{\partial w_\alpha}{\partial y_\gamma}\frac{\partial\delta y_\gamma}{\partial y_\beta}.$$

Combining all these formulas we obtain finally

$$\begin{aligned}\lim_{\epsilon\to 0}\frac{d}{d\epsilon}\int_{\mathcal{B}\setminus\mathcal{S}_\epsilon} W(e(w(y,\epsilon)))\,d^2y &= \lim_{\epsilon\to 0}\int_{\mathcal{B}\setminus\mathcal{S}_0}\Big(\sigma_{\alpha\beta}\frac{\partial\delta w_\alpha}{\partial y_\beta} + \mu_{\alpha\beta}\frac{\partial\delta y_\alpha}{\partial y_\beta}\Big)J\,d^2x \\ &= \lim_{\epsilon\to 0}\int_{\mathcal{B}\setminus\mathcal{S}_\epsilon}(\sigma_{\alpha\beta}\delta w_{\alpha,\beta} + \mu_{\alpha\beta}\delta y_{\alpha,\beta})\,d^2x,\end{aligned} \tag{6}$$

where the tensor

$$\mu_{\alpha\beta} = -\sigma_{\gamma\beta} w_{\gamma,\alpha} + W\delta_{\alpha\beta}$$

was first introduced by Eshelby (1956). Since the stress field σ and the displacement gradient $w_{\alpha,\beta}$ are singular at $\partial\mathcal{S}_0$ and $\partial\mathcal{S}_\epsilon$, Gauss' theorem cannot be applied to the right-hand side of (6) directly. To do this properly we replace the region $\mathcal{B} \setminus \mathcal{S}_\epsilon$ by $\mathcal{B}_\epsilon$, whose interior boundary is shown in Fig. 5. It turns out that formula (6) remains valid for the integral taken over $\mathcal{B}_\epsilon$. Applying Gauss' theorem and letting ϵ approach zero, we obtain

$$\lim_{\epsilon\to 0} \frac{d}{d\epsilon} \int_{\mathcal{B}\setminus\mathcal{S}_\epsilon} W(e(w))\, d^2x = \int_{\mathcal{B}\setminus\mathcal{S}_0} (-\check{\sigma}_{\alpha\beta,\beta}\delta w_\alpha - \check{\mu}_{\alpha\beta,\beta}\delta y_\alpha)\, d^2x + \int_{\mathcal{S}_0} [(-\check{\sigma}^+_{\alpha\beta}\delta w^+_\alpha$$
$$+ \check{\sigma}^-_{\alpha\beta}\delta w^-_\alpha)n_\beta + (-\check{\mu}^+_{\alpha\beta} + \check{\mu}^-_{\alpha\beta})n_\beta \delta y_\alpha]\, ds + \int_{\partial_\tau} \check{\sigma}_{\alpha\beta}\delta w_\alpha n_\beta\, ds - G(\phi)\delta l. \quad (7)$$

In the last term of (7) $G(\phi)$ is given by

$$G(\phi) = \nu_\alpha \lim_{\epsilon\to 0} \int_{\Gamma_\epsilon} \mu_{\alpha\beta}\kappa_\beta ds = \nu_\alpha \lim_{\epsilon\to 0} \int_{\Gamma_\epsilon} (-\sigma_{\gamma\beta} w_{\gamma,\alpha}\kappa_\beta + W\kappa_\alpha)\, ds, \quad (8)$$

with ν being the unit vector pointing to the direction of crack extension, Γ_ϵ the contour of radius $\epsilon/2$ surrounding the crack tip $\partial\mathcal{S}_\epsilon$, κ the outward unit normal vector on Γ_ϵ, and $\delta l = \nu_\alpha \lim_{\epsilon\to 0} \delta y_\alpha|_{\partial\mathcal{S}_\epsilon}$ the virtual crack extension length. Note that, away from the crack tip, the stress field σ, the displacement gradient $w_{\alpha,\beta}$ and the energy density W approach the corresponding quantities calculated for $w = \check{w}$ in the limit $\epsilon \to 0$. Note also that the integral along the contour surrounding the point $\partial\mathcal{S}_0$ tends to zero as $\epsilon \to 0$, because the stress field and the displacement gradients have in its neighborhood the corner singularity, which turns out to be weaker than the square root singularity at the crack tip. When deriving (7) the following asymptotic property is tacitly used

$$\lim_{\epsilon\to 0} \int_{\Gamma_\epsilon} \sigma_{\alpha\beta}\kappa_\beta\, ds = 0.$$

This is due to the square root singularity of the stress field. The integral (8) describes the elastic energy release per unit crack extension length which is expected to be a function of the kink angle ϕ.

Under the assumption that $\Theta([[w]], n) = \Theta_0 = \text{const}$ the variation of the energy of crack is given by

$$\lim_{\epsilon\to 0} \frac{d}{d\epsilon} \int_{\mathcal{S}_\epsilon} \Theta_0\, ds = \Theta_0 \delta l \quad (9)$$

The variation of the external work is equal to

$$\lim_{\epsilon\to 0} \frac{d}{d\epsilon} \int_{\partial_\tau} \tau_\alpha w_\alpha\, ds = \int_{\partial_\tau} \tau_\alpha \delta w_\alpha\, ds. \quad (10)$$

Combining formulae (7)-(10), one transforms the inequality (5) to

$$\int_{\mathcal{B}\setminus\mathcal{S}_0} (-\check{\sigma}_{\alpha\beta,\beta}\delta w_\alpha - \check{\mu}_{\alpha\beta,\beta}\delta y_\alpha)\, d^2x + \int_{\mathcal{S}_0} [(-\check{\sigma}^+_{\alpha\beta}\delta w^+_\alpha + \check{\sigma}^-_{\alpha\beta}\delta w^-_\alpha)n_\beta$$
$$+ (-\check{\mu}^+_{\alpha\beta} + \check{\mu}^-_{\alpha\beta})n_\beta\delta y_\alpha]\, ds + \int_{\partial_\tau} (\check{\sigma}_{\alpha\beta}n_\beta - \tau_\alpha)\delta w_\alpha\, ds - [J(\phi) - \Theta_0]\delta l \geq 0. \quad (11)$$

We shall now analyze the inequality (11). It is obvious that the variations δw and δy in the region $\mathcal{B}\setminus\mathcal{S}_0$ as well as δw on ∂_τ can be chosen arbitrarily. On the contrary, $\delta w^\pm$ and δy on $\mathcal{S}_0$ should satisfy some constraints. When the crack faces are not in contact with each other, then the variations $\delta w^\pm$ on $\mathcal{S}_0$ can obviously have arbitrary values. If this is not the case, the constraint $[[\delta w_\alpha]]n_\alpha \geq 0$ must be obeyed. Since $\mathcal{S}_\epsilon \supseteq \mathcal{S}_0$, δy should satisfy the constraints

$$\delta y_\alpha n_\alpha = 0 \quad \text{on } \mathcal{S}_0, \quad \delta l \geq 0 \quad \text{at } \partial\mathcal{S}_0.$$

Taking all these constraints into account, one can show that the variational inequality (11) leads to

$$\begin{gathered}
\check{\sigma}_{\alpha\beta,\beta} = 0, \quad \check{\sigma}_{\alpha\beta} = \frac{\partial W}{\partial \check{e}_{\alpha\beta}} \quad \text{in} \quad \mathcal{B}\setminus\mathcal{S}_0, \\
\check{w}_\alpha = 0 \quad \text{on } \partial_w, \quad \check{\sigma}_{\alpha\beta} n_\beta = \tau_\alpha \quad \text{on } \partial_\tau, \\
(\check{w}_\alpha^+ - \check{w}_\alpha^-) n_\alpha \geq 0 \quad \text{on } \mathcal{S}_0, \\
\check{\sigma}_{\alpha\beta}^+ n_\beta = \check{\sigma}_{\alpha\beta}^- n_\beta = -p n_\alpha, \quad p \geq 0 \quad \text{on } \mathcal{S}_0, \\
(\check{w}_\alpha^+ - \check{w}_\alpha^-) n_\alpha > 0 \Rightarrow p = 0, \\
\max_\phi G(\phi) \leq \Theta_0 \quad \text{at } \partial\mathcal{S}_0.
\end{gathered} \tag{12}$$

Additionally, we also get the relations

$$\begin{gathered}
\check{\mu}_{\alpha\beta,\beta} = 0, \quad \check{\mu}_{\alpha\beta} = -\check{\sigma}_{\gamma\beta}\check{w}_{\gamma,\alpha} + \check{W}\delta_{\alpha\beta} \quad \text{in} \quad \mathcal{B}\setminus\mathcal{S}_0, \\
(-\check{\mu}_{\alpha\beta}^+ + \check{\mu}_{\alpha\beta}^-) n_\beta \nu_\alpha = 0 \quad \text{on } \mathcal{S}_0,
\end{gathered}$$

with ν being the tangent vector to the curve $\mathcal{S}_0$. However, it is easy to see that these equations are satisfied identically by virtue of other equations in (12). This is due to the invariant properties of the energy functional with respect to the group of parametrizations leaving the curve of discontinuity unchanged.

Exercise 1. Prove this statement.

Thus, Eqs. (12) are necessary conditions for the displacement field of the cracked body $\check{w}$ to be in safe equilibrium. The difference between equilibrium and safe equilibrium reduces to the last condition, $\max_\phi G(\phi) \leq \Theta_0$, called the maximum energy release rate criterion. The calculation of $G(\phi)$ should be done with care. We cannot for example replace in (8) the stress field σ by the stress field $\check{\sigma}$ corresponding to $\check{w}$, because as $\epsilon \to 0$ the stress field σ, in general, does not tend to $\check{\sigma}$, as we shall see below. Let us consider the one-parameter family $w'(.,\epsilon)$ of solutions of static boundary-value problems, in which the crack grows along the path $\mathcal{S}_\epsilon$ under the same external boundary conditions. For simplicity we calculate $G(\phi)$ for isotropic materials. Using standard asymptotic methods, one can show that the singular stress and displacement fields near the extended crack tip $\partial\mathcal{S}_\epsilon$ are given by

$$\begin{aligned}
\sigma'_{\alpha'\beta'}(\epsilon) &= K'_I(\epsilon)\frac{f_{I\alpha'\beta'}(\theta')}{\sqrt{2\pi r'}} + K'_{II}(\epsilon)\frac{f_{II\alpha'\beta'}(\theta')}{\sqrt{2\pi r'}} + O(1), \\
w'_{\alpha'}(\epsilon) &= K'_I(\epsilon)\sqrt{\frac{r'}{2\pi}}v_{I\alpha'}(\theta') + K'_{II}(\epsilon)\sqrt{\frac{r'}{2\pi}}v_{II\alpha'}(\theta') + O(r'),
\end{aligned} \tag{13}$$

where the indices I, II denote the two local plane-strain modes of crack deformation, r', θ' correspond to the polar coordinates of the shifted and rotated coordinate system, and the indices with prime denote the projections onto the corresponding rotated coordinate axes (Fig. 4). The radius of domain, where Eq. (13) applies, is of the order ϵ, and tends to zero together with ϵ. However, the stress intensity factors do not approach those prior to the crack extension as $\epsilon \to 0$. We denote the limiting values of $K'_I(\epsilon), K'_{II}(\epsilon)$ by $K^\star_I, K^\star_{II}$, which are related to the stress intensity factors before the crack kinking, K_I, K_{II}, by

$$\begin{aligned} K^\star_I &= F_{I,I}(\phi)K_I + F_{I,II}(\phi)K_{II}, \\ K^\star_{II} &= F_{II,I}(\phi)K_I + F_{II,II}(\phi)K_{II}. \end{aligned} \tag{14}$$

where $F(\phi)$ is the 2×2 matrix, whose elements are universal functions of ϕ (see Amestoy and Leblond (1992)), ϕ being the kink angle.

We rewrite integral (8) in the local rotated coordinate system x' shown in Fig. 4

$$G(\phi) = \lim_{\epsilon \to 0} \int_\Gamma (-\sigma'_{\alpha'\beta'} w'_{\alpha',1'} \kappa_{\alpha'} + W' \kappa_{1'})\, ds. \tag{15}$$

Using the asymptotic formulas (13) with the well-known functions $f_{\alpha'\beta'}(\theta')$ describing the angular distributions of the stress field, and with $v_{\alpha'}(\theta')$ corresponding to the angular distributions of the displacements (see, e.g., Rice (1968)), we get

$$G(\phi) = \frac{1-\nu^2}{E}[(K^\star_I)^2 + (K^\star_{II})^2] \tag{16}$$

Exercise 2. Check this formula.

Let us now determine the maximum of the energy release rate and the direction at which it is reached. We shall use the formulas (14), in which the approximate expressions of the functions F are given by

$$\begin{aligned} F_{I,I} &= 1 - \frac{3\pi^2}{8}m^2 + (\pi^2 - \frac{5\pi^4}{128})m^4 + (\frac{\pi^2}{9} - \frac{11\pi^4}{72} + \frac{119\pi^6}{15360})m^6 + 5.07790m^8 \\ &\quad -2.88312m^{10} - 0.0925m^{12} + 2.996m^{14} - 4.059m^{16} + 1.63m^{18} + 4.1m^{20}, \\ F_{I,II} &= -\frac{3\pi}{2}m + (\frac{10\pi}{3} + \frac{\pi^3}{16})m^3 + (-2\pi - \frac{133\pi^3}{180} + \frac{59\pi^5}{1280})m^5 + 12.313906m^7 \\ &\quad -7.32433m^9 + 1.5793m^{11} + 4.0216m^{13} - 6.915m^{15} + 4.21m^{17} + 4.56m^{19}, \\ F_{II,I} &= \frac{\pi}{2}m - (\frac{4\pi}{3} + \frac{\pi^3}{48})m^3 + (-\frac{2\pi}{3} + \frac{13\pi^3}{30} - \frac{59\pi^5}{3840})m^5 - 6.176023m^7 \\ &\quad +4.44112m^9 - 1.5340m^{11} - 2.0700m^{13} + 4.684m^{15} - 3.95m^{17} - 1.32m^{19} \\ F_{II,II} &= 1 - (4 + \frac{3\pi^2}{8})m^2 + (\frac{8}{3} + \frac{29\pi^2}{18} - \frac{5\pi^4}{128})m^4 + (-\frac{32}{15} - \frac{4\pi^2}{9} - \frac{1159\pi^4}{7200} \\ &\quad + \frac{119\pi^6}{15360})m^6 + 10.58254m^8 - 4.78511m^{10} - 1.8804m^{12} + 7.280m^{14} \\ &\quad -7.591m^{16} + 0.25m^{18} + 12.5m^{20}, \end{aligned}$$

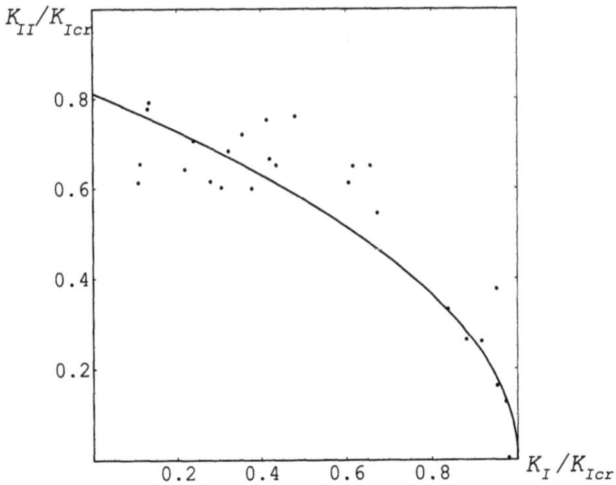

Figure 6. Fracture locus in (K_I, K_{II})-plane: a) solid line: theory, b) points: experimental data.

where $m = \phi/\pi$ (cf. Amestoy and Leblond (1992), Eqn. (66)).

To be in safe equilibrium it is necessary that the stress intensity factors K_I and K_{II} lie inside the curve

$$\max_{\phi} \frac{1-\nu^2}{E}[(K_I^\star)^2 + (K_{II}^\star)^2] = \Theta_0, \tag{17}$$

which is called the fracture locus. In Fig. 6 the graph of the dimensionless fracture locus in accordance with the equation (17) is shown. Finally, in Fig. 7 the dependence of the angle ϕ_{cr}, at which the energy release rate reaches its maximum, on the mode-mixity parameter $\lambda = K_{II}/(K_I + K_{II})$ is shown. In these figures the solid lines correspond to the theory proposed here, points correspond to the experimental data taken from Williams and Ewing (1967) (see also Palaniswamy and Knauss (1978)).

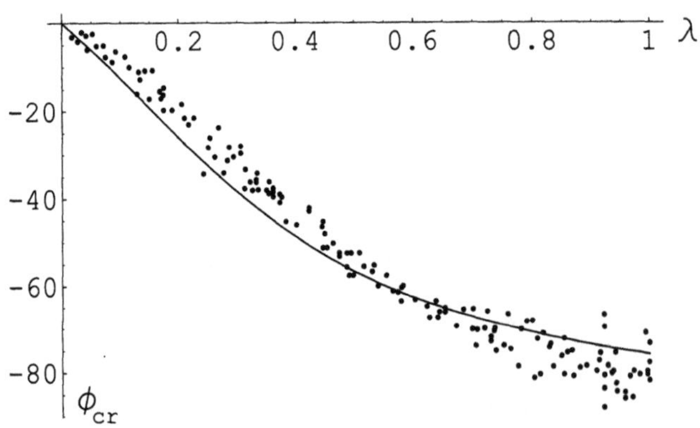

Figure 7. Kink angle ϕ_{cr} as function of λ: a) solid line: theory, b) points: experimental data.

Exercise 3. Generalize this theory to the 3-D case. (*Hint*: see Le et al. (1999))

Note that the above model of cracked body with $\Theta([[w]], n) = \Theta_0 = \text{const}$ is not suitable for predicting safe equilibrium of bodies without pre-existing cracks and for describing the onset of crack nucleation. The reason for that is rather simple: the reduction of elastic energy induced by a newly formed crack of length ϵ is proportional to ϵ^2. At the same time, the increase of surface energy by this crack, under the assumption $\Theta([[w]], n) = \Theta_0 = \text{const}$, is of the amount $\Theta_0\epsilon$ which is much larger than the decrease in elastic energy for small ϵ. Thus, it is not preferable energetically for the body to have small cracks nucleated. The situation changes if we assume that the surface energy density depends on $[[w]]$ (see the solution of crack problem taking the cohesive force into account in 1-D case by Del Piero and Truskinovsky (1998)).

Exercise 4. Generalize Del Piero–Truskinovsky's solution to the 2-D crack problem with the cohesive force.

2 Crack propagation

Consider now the situation when safe equilibrium is not possible, and the crack will propagate throughout the body until its separation into pieces. Since the trajectory of the crack tip is *a priori* unknown, we are here dealing with the problem of a body with unknown boundary. Our aim is to formulate a mathematical model which enables us to predict the displacement field as well as the trajectory of the crack tip. Again, for the sake of simplicity we restrict ourselves to the 2-D case. Let us consider first the displacement field of a body containing a crack propagating along a *known* trajectory

$$w(x,t), \quad x \in \mathcal{B}_t = \mathcal{B} \setminus \mathcal{S}_t,$$

where $\mathcal{S}_t$ is the discontinuity curve at time t. In general, the trajectory of the crack tip is assumed to be a piecewise smooth curve which may have several branches (see Fig. 3). For simplicity we analyze only the case of a single curve $\mathcal{S}_t$ which is smooth at any time t. We also require that $\mathcal{S}_t \supseteq \mathcal{S}_\tau$, $t > \tau$. The position of the crack tip at time t, $\partial\mathcal{S}_t$, is denoted by $l(t)$. The function $w(x,t)$ and its first derivatives with respect to x and t are assumed to be square integrable in $\mathcal{B}_t$ at any time t. Since the region $\mathcal{B}_t$ is non-smooth, this assumption does not exclude the singular behavior of the velocity field $\dot{w}$ and displacement gradients $w_{\alpha,\beta}$ near the crack tip $l(t)$.

We first study the rate of energy change of the cracked body

$$\frac{d}{dt}(E+T),$$

where E is the elastic energy plus the surface energy

$$E = \int_{\mathcal{B}_t} W(e_{\alpha\beta})\, d^2x + \int_{\mathcal{S}_t} \Theta_0\, ds, \tag{18}$$

and T is the kinetic energy

$$T = \int_{\mathcal{B}_t} \frac{1}{2}\rho \dot{w}_\alpha \dot{w}_\alpha\, d^2x.$$

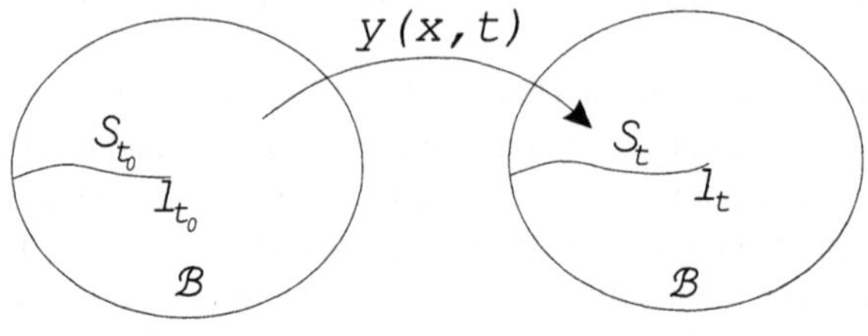

Figure 8. Time-dependent parametrization.

Note that, similar to the case of crack growth, the region and curve of integration $\mathcal{B}_t$ and $\mathcal{S}_t$ are time-dependent. Therefore, in order to calculate the time rates $\dot{E}$ and $\dot{T}$ at $t = t_0$, t_0 being an arbitrary fixed instant of time, we introduce a family of time-dependent parametrizations $t \mapsto y(.,t)$ so that (see Fig. 8)

$$\mathcal{B}_{t_0} \overset{y}{\mapsto} \mathcal{B}_t, \quad \mathcal{S}_{t_0} \overset{y}{\mapsto} \mathcal{S}_t, \quad l(t_0) \overset{y}{\mapsto} l(t),$$
$$y(x,t) = x, \quad \text{when } t = t_0 \text{ or } x \in \partial\mathcal{B}.$$

The quantity $v = \nu_\alpha \dot{y}_\alpha|_{l(t_0)}$, where ν is the tangent vector to $\mathcal{S}_t$ at the point $l(t_0)$, then corresponds to the magnitude of the crack tip velocity $|\dot{l}|$ at $t = t_0$. Note that the functional E in (18) depends upon t in the same way as the functional (2) upon ϵ. Therefore it is easy to see that, at time $t = t_0$, the following equation is valid

$$\dot{E} = \int_{\mathcal{B}_{t_0}} (-\sigma_{\alpha\beta,\beta}\delta_t w_\alpha - \mu_{\alpha\beta,\beta}\dot{y}_\alpha)\, d^2x + \int_{\mathcal{S}_{t_0}} [(-\sigma^+_{\alpha\beta}\delta_t w^+_\alpha + \sigma^-_{\alpha\beta}\delta_t w^-_\alpha)n_\beta$$
$$+ (-\mu^+_{\alpha\beta} + \mu^-_{\alpha\beta})n_\beta \dot{y}_\alpha]\, ds + \int_{\partial_\tau} \sigma_{\alpha\beta} n_\beta \dot{w}_\alpha\, ds - (G - \Theta_0)v|_{l(t_0)}. \tag{19}$$

The elastic energy release rate due to crack propagation is given by

$$G = \nu_\alpha \lim_{|\Gamma|\to 0} \int_\Gamma \mu_{\alpha\beta}\kappa_\beta\, ds = \nu_\alpha \lim_{|\Gamma|\to 0} \int_\Gamma (-\sigma_{\gamma\beta} w_{\gamma,\alpha}\kappa_\beta + W\kappa_\alpha)\, ds,$$

where, as before, the curve Γ surrounds the point $l(t_0)$ and κ is the unit outward normal vector on Γ. The symbol $\delta_t w$ denotes partial derivative of the composite function $w(y(x,t),t)$ with respect to t where x is held constant

$$\delta_t w = \frac{\partial}{\partial t} w(y(x,t),t)\Big|_{x=\text{const},t=t_0}$$

It should be emphasized that $\delta_t w$ and the material velocity $\dot{w}$ do not coincide. Let us now occupy ourselves with the calculation of $\dot{T}$

$$\dot{T} = \frac{d}{dt}\Big|_{t=t_0} \int_{\mathcal{B}_{t_0}} \frac{1}{2}\rho \dot{w}_\alpha \dot{w}_\alpha \det y_{\beta,\gamma}\, d^2x = \int_{\mathcal{B}_{t_0}} (\rho \dot{w}_\alpha \delta_t \dot{w}_\alpha + \frac{1}{2}\rho \dot{w}_\alpha \dot{w}_\alpha \dot{y}_{\beta,\beta})\, d^2x. \tag{20}$$

It is easy to see that

$$\delta_t w_\alpha = \dot{w}_\alpha + w_{\alpha,\beta}\dot{y}_\beta, \quad \delta_t \dot{w}_\alpha = \ddot{w}_\alpha + \dot{w}_{\alpha,\beta}\dot{y}_\beta, \tag{21}$$

where $\ddot{w}$ is the material acceleration. Substituting $\delta_t \dot{w}$ from (21) into Eq. (20) and using Gauss' theorem one obtains

$$\dot{T} = \int_{\mathcal{B}_{t_0}} \rho\ddot{w}_\alpha(\delta_t w_\alpha - w_{\alpha,\beta}\dot{y}_\beta)\, d^2x - (Qv)|_{l(t_0)}, \tag{22}$$

where

$$Q = \nu_\beta \lim_{|\Gamma|\to 0} \int_\Gamma \frac{1}{2}\rho\dot{w}_\alpha\dot{w}_\alpha\kappa_\beta\, ds. \tag{23}$$

The quantity Q can be interpreted as the kinetic energy release rate due to crack propagation.

Since the time t_0 is arbitrary, one obtains from (19)–(23) the rate of energy change for the cracked body in the form

$$\dot{E} + \dot{T} = \int_{\mathcal{B}_t} [(\rho\ddot{w}_\alpha - \sigma_{\alpha\beta,\beta})\delta_t w_\alpha + (-\mu_{\alpha\beta,\beta} - \rho\ddot{w}_\beta w_{\beta,\alpha})\dot{y}_\alpha]\, d^2x + \int_{\mathcal{S}_t} [(-\delta_t w^+_\alpha \sigma^+_{\alpha\beta}$$
$$+ \delta_t w^-_\alpha \sigma^-_{\alpha\beta})n_\beta + \dot{y}_\alpha(-\mu^+_{\alpha\beta} + \mu^-_{\alpha\beta})n_\beta]\, ds + \int_{\partial_\tau} \dot{w}_\alpha\sigma_{\alpha\beta}n_\beta\, ds - (G_d - \Theta_0)v|_{l(t)}, \tag{24}$$

where G_d denotes the total energy release rate to be calculated by

$$G_d = \nu_\alpha \lim_{|\Gamma|\to 0} \int_\Gamma [-\sigma_{\gamma\alpha}w_{\gamma,\beta}\kappa_\beta + (W + \frac{1}{2}\rho\dot{w}_\beta\dot{w}_\beta)\kappa_\alpha]\, ds.$$

We write down the equations of motion and the boundary conditions for the cracked body in the form:

$$\begin{gathered}
\rho\ddot{w}_\alpha = \sigma_{\alpha\beta,\beta}, \quad \sigma_{\alpha\beta} = \frac{\partial W}{\partial e_{\alpha\beta}} \quad \text{in } \mathcal{B}_t, \\
w = 0 \quad \text{on } \partial_w, \quad \sigma_{\alpha\beta}n_\beta = \tau_\alpha \quad \text{on } \partial_\tau, \\
(w^+_\alpha - w^-_\alpha)n_\alpha \geq 0 \quad \text{on } \mathcal{S}_t, \\
\sigma^+_{\alpha\beta}n_\beta = \sigma^-_{\alpha\beta}n_\beta = -pn_\alpha, \quad p \geq 0 \quad \text{on } \mathcal{S}_t, \\
(\dot{w}^+_\alpha - \dot{w}^-_\alpha)n_\alpha > 0 \Rightarrow p = 0.
\end{gathered} \tag{25}$$

Moreover, if $p > 0$ then $(\dot{w}^+_\alpha - \dot{w}^-_\alpha)n_\alpha = 0$. One can also prove that the following equations

$$\begin{gathered}
\mu_{\alpha\beta,\beta} = -\rho\ddot{w}_\beta w_{\beta,\alpha}, \quad \mu_{\alpha\beta} = -\sigma_{\gamma\beta}w_{\gamma,\alpha} + W\delta_{\alpha\beta} \quad \text{in } \mathcal{B}_t, \\
(-\mu^+_{\alpha\beta} + \mu^-_{\alpha\beta})n_\beta\nu_\alpha = 0 \quad \text{on } \mathcal{S}_t
\end{gathered}$$

are satisfied identically by virtue of other equations (see, e.g. Le (1990)). Taking all these equations into account we obtain

$$\dot{E} + \dot{T} = \int_{\partial_\tau} \tau_\alpha\dot{w}_\alpha\, ds - (G_d - \Theta_0)v|_{l(t)}. \tag{26}$$

Exercise 5. Derive the rate of energy change for a kink crack.

Now we formulate the equation of energy balance in dynamic fracture mechanics: the time rate of the energy change of a cracked body is equal to the power of the external forces minus the energy dissipation rate due to the crack resistance force r acting on the crack tip

$$\frac{d}{dt}(E+T) = L - D,$$

where

$$L = \int_{\partial_\tau} \tau_\alpha \dot{w}_\alpha \, ds, \quad D = (rv)|_{l(t)}.$$

Combining the equation of energy balance with (26) we get

$$D = (G_d - \Theta_0)v, \quad r(v) = G_d - \Theta_0 \quad \text{when } v > 0. \tag{27}$$

We also require the energy dissipation rate $D = (G_d - \Theta_0)v$ to be always non-negative. Thus, if $G_d < \Theta_0$, then $v = 0$, and the crack stops its propagation.

Note that all these equations are satisfied for the cracked body with a *known* trajectory $l(t)$ of the crack tip. How do we find the trajectory of the crack tip? If we know the velocity $\dot{l}(t)$ as function of $l(t)$ at each instant of time

$$\dot{l} = F(l),$$

then we can integrate this equation to obtain the trajectory of the crack tip. Following Slepyan (1993) we formulate the criterion of maximum energy dissipation rate: the true crack tip velocity $\breve{v}$ maximizes the energy dissipation rate (27)

$$D(l, \breve{v}) = \max_v D(l, v).$$

For a crack propagating along the x_1-axis under the dynamic mode I we obtain G_d in the form (see, e.g., Freund (1990))

$$G_d = \lim_{|\Gamma|\to 0} \int_\Gamma [-\sigma_{\gamma\beta} w_{\gamma,1} \kappa_\beta + (W + \frac{1}{2}\rho \dot{w}_\beta \dot{w}_\beta)\kappa_1] \, ds = \frac{1-\nu^2}{E} A(v) K_I^2, \tag{28}$$

where

$$A(v) = \frac{v^2 \alpha_d}{(1-\nu)c_s^2 D}, \quad \alpha_d = \sqrt{1 - v^2/c_d^2}, \quad \alpha_s = \sqrt{1 - v^2/c_s^2},$$
$$D = 4\alpha_d \alpha_s - (1+\alpha_s^2)^2,$$

$c_d = \sqrt{(\lambda + 2\mu)/\rho}$ and $c_s = \sqrt{\mu/\rho}$ being the dilatation and the shear wave speed, respectively. In the particular case of a semi-infinite crack propagating in an unbounded elastic body under a constant loading one can show that the dynamic stress intensity factor K_I has the form (see Freund (1990))

$$K_I(l, v) = k(1/v) K_I(l, 0). \tag{29}$$

Thus, the dynamic stress intensity factor is equal to the universal function $k(1/v)$ multiplied by the instantaneous equilibrium stress intensity factor which does not depend on v. The function $k(h)$ is given by

$$k(h) = \frac{(1 - c/h)S_-(h)}{\sqrt{1 - a/h}},$$

where

$$a = 1/c_d, \quad b = 1/c_s, \quad c = 1/c_R,$$

c_R being the Rayleigh wave speed, and

$$S_-(\zeta) = \exp\left\{-\frac{1}{\pi}\int_a^b \arctan\left[\frac{4\eta^2\sqrt{(\eta^2 - a^2)(b^2 - \eta^2)}}{(b^2 - 2\eta^2)^2}\right]\frac{d\eta}{\eta - \zeta}\right\}.$$

From (27)-(29) we obtain the normalized energy dissipation rate in the form

$$M(l, v) = \frac{D(l, v)}{G(l)} = N(v) - \gamma(l)v, \quad N(v) = A(v)k^2(v)v,$$

where $\gamma(l) = \Theta_0/G(l)$ with $G(l)$ the instantaneous elastic energy release rate. The graph of $N(v_*)$, where $v_* = v/c_R$ is the dimensionless crack tip speed, is shown in Fig. 9 for Poisson's ratio $\nu = 0.3$.

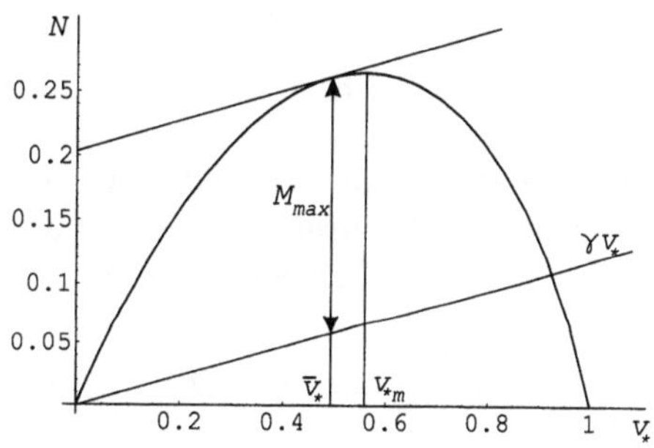

Figure 9. Maximum energy dissipation rate.

From this figure we can see that $N(v)$ is a concave function of v. The maximum of M is achieved at $\check{v}$ for which

$$\frac{dN(\check{v})}{dv} = \gamma(l).$$

Since $\gamma(l) > 0$, the limit crack tip speed is equal to v_m which is the root of the equation

$$\frac{dN(v_m)}{dv} = 0.$$

For $\nu = 0.3$ the limit crack tip speed is $v_m = 0.558c_R$, or $v_m = 0.517c_s$. This theoretical value is close to the experimental results (see Slepyan (1993)).

3 Safe equilibrium at large strain

Up to now we have assumed that the displacement gradient in a cracked body is small everywhere so that linearized elasticity can be used. However, the unbounded displacement gradient near the crack tip are inconsistent with this assumption, and the correctness of the detailed predictions of the linear theory is subject to question. In this Section we want to study the finite anti-plane shear deformation near the tip of a crack within nonlinear elasticity.

Suppose that an isotropic, homogeneous, incompressible elastic body occupies the region $\mathcal{B}$ in the undeformed configuration, and consider the deformation

$$f(x) = x + w(x), \quad x \in \mathcal{B},$$

where x and $x + w(x)$ are the places occupied by the same material point in the undeformed and deformed configuration, respectively, and w is the displacement vector. Let

$$F = \nabla f = I + \nabla w$$

be the deformation gradient. Incompressibility requires that

$$J = \det F = 1.$$

Let $C = F^T F$ denote the right Cauchy-Green deformation tensor. Then

$$I_1 = \mathrm{tr} C, \quad I_2 = \det C \,\mathrm{tr} C^{-1}$$

are two of the three principal invariants of C. The third principal invariant $I_3 = \det C$ is equal to 1 due to incompressibility.

The mechanical response of the material is governed by the stored energy per unit undeformed volume W, which, for incompressible materials, depends only on the first two principal invariants I_1, I_2: $W = W(I_1, I_2)$. The Piola-Kirchhoff stress tensor is given by

$$P = 2\left[\frac{\partial W}{\partial I_1} F + \frac{\partial W}{\partial I_2}(I_2 - C^{-1})F^{-T}\right] - pF^{-T}, \tag{30}$$

with p being the unknown pressure. The equilibrium equations read

$$\mathrm{div} P = 0.$$

If ∂_0 is a traction-free portion of the boundary, P satisfies

$$Pn = 0 \quad \text{on } \partial_0.$$

Suppose now that the body is of cylindrical shape and choose rectangular cartesian coordinates x_1, x_2, x_3 with the x_3-axes parallel to the generators of the cylinder. The cross section of the body with the x_1, x_2-plane occupies the 2-D region $\mathcal{B} \setminus \mathcal{S}_0$. The curve of discontinuity $\mathcal{S}_0$, which corresponds to the cross section of the crack, is assumed to be located on the negative x_1-axis (see Fig. 10). Consider the anti-plane shear

$$f_1 = x_1, \quad f_2 = x_2, \quad f_3 = x_3 + u(x_1, x_2), \quad x_\alpha \in \mathcal{B} \setminus \mathcal{S}_0. \tag{31}$$

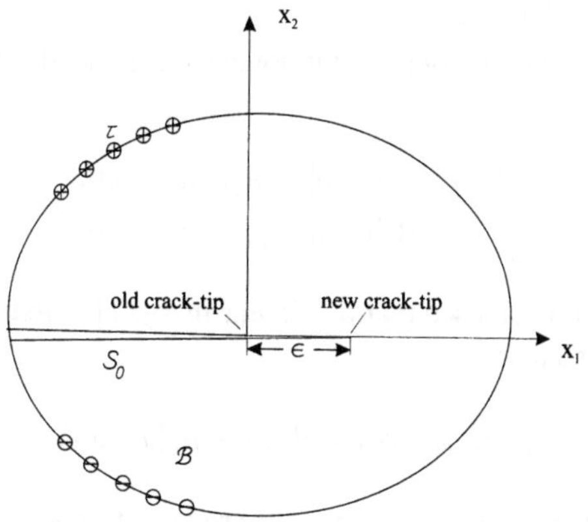

Figure 10. A virtually growing crack under the anti-plane shear.

It turns out that the governing equations written in terms of two function u and p are not consistent unless the stored energy density $W(I_1, I_2)$ is suitably restricted (see Knowles (1977)). One class of incompressible materials for which the governing equations may be consistently reduced to a single equation for u is that for which $W(I_1, I_2)$ is independent of the second invariant

$$W = W(I_1).$$

The analysis will be provided for this special class of materials. We assume that $W(I_1)$ is twice differentiable for $I_1 \geq 3$, vanishes in the undeformed state, and satisfies the Baker-Ericksen inequality

$$W' > 0 \quad \text{for } I_1 \geq 3.$$

The constitutive law takes the simpler form

$$P = 2W'(I_1)F - pF^{-T}. \tag{32}$$

With (31) the tensors F and C are found to be

$$F = \begin{pmatrix} 1 & 0 & 0 \\ 0 & 1 & 0 \\ u_{,1} & u_{,2} & 1 \end{pmatrix}, \quad C = \begin{pmatrix} 1+u_{,1}^2 & u_{,1}u_{,2} & u_{,1} \\ u_{,1}u_{,2} & 1+u_{,2}^2 & u_{,2} \\ u_{,1} & u_{,2} & 1 \end{pmatrix},$$

It can be seen that the incompressibility condition is automatically satisfied and that the first principal invariant is given by

$$I_1 = 3 + u_{,\alpha}u_{,\alpha}.$$

We use the constitutive law (32) to calculate the components of the first Piola-Kirchhoff stress tensor

$$\begin{aligned} P_{3\alpha} = 2W'(I_1)u_{,\alpha}, \quad P_{\alpha 3} &= pu_{,\alpha}, \\ P_{33} &= 2W'(I_1) - p, \\ P_{\alpha\beta} &= [2W'(I_1) - p]\delta_{\alpha\beta}, \end{aligned} \tag{33}$$

where $\delta_{\alpha\beta}$ is the Kronecker delta.

Rewriting the equilibrium equations in component form and substituting (33) into them, we obtain

$$[2W'(I_1) - p]_{,\alpha} + p_{,3}u_{,\alpha} = 0,$$
$$[2W'(I_1)u_{,\alpha}]_{,\alpha} - p_{,3} = 0.$$

From the second equation one can see that p is linear in x_3. The first equation then requires that $p_{,\alpha}$ be independent of x_3, so that

$$p(x_1, x_2, x_3) = d_0 x_3 + p_1(x_1, x_2),$$

where the constant d_0 and the function $p_1(x_1, x_2)$ are as yet undetermined. Substitution of this into the first equilibrium equation gives

$$p_1 = 2W'(I_1) + d_0 u + d_1,$$

where d_1 is an arbitrary constant. The only remaining equilibrium equation is

$$[2W'(I_1)u_{,\alpha}]_{,\alpha} - d_0 = 0.$$

Taking into account the traction-free boundary conditions on the crack faces, we have

$$P_{12} = P_{22} = P_{32} = 0 \quad \text{for } x_2 = \pm 0, x_1 < 0.$$

Thus, the first boundary condition is automatically satisfied, while the second requires that

$$d_0 = d_1 = 0.$$

Now the differential equation for u reads

$$[2W'(I_1)u_{,\alpha}]_{,\alpha} = 0. \tag{34}$$

Since $d_0 = d_1 = 0$ one can reduce the first Piola-Kirchhoff stress components to the following form

$$P_{\alpha 3} = P_{3\alpha} = 2W'(I_1)u_{,\alpha}, \quad P_{\alpha\beta} = 0, \quad P_{33} = 0.$$

The traction-free boundary conditions lead to

$$u_{,2} = 0 \quad \text{for } x_2 = \pm 0, x_1 < 0. \tag{35}$$

We shall not specify the boundary conditions at the exterior boundary of the body, since the main interest will be focused at obtaining the consequences of the local crack problem (34) and (35).

Let us further specify the stored energy introduced by Knowles (1977)

$$W(I_1) = \frac{\mu}{2b}\{[1 + \frac{b}{n}(I_1 - 3)]^n - 1\}, \quad I_1 \geq 3, \tag{36}$$

where μ is the infinitesimal shear modulus, n is a "hardening" parameter, and b is a third material constant. It will be assumed throughout that $\mu > 0, n > 0$, and $b > 0$. For the elastic materials defined by (32) the shear stresses $P_{3\alpha}$ are

$$P_{3\alpha} = P_{\alpha 3} = \mu \left[1 + \frac{b}{n}|\nabla u|^2\right]^{n-1} u_{,\alpha}.$$

The equilibrium equation becomes

$$[\mu(1 + \frac{b}{n}|\nabla u|^2)^{n-1} u_{,\alpha}]_{,\alpha} = 0. \tag{37}$$

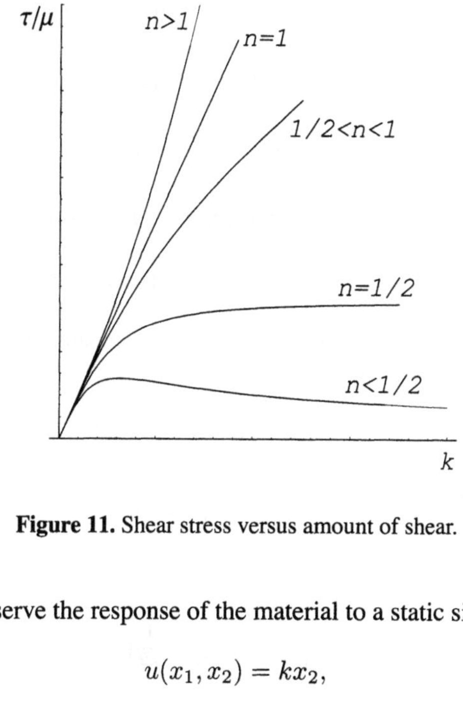

Figure 11. Shear stress versus amount of shear.

It is interesting to observe the response of the material to a static simple shear

$$u(x_1, x_2) = kx_2,$$

with the constant k being the amount of shear. The shear stress $\tau = P_{32}$ is found to be

$$\tau = \mu(1 + \frac{b}{n}k^2)^{n-1}k.$$

For $n > 1$, τ is a monotone increasing function of k, as is $d\tau/dk$, so that materials with $n > 1$ can be classified as "hard" in shear. For $1/2 \leq n < 1$, τ increases with k, but $d\tau/dk$ decreases; as $k \to \infty$, $\tau \to \infty$ provided $n > 1/2$, while for $n = 1/2$, $\tau \to (2b)^{-1/2}\mu$ as $k \to \infty$, corresponding to a finite "ultimate stress" in shear. The materials with $1/2 \leq n < 1$ will be called "soft" in shear. Finally in the range $0 < n < 1/2$, τ rises from zero at $k = 0$ to a maximum and then tends to zero as $k \to \infty$, demonstrating the softening behavior. Various curves of normalized shear stress τ/μ *vs.* amount of shear k in simple shear are illustrated in Fig. 11.

One can show that the governing equation (37) is elliptic if the curve of shear stress *vs.* amount of shear in simple shear has a positive slope at k, i.e., if $n \geq 1/2$. We shall further restrict ourselves to the range $n > 1/2$.

In the polar coordinates r, ϑ we assume that

$$u \sim r^m v(\vartheta) \quad \text{as } r \to 0,\ -\pi \leq \vartheta \leq +\pi,$$

where m and $v(\vartheta)$ are unknowns. Since u remains bounded as $r \to 0$, m must be non-negative. One seeks the smallest value of $m \in (0,1)$ for which the differential equation (37) and the boundary condition (35) are asymptotically satisfied. The function $v(\vartheta) \neq 0$ is assumed to be twice continuously differentiable with respect to ϑ.

Using the rule of differentiation, we calculate the partial derivatives of u

$$u_{,1} = u_{,r} \cos\vartheta - \frac{1}{r} u_{,\vartheta} \sin\vartheta \sim r^{m-1}(mv\cos\vartheta - v'\sin\vartheta),$$
$$u_{,2} = u_{,r} \sin\vartheta + \frac{1}{r} u_{,\vartheta} \cos\vartheta \sim r^{m-1}(mv\sin\vartheta + v'\cos\vartheta),$$

where the prime denotes the derivative with respect to ϑ. Therefore

$$|\nabla u|^2 = u_{,1}^2 + u_{,2}^2 \sim r^{2(m-1)}(m^2 v^2 + v'^2).$$

On the right-hand side of (37) one can neglect 1 as small compared with $b|\nabla u|^2/n$. Dividing the remaining terms by $\mu b/n$, we obtain

$$[r^{2(m-1)(n-1)} q^{n-1} u_{,\alpha}]_{,\alpha} = 0,$$

where the function q is given by

$$q = m^2 v^2 + v'^2.$$

Making use of the formula

$$(f u_{,\alpha})_{,\alpha} = \frac{1}{r}(r f u_{,r})_{,r} + \frac{1}{r^2}(f u_{,\vartheta})_{,\vartheta},$$

we derive the following quasi-linear differential equation for v

$$(q^{n-1}v')' + m[(m-1)(2n-1)+1]q^{n-1}v = 0. \tag{38}$$

The boundary condition (35) clearly leads to

$$v'(\pm\pi) = 0. \tag{39}$$

Thus, we obtain a nonlinear eigenvalue problem. The unique eigenvalue $m \in (0,1)$ is

$$m = 1 - \frac{1}{2n} \tag{40}$$

corresponding to the eigenfunction

$$v(\vartheta) = A \sin\frac{\vartheta}{2}\left[1 - \frac{2k^2\cos^2(\vartheta/2)}{1+\omega(\vartheta,n)}\right]^{1/2}[\omega(\vartheta,n)+k\cos\vartheta]^{k/2}, \tag{41}$$

where A is an arbitrary constant having the meaning of the "stress intensity factor" and

$$\omega(\vartheta,n) = [1-k^2\sin^2\vartheta]^{1/2}, \quad k = 1-1/n.$$

It must be emphasized that A cannot be determined from a purely local analysis. One can directly check the validity of this solution by using the following formulas:

$$v' = \frac{\cos\vartheta+\omega(\vartheta,n)}{2\sin\vartheta}v, \quad v'' = -\frac{1}{4}[1+\frac{2k^2\cos\vartheta}{\omega(\vartheta,n)}+k^2]v,$$
$$q = \frac{1}{4}(1-k^2)(1+k)[\omega(\vartheta,n)+k\cos\vartheta]^{k-1}.$$

Note that $v(\vartheta,t)$ possesses continuous derivatives of all orders with respect to $\vartheta \in [-\pi,\pi]$, is odd in ϑ, and, if $A>0$, v is monotone increasing on $(-\pi,\pi)$. The dependence of the normalized angular distribution of the displacement v/A on ϑ for various values of n is shown in Fig. 12.

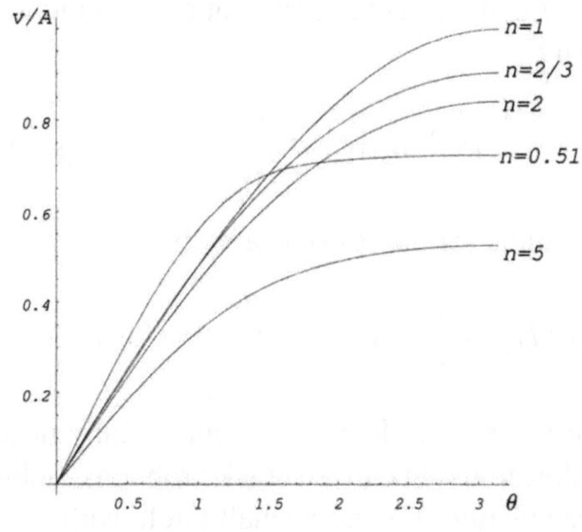

Figure 12. Normalized angular distribution of the displacement.

Having the displacement field $\breve{w}$, let us derive the necessary condition for $\breve{w}$ to be in safe equilibrium. We consider a one parameter family of admissible displacements $\epsilon \mapsto w(.,\epsilon)$, whose discontinuity curves describe a virtually growing crack S_ϵ. For simplicity we assume that the crack grows in the x_1-direction (see Fig. 10). The energy of the cracked body then becomes a function of ϵ

$$I_c[w(\epsilon)] = \int_{B\backslash S_\epsilon} W(I_1(w))\,d^2x + \int_{S_\epsilon}\Theta_0\,ds - \int_{\partial_\tau}\tau w\,ds$$

In order for $I_c[w(\epsilon)]$ to have a local minimum at $\check{w} = w(0)$, its first order derivative with respect to ϵ at $\epsilon = 0$ must be non-negative

$$\lim_{\epsilon \to 0} \frac{d}{d\epsilon} I_c[w(\epsilon)] \geq 0.$$

Applying the calculus of variation, one can show that, for $\check{w}$ to be in safe equilibrium, the equation (37) in $\mathcal{B} \setminus \mathcal{S}_0$ as well as the boundary conditions on $\partial\mathcal{B}$ and the traction-free boundary condition on $\mathcal{S}_0$ must be satisfied. The following additional condition

$$G - \Theta_0 \leq 0 \tag{42}$$

must also be fulfilled, where G is the nonlinear energy release rate which, in our case, is given by

$$G = \lim_{|\Gamma| \to 0} \int_\Gamma (-u_{,1} P_{3\alpha} \kappa_\alpha + W \kappa_1)\, ds. \tag{43}$$

Exercise 6. Derive the necessary condition (42).

To calculate G we need the asymptotic formulas for the shear amount, the shear stresses and the stored energy density near the crack tip. The asymptotic formula of the shear amounts $u_{,\alpha}$ reads

$$u_{,\alpha} \sim [(1 - \frac{1}{2n}) v c_\alpha - v' \varepsilon_{\alpha\beta} c_\beta] r^{-1/2n},$$

where $c_1 = \cos\vartheta, c_2 = \sin\vartheta$ and $\varepsilon_{\alpha\beta}$ is the permutation symbol. Therefore the shear stresses $P_{3\alpha}$ are asymptotically given by

$$P_{3\alpha} \sim \mu \left(\frac{b}{n}\right)^{n-1} p^{n-1}(\vartheta) \left[(1 - \frac{1}{2n}) v c_\alpha - v' \varepsilon_{\alpha\beta} c_\beta\right] r^{-(1-1/2n)}$$

as $r \to 0$. The asymptotic formula for the stored energy is

$$W(I_1) \sim \frac{\mu}{2b} p^n(\vartheta) \left(\frac{b}{n}\right)^n r^{-1} \quad \text{as } r \to 0.$$

Thus, the stored energy density behaves like r^{-1} asymptotically near the crack tip. This is the common result for both nonlinear and linearized crack problems. Substituting all these formulas into the integral (43) and calculating it along a small circle with $\kappa_1 = \cos\theta, \kappa_2 = \sin\vartheta$, one finds that

$$G = \mu A^{2n} \frac{\pi}{2} b^{n-1} \frac{n(2n-1)^{2n-1}(2n^2 - 2n + 1)}{(4n^4)^n}.$$

Exercise 7. Check this formula.

Substituting this formula into (42), we get the necessary condition for safe equilibrium in terms of the stress intensity factor A.

Exercise 8. Generalize this theory to the plane strain case for Ogden's materials. (*Hint*: see Le and Stumpf (1993))

4 Nonlinear dynamic crack problem

For the same reason as in statics the linearized dynamic elasticity of cracked body leads to the unbounded displacement gradient and velocity near the crack tip which is inconsistent with its main assumptions. This is the motivation for us to study in this Section the nonlinear dynamic crack problem. Let an undeformed configuration of an elastic body occupy the region $\mathcal{B}$. A motion of the body is described by the time-dependent displacement field $w(x,t)$. In the absence of body forces the equations of motion read

$$\rho \ddot{w} = \text{div} P. \tag{44}$$

Here ρ is the mass density and P is the first Piola-Kirchhoff stress tensor. If ∂_0 is a traction-free portion of the boundary, P satisfies the boundary condition

$$Pn = 0 \quad \text{on } \partial_0.$$

Combining Eq. (44) with the constitutive equation (30) expressing P as function of the deformation gradient $F = I + \nabla w$, we get three equations in terms of displacement w. We assume that the elastic material is incompressible and isotropic and that the stored energy density W depends on F only through the first invariant of the strain tensor in accordance with (36).

For simplicity of the analysis we again restrict ourselves to the anti-plane shear motion of a cracked body. Suppose that the body is of cylindrical shape and choose rectangular cartesian coordinates x_1, x_2, x_3 with the x_3-axes parallel to the generator of the cylinder. The cross section of the body with the x_1, x_2-plane occupies the 2-D region $\mathcal{B} \backslash \mathcal{S}_t$ at time t. The crack $\mathcal{S}_t$ propagates in the x_1-direction, with $s(t)$ being the position of the crack-tip at time t. An anti-plane shear motion is then characterized by the "out-of-plane" displacement $w_3 = w(x_1, x_2, t)$. Repeating the procedure provided in the previous Section with some minor changes we can reduce the equations of motion to a single differential equation for w

$$\rho \ddot{w} = [2W'(I_1) w_{,\alpha}]_{,\alpha}. \tag{45}$$

The first Piola-Kirchhoff stress components are given by

$$T_{\alpha 3} = T_{3\alpha} = 2W'(I_1) w_{,\alpha}, \quad T_{\alpha\beta} = 0, \quad T_{33} = 0.$$

The traction-free boundary condition reads

$$w_{,2} = 0 \quad \text{for } x_2 = \pm 0, x_1 < s(t). \tag{46}$$

Just as in the static case we shall not impose the boundary conditions at the exterior boundary of the cross section, since the main interest will be focused at obtaining the consequences of the local crack problem (45)–(46).

It is interesting to study the quasi-linear second order partial differential equation (45) which can be rewritten in the form

$$\rho \ddot{w} = [2W'(3 + |\nabla w|^2) w_{,\alpha}]_{,\alpha} \tag{47}$$

where $|\nabla w|^2 = w_{,\alpha} w_{,\alpha}$. One can easily show that (47) is everywhere hyperbolic according to the standard definition if the magnitude $k = |\nabla w|$ of the local displacement gradient satisfies the inequality

$$W'(3+k^2)[W'(3+k^2) + 2k^2 W''(3+k^2)] > 0.$$

This inequality means that the curve of shear stress *vs.* amount of shear in simple shear has positive slope at k. Thus, for the material whose stored energy is given by Eq. (36), Eq. (47) is always hyperbolic if $n \geq 1/2$. Correspondingly, the equation without the inertial term is elliptic for this range of the hardening parameter n. We shall further restrict ourselves to the range $n > 1/2$.

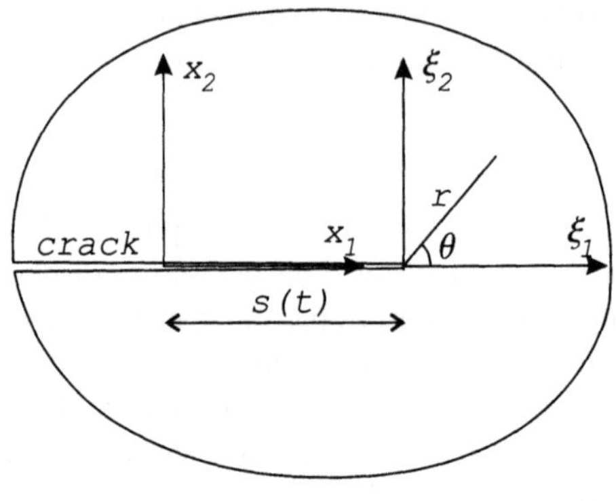

Figure 13. Crack propagating along x_1-axes and co-moving coordinates.

In order to fix a boundary for the local dynamic crack problem (45)–(46) we introduce a co-moving rectangular coordinate system ξ_1, ξ_2 by

$$\xi_1 = x_1 - s(t), \quad \xi_2 = x_2.$$

We also let r and θ denote the polar coordinates associate with ξ_1, ξ_2: $r = \sqrt{\xi_1^2 + \xi_2^2}$ and $\tan\theta = \xi_2/\xi_1$ (see Fig. 13). The displacement w is then viewed as a function of ξ_α and t, that is $w(x_1, x_2, t) = u(\xi_1, \xi_2, t)$. Using the rule of differentiation we obtain for the velocity and acceleration

$$\dot{w} = -u_{|1}\dot{s} + u_{|t}, \quad \ddot{w} = u_{|11}\dot{s}^2 - u_{|1}\ddot{s} - 2u_{|1t}\dot{s} + u_{|tt},$$

where a subscript preceded by a vertical bar indicates partial differentiation of u with respect to the corresponding variables ξ_1, ξ_2, or t. Thus, under the transformation of coordinates the partial differential equation (45) and the boundary condition (46) become

$$\rho_0(u_{|11}\dot{s}^2 - u_{|1}\ddot{s} - 2u_{|1t}\dot{s} + u_{|tt}) = [\mu(1 + \frac{b}{n}|\nabla u|^2)^{n-1} u_{|\alpha}]_{|\alpha}, \tag{48}$$

and

$$u_{|2} = 0 \quad \text{for } \xi_2 = \pm 0, \xi_1 < 0, \tag{49}$$

where, now, $|\nabla u|^2 = u_{|\alpha} u_{|\alpha}$.

In terms of the local polar coordinates r, θ it is assumed that

$$u \sim r^m v(\theta, t) \quad \text{as } r \to 0, \ -\pi \le \theta \le +\pi, \tag{50}$$

for some nonnegative exponent m and some as yet undetermined function $v(\theta, t)$. Since m is nonnegative, (50) ensures that u remains bounded as $r \to 0$. One seeks the smallest value of $m \ge 0$ for which (50) is asymptotically consistent with the differential equation (48) and the boundary condition (49). Since we are interested in the singular fields as $r \to 0$, we look for m that lies in the interval $0 \le m < 1$. Concerning the function $v(\theta, t) \ne 0$, we assume that it is twice continuously differentiable with respect to θ and t except at some points θ_i, which correspond to radial lines of discontinuity moving with the crack and inclined at the angle θ_i to the crack line.

Using the rule of differentiation, we calculate the partial derivatives of u

$$\begin{aligned} u_{|1} &= u_{|r} \cos\theta - \frac{1}{r} u_{|\theta} \sin\theta \sim r^{m-1}(mv\cos\theta - v'\sin\theta), \\ u_{|2} &= u_{|r} \sin\theta + \frac{1}{r} u_{|\theta} \cos\theta \sim r^{m-1}(mv\sin\theta + v'\cos\theta), \end{aligned} \tag{51}$$

where the prime denotes the derivative with respect to θ. With (51) it is easily seen that

$$|\nabla u|^2 = u_{|11}^2 + u_{|22}^2 \sim r^{2(m-1)}(m^2 v^2 + v'^2).$$

By taking the derivatives of (51) we get the estimation $u_{|\alpha\beta} \sim r^{m-2} f_{\alpha\beta}(\theta)$. Consequently, the leading terms in (48) are quite different in the three following cases: i) $n > 1$, ii) $n = 1$, and iii) $1/2 < n < 1$. Therefore the asymptotic analysis should be made separately for these cases.

In case i) for which $n > 1$ all terms in the left-hand side of (48) are small compared with the leading terms in its right-hand side as $r \to 0$ and can be removed. Without these inertial terms Eq. (48) becomes elliptic and does not admit non-smooth solutions. In the right-hand side of (48) one can neglect 1 as small compared with $b/n|\nabla u|^2$. Dividing the remaining terms by $\mu b/n$ we obtain the quasi-linear differential equation (38) which is subjected to the boundary conditions (39). They constitute a nonlinear eigenvalue problem obtained in the previous Section. The unique eigenvalue $m \in (0,1)$ is given by (40) while the corresponding eigenfunction by (41) with the difference that now $A(t)$, called "dynamic stress intensity factor", depends on time t. It must be emphasized that $A(t)$ cannot be determined from a purely local analysis. The dependence of the normalized angular distribution of the displacement v/A on θ for various values of n was shown on Fig. 12. Note that this normalized angular distribution coincides with that of the equilibrium crack.

In case ii) with $n = 1$ Eq. (48) reduces to

$$\rho_0(u_{|11}\dot{s}^2 - u_{|1}\ddot{s} - 2u_{|1t}\dot{s} + u_{|tt}) = \mu u_{|\alpha\alpha}. \tag{52}$$

Under the assumption (50) all terms in the left-hand side of (52) except the first one can be neglected as $r \to 0$ giving

$$(1 - \dot{s}^2/c_s^2)u_{|11} + u_{|22} = 0, \tag{53}$$

where $c_s = \sqrt{\mu/\rho_0}$. This equation is elliptic if $|\dot{s}/c_s| < 1$ (sub-sonic crack propagation). Thus, non-smooth solutions do not exist in the case of sub-sonic crack propagation. Equation (53) can be transformed to the Laplace equation by stretching the coordinate ξ_2 by the factor $\alpha_s = \sqrt{1-\dot{s}^2/c_s^2}$. The well-known solution of (53) and (49) can be written in the form (see Freund (1990))

$$u \sim A(t) r_s^{1/2} \sin \frac{1}{2}\theta_s, \tag{54}$$

where

$$r_s = \sqrt{\xi_1^2 + \alpha_s^2 \xi_2^2}, \quad \tan\theta_s = \frac{\alpha_s \xi_2}{\xi_1}.$$

The polar coordinates r, θ are related to r_s, θ_s by

$$r_s = r\sqrt{1-(\dot{s}\sin\theta/c_s)^2}, \quad \tan\theta_s = \alpha_s \tan\theta.$$

Thus, the singularity degree $m = 1/2$ remains the same as that for the equilibrium crack, but the angular distribution of the displacement turns out to be different due to the inertial effects. Note that the dynamic angular distribution reduces to the static one as $\dot{s} \to 0$.

Finally, in case iii) for which $1/2 < n < 1$ all terms in the right-hand side as well as those in the left-hand side of (48) except the first one can be neglected in the limit $r \to 0$, provided $\dot{s} \neq 0$. Thus, for $\dot{s} \neq 0$ this equation reduces to

$$u_{|11} = 0. \tag{55}$$

If non-smooth solutions are admitted, then one can use the jump condition to show that the derivative $u_{|1}$ should be continuous across the lines of discontinuity. It is then straightforward to see that a non-trivial solution of the type (50) within the range $m \in [0,1)$ does not exist for (55) and (48). This can also be justified by the finiteness of the total energy of the body. Thus, the dynamic fields near the moving crack tip for this subclass of materials exhibit no singular behavior. The smallest $m > 1$ for which a nontrivial solution of (48) and (49) exists is equal to 3/2. Indeed, for $m > 1$ the term $|\nabla u|^2$ is small compared with 1 in the right-hand side of (48) and can be neglected. Keeping the principal terms in (48) we arrive again at the linear equation (53) which gives the smallest eigenvalue

$$m = 3/2 > 1,$$

corresponding to the regular displacement field

$$w \sim A(t) r_s^{3/2} \sin \frac{3}{2}\theta_s.$$

It should be noted that, at the same time, the fields near the crack with $\dot{s} = 0$ remain singular. The major term in the left-hand side of (48), $-u_{|1}\ddot{s}$, has the singularity of the type r^{m-1}, provided the crack-tip acceleration does not vanish: $\ddot{s} \neq 0$. The major term in the right-hand side of (48), $(|\nabla u|^{2(n-1)} u_{|\alpha})_{|\alpha}$, has the singularity of the type $r^{(m-1)(2n-1)-1}$ which is stronger than its counterpart in the left-hand side, provided $1/2 < n < 1$. Therefore after neglecting all terms in the left-hand side of (48) and keeping the major terms in its right-hand side, we arrive again at Eq. (38). Thus, for $n \in (1/2, 1)$ formulas (40)–(41) still hold true for the displacement field of the crack with the vanishing crack-tip velocity or of the equilibrium crack.

The dynamic energy release rate, in our case, is calculated by

$$G_d = \lim_{|\Gamma|\to 0} \int_\Gamma [-w_{,1} P_{3\alpha}\kappa_\alpha + (W + \frac{1}{2}\rho\dot{w}^2)\kappa_1]\, ds. \tag{56}$$

Here Γ denotes the contour surrounding the instantaneous crack tip and being shrunk onto it as $|\Gamma| \to 0$, κ_α is the outward unit normal vector on Γ, and ds the element of length. Referring to the co-moving coordinates ξ_1, ξ_2 (56) takes the form

$$G_d = \lim_{|\Gamma|\to 0} \int_\Gamma \{-u_{|1} P_{3\alpha}\kappa_\alpha + [W + \frac{1}{2}\rho(-u_{|1}\dot{s} + u_{|t})^2]\kappa_1\}\, ds. \tag{57}$$

Now Γ is the contour in the ξ_1, ξ_2-plane. In order to calculate this integral we need the asymptotic formulas near the crack tip for the shear amount, the shear stresses and the kinetic and stored energy densities. All these quantities will be calculated separately in the three above-mentioned cases.

Exercise 9. Derive formula (56).

In case i) the asymptotic formulae as $r \to 0$ of the shear amounts $w_{|\alpha}$ and the shear stresses $P_{3\alpha}$ can be obtained from (51), (38), and (39). With the help of (51) it is easily shown that as $r \to 0$

$$w_{|\alpha} \sim [(1 - \frac{1}{2n})vc_\alpha - v'\varepsilon_{\alpha\beta}c_\beta]r^{-1/2n}, \tag{58}$$

where $c_1 = \cos\theta, c_2 = \sin\theta$ and $\varepsilon_{\alpha\beta}$ is the permutation symbol. Therefore the shear stresses $P_{3\alpha}$ are asymptotically given by

$$P_{3\alpha} \sim \mu \left(\frac{b}{n}\right)^{n-1} p^{n-1}(\theta) \left[(1 - \frac{1}{2n})vc_\alpha - v'\varepsilon_{\alpha\beta}c_\beta\right] r^{-(1-1/2n)}$$

as $r \to 0$. The asymptotic formula for the stored energy reads

$$W(I_1) \sim \frac{\mu}{2b} p^n(\theta) \left(\frac{b}{n}\right)^n r^{-1} \quad \text{as } r \to 0.$$

It is also clear from (58) that the main term of the kinetic energy in (57) has the singularity of the type $r^{-1/n}$ which is weaker than the standard r^{-1} for $n > 1$ and therefore gives no contribution to the energy release rate. Neglecting the kinetic energy in (57) and calculating this integral along a small circle with $\kappa_1 = \cos\theta, \kappa_2 = \sin\theta$, one finds that

$$G_d = \mu A^{2n}(t) \frac{\pi}{2} b^{n-1} \frac{n(2n-1)^{2n-1}(2n^2 - 2n + 1)}{(4n^4)^n}. \tag{59}$$

Thus, the inertial effects influence the dynamic driving force only through the dynamic stress intensity factor. If we substitute the static stress intensity factor A instead of $A(t)$ in (59), we obtain the static energy release rate G.

In case ii) the shear amounts $w_{|\alpha}$ are asymptotically given by

$$w_{|1} \sim -\frac{A(t)}{2r_s^{1/2}} \sin\frac{1}{2}\theta_s, \quad w_{|2} \sim \frac{A(t)\alpha_s}{2r_s^{1/2}} \cos\frac{1}{2}\theta_s$$

as $r \to 0$. The shear stresses are expressed by

$$P_{31} \sim -\mu \frac{A(t)}{2r_s^{1/2}} \sin \frac{1}{2}\theta_s, \quad P_{32} \sim \mu \frac{A(t)\alpha_s}{2r_s^{1/2}} \cos \frac{1}{2}\theta_s,$$

while the stored and kinetic energy densities are given by

$$W \sim \frac{\mu A^2(t)}{8r_s}(\alpha_s^2 \cos^2 \frac{1}{2}\theta_s + \sin^2 \frac{1}{2}\theta_s), \quad K \sim \frac{\mu A^2(t)}{8r_s}\frac{\dot{s}^2}{c_s^2} \sin^2 \frac{1}{2}\theta_s$$

as $r \to 0$. Substituting these formulas into (57) and calculating the integral we obtain

$$G_d = \mu \frac{\pi}{4} \alpha_s A^2(t).$$

This formula differs from that for the equilibrium crack

$$G = \mu \frac{\pi}{4} A^2$$

by the factor $\alpha_s = \sqrt{1 - \dot{s}^2/c_s^2}$ depending on the crack-tip velocity $\dot{s}$.

In case iii) the leading term of the displacement field near the moving crack tip with $\dot{s} \neq 0$ should be of the type (50) with $m \geq 1$. Thus, the stresses and the displacement gradients exhibit no singular behavior, and the integral (57) should vanish. However, the corresponding fields for the accelerating crack with $\dot{s} = 0, \ddot{s} \neq 0$ turn out to be singular, and the energy release rate does not vanish for the crack with $\dot{s} = 0$ and is given by (59).

Exercise 10. Generalize the theory to the plane strain case.

According to the formula (59) one can see that, for materials with the hardening parameter $n > 1$, the energy release rate increases for the increasing stress intensity factor $A(t)$, and this might lead to the unstable crack propagation. Thus, the subclass of hard materials may exhibit the regime of unstable crack propagation. In case $n < 1$ the crack propagation can only be slow and non-monotonic, since the energy release rate becomes finite only for the equilibrium crack or crack propagating with an infinitesimally small speed. As the crack tip velocity becomes finite, the singularity disappears, the energy release rate vanishes, and the crack stops its propagation, waiting until the energy release rate becomes critical again due to the loading process. This qualitative feature agrees well with experimental observations of non-monotonic crack propagation with numerous intermediate crack arrests in polymers by Chudnovsky et al. (1995). Thus, the subclass of soft materials with $n < 1$ exhibits the regime of non-monotonic crack propagation. The transition between these two regimes is possible, if one imagines that the hardening parameter n depends on the temperature. As the temperature crosses a critical temperature T_c this parameter crosses 1, thus changing suddenly the regime of crack propagation. The material with $n = 1$ represents an exceptional case, in which the inertial terms affect both the dynamic stress intensity factor and the local fields. The boundary-value problem becomes linear, and many exact solution can be found in particular situations (see Freund (1990)).

References

Amestoy, M., and Leblond, J. (1992). Crack paths in plane situations - II. Detailed form of the expansion of the stress intensity factors. *Int. J. Solids Struc.* 29:465–501.

Chudnovsky, A., Shulkin, Y., Baron, D., and Lin, K. P. (1995). New method of lifetime prediction for brittle fracture of polyethylene. *Journal of Applied Polymer Science* 56:1465–1478.

De Giorgi, E., and Ambrosio, L. (1988). Un nuovo tipo di funzionale del calcolo delle variazioni. *Atti Academia Nazionale Lincei* 82:199–210.

Del Piero, G., and Truskinovsky, L. (1998). A one-dimensional model for localized and distributed failure. *Journal de Physique IV* 8:95–102.

Eshelby, J. D. (1956). The continuum theory of lattice defects. In *Progress in solids state physics*, volume 3. New York: Academic Press.

Freund, L. B. (1990). *Dynamic Fracture Mechanics*. Cambridge: Cambridge University Press.

Griffith, A. A. (1920). The phenomenon of rupture and flow in solids. *Phil. Trans. Roy. Soc.* A221:163–198.

Griffith, A. A. (1924). The theory of rupture. In *Proc. 1st Int. Congr. Appl. Mech., Delft, 1924*, 55–63.

Knowles, J. K. (1977). The finite anti-plane shear field near the tip of a crack for a class of incompressible elastic solids. *Int.J.Fracture* 13:611–639.

Le, K. C., and Stumpf, H. (1993). The singular elastoplastic field due to a crack in rubberlike materials. *J. Elasticity* 32:183–222.

Le, K. C., Schütte, H., and Stumpf, H. (1999). Determination of the driving force acting on a kinked crack. *Arch. Appl. Mech.* 69:337–344.

Le, K. C. (1989). Variational inequalities in brittle fracture mechanics. In Brandt, A. M., and Marshall, I. H., eds., *Brittle Matrix Composites 2*, 144–151. London: Elsevier Applied Science.

Le, K. C. (1990). Variational principles of the nonlinear theory of brittle fracture. *Applied Mathematics and Mechanics (PMM)* 54:658–665.

Palaniswamy, K., and Knauss, W. G. (1978). On the problem of crack extension in brittle solids under general loading. In Nemat-Nasser, S., ed., *Mechanics Today*, volume 4. Oxford: Pergamon Press. 87–148.

Rice, J. R. (1968). A path independent integral and the approximate analysis of strain concentration by notches and cracks. *J. Appl. Mech.* 35:379–388.

Slepyan, L. I. (1993). Principle of maximum energy dissipation rate in crack dynamics. *J. Mech. Phys. Solids* 41:1019–1033.

Williams, J. G., and Ewing, P. D. (1967). Fracture under complex stress – the angled crack problem. *Int. J. Fract. Mech.* 8:441–446.

References

[illegible]

Griffith Theory Revisited

Jean-Jacques Marigo*

Institut Galilée, Université Paris-Nord, France

1 The Formulation

Throughout the section, Ω denotes a bounded connected open domain of $\mathbb{R}^N$, $1 \leq N \leq 3$, with smooth boundary $\partial\Omega$ the surface measure of which is finite and such that Ω is the interior of $\overline{\Omega}$. As such, Ω represents the crack-free reference configuration of an elastic body.

1.1 The Ingredients

The family of possible cracks and their surface energy. In classical fracture mechanics, the family of possible cracked configurations are few, restricted in general to a few number of smooth surfaces; by contrast, our family is rather extended, since it is made up of all closed subdomains of $\overline{\Omega}$, independently of their shape, subject only to the condition that their dimension is not greater than $N-1$ so that they do not consume too much surface energy. Indeed, an energy is assigned to each member of the family. In the spirit of Griffith, if $k(x)$ represents the energy required to create an "infinitesimal" crack at the point x of $\overline{\Omega}$, then the surface energy associated to the crack $\Gamma \subset \overline{\Omega}$ is given by

$$E_s(\Gamma) = \int_\Gamma k(x) d\mathcal{H}^{N-1}(x), \tag{1}$$

where $\mathcal{H}^{N-1}$ denotes the $N-1$-dimensional Hausdorff measure (a "surface" measure which amounts to the usual surface measure for smooth enough hypersurfaces).

Remark 1. *The energy density $k(x)$, the toughness of the given material, is assumed to be strictly positive (***and even bounded away from zero***), so that energy is created by introducing additional cracks. Furthermore $k(x)$ can take the value ∞, which corresponds to a material that would not be breakable at the given point x. This will prove useful when analyzing debonding in a composite material, see Section 3. We could also introduce a possible dependence of k upon the normal vector to the crack. This is necessary for anisotropic materials.*

* I thank Gilles Francfort and Blaise Bourdin for their help during the preparation of this document.

The type of loading. For reasons that will be clear in Section 5 we consider only the cases where the displacements are prescribed (at a value U) on a part $\partial_d\Omega$ of the boundary, while the remaining part of the boundary and the lips of the cracks are free (of forces), and the body forces vanish.

Remark 2. *A no-contact condition on the "lips" of the crack is thus assumed. Moreover, the crack can be partially located at the boundary where the displacements are prescribed. In such a case the "inner trace" v^- of an admissible displacement v does not satisfy the boundary condition.*

In conclusion, the kinematically admissible set of displacement fields is a function of both the loading U and the crack Γ :

$$\mathcal{C}(\Gamma, U) = \{v \in W^{1,2}(\Omega \setminus \Gamma; \mathbb{R}^N) \; : \; v = U \text{ on } \partial_d\Omega \setminus \Gamma\}. \tag{2}$$

The material behavior. In all that follows the material is assumed to behave elastically and to only undergo infinitesimal transformations. If $\varepsilon(v)$ denotes the symmetric gradient of v, an elastic energy density $W(x, \varepsilon(v)(x))$ is thus given at each point x of $\Omega \setminus \Gamma$ by

$$W(x, \xi) = \frac{1}{2} A(x)\xi \cdot \xi, \tag{3}$$

where $A(x)$ denotes the elasticity tensor at x. $A(x)$ is symmetric and definite positive, *i.e.* such that $0 < \alpha I \leq A(x) \leq \beta I$ for a.e. $x \in \Omega$.

We accordingly define the bulk energy as

$$E_d(\Gamma, U) = \inf_{v \in \mathcal{C}(\Gamma, U)} \int_{\Omega \setminus \Gamma} W(x, \varepsilon(v)(x)) dx. \tag{4}$$

The total energy of the body for a given crack Γ and a given loading U is then given by

$$E(\Gamma, U) = E_d(\Gamma, U) + E_s(\Gamma). \tag{5}$$

1.2 The Evolution Law

We now propose to follow the response of the body to a time-dependent loading. To this effect a time-parameterized loading $U(t)$ is applied to $\partial_d\Omega$. Assume that a (maybe empty) initial crack Γ_0 is present in the body at the onset of the loading process. Our goal is to determine the evolution of the crack(s) during the loading, *i.e.*, to obtain the time-parameterized mapping $t \mapsto \Gamma(t)$. (In the sequel $\Gamma(t)$ will be called the *cracked state* at time t.) The basic idea is as follows. At a given time t, and for the corresponding loading $U(t)$ the crack state $\Gamma(t)$ will be the closed subset of $\overline{\Omega}$ which minimizes $E(\Gamma, U(t))$ among all cracks Γ which contain all previous $\Gamma(s)$, $s < t$.

There are two important features in this formulation. On the one hand, the driving principle is global energy minimization. On the other hand, the geometry and size of the crack is limited by its predecessors; this is an attempt at expressing the irreversibility of the cracking process, and the absence of healing.

If the conceptual principle is fairly clear, its precise formulation is not so easily laid out. We proceed in two steps, the first of which addresses time-discretized evolutions, while the second pertains to time-continuous evolutions, albeit for a specific subclass of all possible loadings.

Law 1. Discrete evolutions:

Assume that Γ_0 is given. Let U_i, $1 \leq i \leq p$ be a sequence of loadings. Then, the corresponding cracked states Γ_i have to satisfy

$$\Gamma_i \supset \Gamma_{i-1}, \qquad E(\Gamma_i, U_i) \leq E(\Gamma, U_i) \quad \textit{for every } \Gamma \supset \Gamma_{i-1}. \tag{6}$$

As such, the evolution is dicretization-dependent. The real evolution should be constructed as a limit of the discrete evolution as the time-step tends to 0. It has been recently shown in Francfort and Larsen (2002) that a precise law can then be derived for all loading scenarios. In the present study we will only investigate *monotonically increasing loadings*. Assume therefore that the loading is of the form

$$U(x,t) = \begin{cases} tU_0(x), & t \geq 0, \\ 0, & t < 0, \end{cases} \tag{7}$$

for all x's on $\partial_d \Omega$.

Law 2. Continuous monotone evolutions:

The cracked state $\Gamma(t)$ have to satisfy

(i) $\Gamma(t) = \Gamma_0$ *for* $t < 0$, *and* $\Gamma(t)$ *does not decrease with* t,

(ii) $E(\Gamma(t), U(t)) \leq E(\Gamma, U(t))$, *for all* $\Gamma \supset \Gamma^-(t) \equiv \cup_{s<t}\Gamma(s)$,

(iii) $E(\Gamma(t), U(t)) \leq E(\Gamma, U(t))$, *for all* $\Gamma \subset \Gamma(t)$.

The above law immediately calls for few comments. The first condition is merely the requirement that the crack increases with time from its original state Γ_0. (We show in the next paragraph that $\Gamma(0) = \Gamma_0$.) That the energy should be minimum among all possible cracks at a given time is expressed in the second condition which can be formally derived from Law 1 upon replacing t_i by t and Γ_{i-1} by $\Gamma^-(t)$. The third condition is a comparison with the energy that the body would have if it were left in a smaller cracked state but submitted to the present load. It is a condition stronger than that introduced in Francfort and Marigo (1998).

Remark 3. *Noting that there does not exist a total order relation between all possible cracked states, Law 2 simply requires that the crack grows and gives to the body the smallest energy among the subfamily of greater or smaller cracked configurations.*

1.3 Some Elementary Properties

The energies E_d and E_s enjoy the following elementary but very useful properties, direct consequences of their definition :

P 1. *E_s is positive and strictly monotonically increasing in Γ : If $\Gamma_1 \subset \Gamma_2$, then $0 \leq E_s(\Gamma_1) \leq E_s(\Gamma_2)$. Moreover if $E_s(\Gamma_1) = E_s(\Gamma_2) < +\infty$, then $\Gamma_1 = \Gamma_2$ (up to a set of zero $\mathcal{H}^{N-1}$-measure).*

Proof. It is a direct consequence of the assumed strict positivity of the toughness. □

P 2. *E_d is positive and monotonically decreasing in Γ for any fixed U :*

$$E_d(\Gamma_1, U) \geq E_d(\Gamma_2, U) \geq 0 \quad if \quad \Gamma_1 \subset \Gamma_2.$$

Proof. If $\Gamma_1 \subset \Gamma_2$ then $\mathcal{C}(\Gamma_1) \subset \mathcal{C}(\Gamma_2)$, and since the bulk energy $\mathcal{E}_d(\Gamma, U)$ is the infimum on $\mathcal{C}(\Gamma)$, the result follows. □

P 3. *E_d is homogeneous of degree 2 in U :*

$$E_d(\Gamma, tU) = t^2 E_d(\Gamma, U).$$

Proof. It is a direct consequence of the infinitesimal linear elasticity assumption. □

We can immediately deduce that $\Gamma(0) = \Gamma_0$. Indeed, by P3, and *(i)* and *(iii)* of Law 2 we have $E(\Gamma(0), U(0)) = E_s(\Gamma(0)) \leq E_s(\Gamma_0)$. But, by *(i)* of Law 2 we have also $\Gamma_0 \subset \Gamma(0)$ and we can conclude owing to P1.

Another interesting property for the bulk energy is that it is bounded from below as a function of the crack. Let us introduce its greatest lower bound :

$$E_d^{min}(U_0) = \inf_{\Gamma \,:\, E_s(\Gamma) < +\infty} E_d(\Gamma, U_0) \geq 0. \tag{8}$$

The value of $E_d^{min}(U_0)$ depends both on the loading, on the geometry and on the material properties. For instance, we have the following

Proposition 1. *If all the materials as well as all the interfaces are breakable,* i.e. *if $k(x) \leq k_M < \infty$ for all $x \in \overline{\Omega}$, then $E_d^{min}(U_0) = 0$.*

Proof. Indeed, take $\Gamma = \partial_d \Omega$ (its surface energy is finite by hypothesis on the smoothness of $\partial\Omega$ and on the finitude of the toughness), then $v = 0$ becomes an admissible displacement field. Therefore we deduce from (4) and P2 that $E_d(\Gamma, U_0) = 0$. □

In other contexts the lowest bulk energy can be strictly positive. It is the case in the pull-out problem treated in Section 3 where the interface alone is breakable, and the fiber and the matrix are clamped at one end. Moreover it could happen that this infimum is reached by no cracked state.

1.4 Elementary Examples

As a first test of the proposed formulation, three examples for which exact solutions can be computed are presented below: that of a one-dimensional bar under tension for which Law 2 is used, that of a three-dimensional cylinder under uniaxial traction (controlled by a hard device) and that of the tearing of a reinforcement in a cylindrical domain for which Law 1 is used.

Tension of a bar. Assume that $\Omega = (0, L)$, that the bar is homogeneous with stiffness ES and fracture toughness kS (S representing the area of the cross-section of the "real" three-dimensional body). Assume also that the bar is initially uncracked, $\Gamma_0 = \emptyset$, fixed at one end and submitted to an increasing displacement at the other end :

$$u(0-) = 0 \qquad u(L+) = t. \tag{9}$$

Accordingly, the set of admissible displacements reads as

$$\mathcal{C}(\Gamma, t) = \{v \in W^{1,2}([0, L] \setminus \Gamma) \; : \; v(0-) = 0, \; v(L+) = t\}, \tag{10}$$

while the possible cracked configurations of the bar with finite energy are

$$\Gamma = \emptyset \qquad \text{or} \qquad \Gamma = \{x_1, \cdots, x_n\} \quad \text{with} \quad n \geq 1, \quad 0 \leq x_1 < \cdots < x_n \leq L. \tag{11}$$

Recalling that the bulk energy is given by

$$E_d(\Gamma, t) = \inf_{v \in \mathcal{C}(\Gamma, t)} \frac{1}{2} \int_0^L ESv'(x)^2 \, dx, \tag{12}$$

an easy computation would show that the displacement field, say $u(\emptyset, t)$, which achieves the infimum for the crack free bar is given by

$$u(\emptyset, t)(x) = t\frac{x}{L} \tag{13}$$

and the corresponding energies are

$$E_s(\emptyset) = 0, \qquad E(\emptyset, t) = E_d(\emptyset, t) = \frac{ESt^2}{2L}. \tag{14}$$

On the other hand, when the crack is $\Gamma = \{x_1, \cdots, x_n\}$ with $n \geq 1$, we easily find that the bulk energy vanishes and then

$$E_d(\Gamma, t) = 0 \qquad E(\Gamma, t) = E_s(\Gamma) = nkS. \tag{15}$$

The graphs of the total energy corresponding to these different cracked configurations versus the load parameter are plotted in Figure 1. We deduce then the following result

Proposition 2. *The bar remains crack-free as long as the loading parameter t remains strictly less than $t_f = \sqrt{\frac{2kL}{E}}$. For $t > t_f$, the bar is cut into two pieces at an arbitrary point $x_1 \in [0, L]$. In other words, the (family indexed by x_1 of) crack evolution is*

$$\Gamma(t) = \begin{cases} \emptyset, & \text{when} \quad t < t_f \\ \emptyset \text{ or } \{x_1\} & \text{when} \quad t = t_f \\ \{x_1\} & \text{when} \quad t > t_f \end{cases} \tag{16}$$

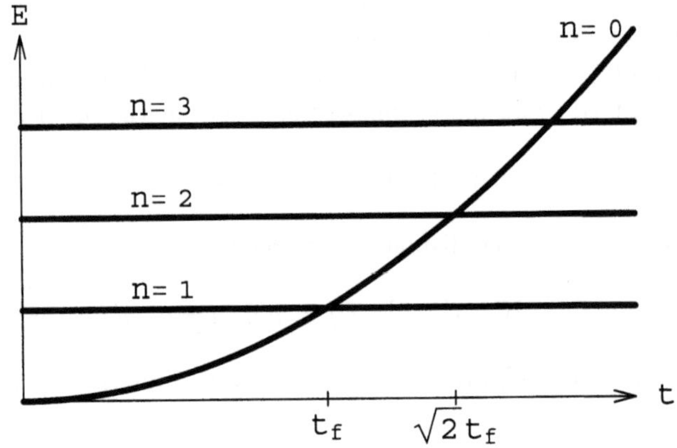

Figure 1. Graphs of the total energy for the different cracked configurations of the bar.

Proof. ▹ The proof is divided into 2 steps.

(i) *Existence.* Let us first verify that the proposed crack evolution $t \mapsto \Gamma(t)$ satisfied the three items of Law 2. $\Gamma(0) = \emptyset$ and $\Gamma(t)$ grows with t, hence item *(i)* is satisfied. It can be easily verified on Figure 1 that, for each t, $\Gamma(t)$ is the cracked state giving the smallest energy among the set of all possible cracked states. Therefore items *(ii)* and *(iii)* are satisfied. □

(ii) *Uniqueness.* Let us show now that there does not exist another solution. Let $t \mapsto \Gamma(t)$ be an evolution verifying items *(i) –(iii)* and $t \mapsto n(t)$ the corresponding evolution of the number of cuts. When $t < t_f$, by choosing $\Gamma = \emptyset$ in item *(iii)* we obtain that $E(\Gamma(t), t) \leq E(\emptyset, t)$, which is only possible if $\Gamma(t) = \emptyset$. Consider now t such that $t_f < t < \sqrt{2}\, t_f$. By choosing once more $\Gamma = \emptyset$ in item *(iii)* , we deduce from Figure 1 that $0 \leq n(t) \leq 1$. But $n(t) = 0$ is impossible by *(ii)* . Indeed, if $n(t) = 0$ we could choose $\Gamma = \{x_1\}$ in *(ii)* and we should obtain $E(\emptyset, t) \leq kS$, *i.e.* a contradiction. Therefore $n(t) = 1$, the bar is cut at one point. But by virtue of *(i)* this point must be the same at each t. So $\Gamma(t) = \{x_1\}$ when $t_f < t < \sqrt{2}\, t_f$. Consider now $t \geq \sqrt{2}\, t_f$. Item *(i)* requires that $\Gamma(t) \supset \{x_1\}$. By choosing $\Gamma = \{x_1\}$, item *(iii)* requires that $E(\Gamma(t), t) \leq kS$ and therefore that $n(t) = 1$. But then *(i)* allows to conclude that $\Gamma(t) = \{x_1\}$. □ ◃

Uniaxial traction of a three-dimensional cylinder. Assume that $\Omega = S \times (0, L)$, with S smooth open connected domain of $\mathbb{R}^2$, that the material is isotropic and homogeneous with Young's modulus E, Poisson's ratio ν, and fracture toughness $k(x) = k$. Assume also that $\Gamma_0 = \emptyset$, and that $u_3 = 0$ on $S \times \{0\}$ while $u_3 = t$ on $S \times \{L\}$. In other words, the boundary conditions look like

$$\begin{cases} u_3 = 0 \; and \; \sigma_{13} = \sigma_{23} = 0 \text{ on } S \times \{0\}, \\ u_3 = t \; and \; \sigma_{13} = \sigma_{23} = 0 \text{ on } S \times \{L\}, \\ \qquad \sigma \cdot n = 0 \qquad \text{on } \partial S \times (0, L). \end{cases} \tag{17}$$

Correspondingly, the set of admissible displacements reads as

$$\mathcal{C}(\Gamma,t)=\{v\in W^{1,2}(\Omega\setminus\Gamma;\mathbb{R}^3)\ :\ v_3=0\text{ in }S\times\{0\}\setminus\Gamma, v_3=t\text{ in }S\times\{L\}\setminus\Gamma\}. \quad (18)$$

An easy computation would show that the displacement field at equilibrium for the crack free cylinder is given (up to an admissible rigid displacement) by

$$u(\emptyset,t)(x)=-\nu t\frac{x_1}{L}e_1-\nu t\frac{x_2}{L}e_2+t\frac{x_3}{L}e_3, \quad (19)$$

the corresponding stress field reads as

$$\sigma(\emptyset,t)(x)=E\frac{t}{L}e_3\otimes e_3. \quad (20)$$

The corresponding energies are

$$E_s(\emptyset)=0,\qquad E(\emptyset,t)=E_d(\emptyset,t)=\frac{1}{2}E\frac{t^2}{L}\mathcal{H}^2(S). \quad (21)$$

Remark 4. *In the absence of crack, the stress field is then uniaxial but the longitudinal extension is controlled. This type of control is essential for the questions of energy minimization, see* Charlotte *et al. (2000). We will call such a test a uniaxial traction controlled by a hard device.*

Let Γ be an arbitrary crack, denote by $P(\Gamma)$ its projection onto S and define

$$\Theta(\Gamma)=\frac{\mathcal{H}^2(P(\Gamma))}{\mathcal{H}^2(S)}. \quad (22)$$

The energy of the body in an arbitrary cracked state is not explicitly computable but an estimate from below will prove sufficient for our purpose. This is the object of the following

Proposition 3. $E(\Gamma,t)\geq(1-\Theta(\Gamma))E(\emptyset,t)+\Theta(\Gamma)k\mathcal{H}^2(S)$. *The equality holds if and only if* $\Gamma=\emptyset$ *or* $\Gamma=S\times\{z\}$ *with* $0\leq z\leq L$.

Proof. The equality is trivial when $\Gamma=\emptyset$. When $\Gamma=S\times\{z\}$, the equality holds because $E_d(\Gamma,t)=0$ and $\Theta(\Gamma)=1$. Let us consider now an arbitrary Γ. Then

$$E_s(\Gamma,t)\geq\Theta(\Gamma)k\mathcal{H}^2(S), \quad (23)$$

the inequality being strict unless $\mathcal{H}^2(P(\Gamma))=\mathcal{H}^2(\Gamma)$, that is unless Γ is transverse. For the strain energy, classical dual methods in linear elasticity give

$$\begin{aligned}E_d(\Gamma,t)&=\inf_{v\in\mathcal{C}(\Gamma,t)}\frac{1}{2}\int_{\Omega_\Gamma}A\varepsilon(v)\cdot\varepsilon(v)dx\\&=\inf_{v\in\mathcal{C}(\Gamma,t)}\sup_{\tau\in[L^2(\Omega_\Gamma)]^9_s}\int_{\Omega_\Gamma}(\tau\cdot\varepsilon(v)-\frac{1}{2}A^{-1}\tau\cdot\tau)dx\\&\geq\sup_{\tau\in[L^2(\Omega_\Gamma)]^9_s}\inf_{v\in\mathcal{C}(\Gamma,t)}\int_{\Omega_\Gamma}(\tau\cdot\varepsilon(v)-\frac{1}{2}A^{-1}\tau\cdot\tau)dx\\&\geq\inf_{v\in\mathcal{C}(\Gamma,t)}\int_{\Omega_\Gamma}(\sigma(t)\cdot\varepsilon(v)-\frac{1}{2}A^{-1}\sigma(t)\cdot\sigma(t))dx, \quad (24)\end{aligned}$$

where

$$\sigma(t) = \begin{cases} 0 & in \quad P(\Gamma) \times (0, L) \\ \sigma(\emptyset, t) & elsewhere \end{cases} \tag{25}$$

is an admissible stress field. Now each element of $\mathcal{C}(\Gamma, t)$ has a third component with a well-defined trace on $(S \setminus P(\Gamma)) \times \{0 \text{ (resp.} L)\}$, namely 0 (resp. t). Thus, since the only non-zero component of $\sigma(t)$ is its 3×3 component,

$$\begin{aligned}\int_{\Omega_\Gamma} \sigma(t) \cdot \varepsilon(v) dx &= \int_{\Omega_\Gamma} \sigma_{33}(t) \cdot \frac{\partial v_3}{\partial x_3} dx \\ &= \int_{(S \setminus P(\Gamma)) \times (0,L)} \sigma_{33}(\emptyset, t) \cdot \frac{\partial u_3(\emptyset, t)}{\partial x_3} dx \\ &= \int_{(S \setminus P(\Gamma)) \times (0,L)} (\sigma(\emptyset, t) \cdot \varepsilon(u(\emptyset, t)) dx.\end{aligned}$$

Thus,

$$\begin{aligned}E_d(\Gamma, t) &\geq \int_{S \setminus P(\Gamma) \times (0,L)} (\sigma(\emptyset, t) \cdot \varepsilon(u(\emptyset, t)) - 1/2 A^{-1} \sigma(\emptyset, t) \cdot \sigma(\emptyset, t)) dx \\ &= (1 - \Theta(\Gamma)) E(\emptyset, t),\end{aligned} \tag{26}$$

which, together with (23), proves the inequality of the Proposition.

Assume that the equalities hold in (23) and (24). Then Γ is transverse, $\sigma(t)$ is the true stress field $\sigma(\Gamma, t)$, and $u(\Gamma, t)_{3,3}$ is equal to 0 in $(P(\Gamma) \times (0, L)) \setminus \Gamma$ and to t/L in $(S \setminus P(\Gamma)) \times (0, L)$. When $t \neq 0$, the continuity of $u(\Gamma, t)_3$ on $\partial P(\Gamma) \times (0, L)$ is then impossible, unless $P(\Gamma) = \emptyset$ or $P(\Gamma) = S$. The former case corresponds to $\Gamma = \emptyset$. In the latter case, we obtain $\sigma(\Gamma, t) = 0$ and $\varepsilon(u(\Gamma, t)) = 0$ in $\overline{\Omega} \setminus \Gamma$. Hence $u(\Gamma, t)$ is a rigid displacement in each connected component of $\overline{\Omega} \setminus \Gamma$. By the boundary conditions there exists at least two connected components, but since Γ is transverse there exists exactly two connected components. Hence Γ is of the form $S \times \{z\}$, $0 \leq z \leq L$. □

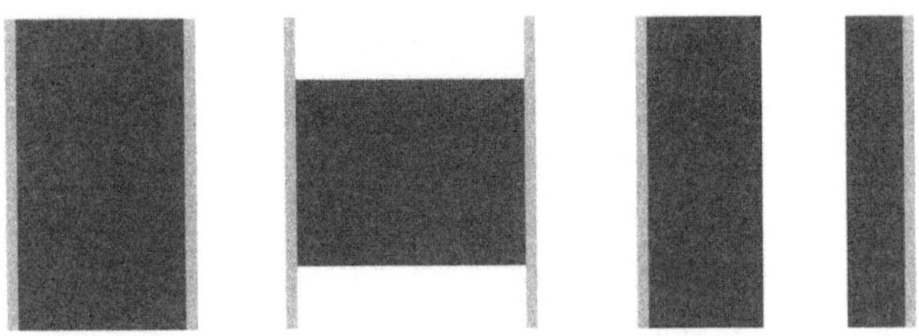

Figure 2. *Failure of the cylinder under uniaxial traction controlled by a hard device.*

We are now in a position to prove the following

Proposition 4. *The cylinder remains crack-free as long as the loading parameter t remains strictly less than $t_f = \sqrt{\frac{2kL}{E}}$. For $t > t_f$, a solution consists in cutting the cylinder into two pieces along an arbitrary transverse section.*

Furthermore these are the unique evolutions which are limits of evolutions satisfying Law 1 when the step of discretization goes to 0.

Proof. Let $z \in [0, L]$ and let $\Gamma_z(t)$ be defined for $t \geq 0$ by

$$\Gamma_z(t) = \begin{cases} \emptyset & \textit{when} \quad t < t_f \\ \emptyset \textit{ or } S \times \{z\} & \textit{when} \quad t = t_f \\ S \times \{z\} & \textit{when} \quad t > t_f \end{cases}. \tag{27}$$

$\Gamma_z(t)$ is a global minimizer of the energy:

$$E(\Gamma_z(t), t) \leq E(\Gamma, t), \qquad \forall \Gamma. \tag{28}$$

Indeed, $E(\Gamma_z(t), t) = E(\emptyset, t) \leq k\mathcal{H}^2(S)$ when $t \leq t_f$ while $E(\Gamma_z(t), t) = k\mathcal{H}^2(S) \leq E(\emptyset, t)$ when $t \geq t_f$. Thus, by virtue of Proposition 3,

$$E(\Gamma, t) \geq (1 - \Theta(\Gamma))E(\emptyset, t) + \Theta(\Gamma)k\mathcal{H}^2(S) \geq \min\{E(\emptyset, t), k\mathcal{H}^2(S)\} = E(\Gamma_z(t), t)$$

and the first inequality is strict unless $\Gamma = \emptyset$ or $\Gamma = S \times \{z'\}$ for some $z' \in [0, L]$.

Let us consider an arbitrary discretization of the loading: $0 = t_0 < t_1 < \cdots < t_n < \cdots$. It is clear that $\{\Gamma_z(t_n)\}_{n \in \mathbb{N}}$ satisfies Law 1 because $\Gamma_z(t_n)$ is increasing with n and an energy minimizer.

Let $\{\Gamma_n\}_{n \in \mathbb{N}}$ be another solution. Since $\Gamma_0 = \emptyset$, Law 1 requires that Γ_1 is a global minimizer of the energy at t_1. Hence $\Gamma_1 = \emptyset$ if $t_1 < t_f$. By repeating the same argument, as long as $t_i < t_f$ Law 1 yields $\Gamma_i = \emptyset$. Let n such that $t_{n-1} < t_f$ and $t_n > t_f$. By Law 1 Γ_n must be a global minimizer of the energy at t_n. Hence $\Gamma_n = S \times \{z\}$ for some $z \in [0, L]$. Since $\Gamma_{n+1} \supset \Gamma_n$, since $k\mathcal{H}^2(S) = E(\Gamma_n, t_{n+1})$ and since $k\mathcal{H}^2(S)$ is the minimum of the energy at t_{n+1}, Law 1 requires that $\Gamma_{n+1} = \Gamma_n$. By induction, $\Gamma_i = \Gamma_n$ for all $i \geq n$. We have then proved that $\Gamma_i = \Gamma_z(t_i)$ for all i whatever the load-discretization. (The value of z is of course dependent on the discretization, but that indetermination is by essence due to the non uniqueness of the solution.) □

Tearing of a reinforcement. A three-dimensional cylinder $\Omega = S \times (0, L)$ is investigated; its cross-section S is an annulus of inner (resp. outer) radius R_i (resp. R_o); it is made of a homogeneous, isotropic elastic material with Lamé coefficients λ, μ and toughness k. A rigid, unbreakable reinforcement is bonded to the inner surface C_i of the cylinder. Assume that the cylinder is initially crack-free ($\Gamma(t) = \emptyset, t < 0$), and that the following load is applied:

$$u_1 = u_2 = \sigma_{33} = 0 \text{ on } S \times \{0\}, S \times \{L\}; \; u = 0 \text{ on } \Sigma_o; u = te_3 \text{ on } C_i, \tag{29}$$

where Σ_o denotes the outer surface of the cylinder. Therefore the set of kinematically admissible displacement fields reads as

$$\begin{aligned}\mathcal{C}(\Gamma,t) = \{\ v \in W^{1,2}(\Omega \setminus \Gamma; \mathbb{R}^3)\ :\\ v_1 = v_2 = 0 \text{ on } (S \times \{0,L\}) \setminus \Gamma,\\ v = 0 \text{ on } \Sigma_o \setminus \Gamma, v = te_3 \text{ on } \mathcal{C}_i \setminus \Gamma\},\end{aligned} \tag{30}$$

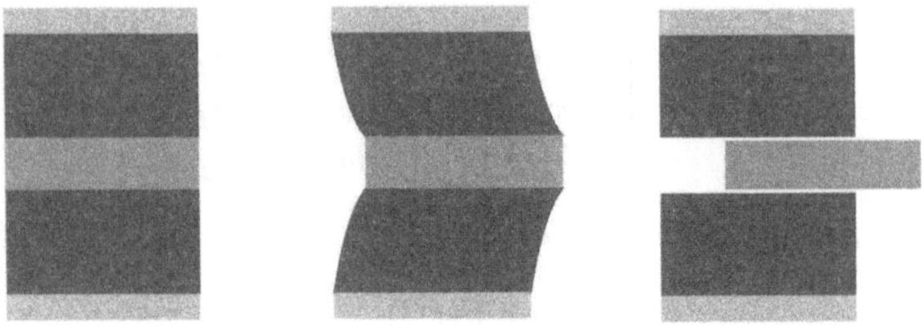

Figure 3. *Tearing of a reinforcement.*

The result is the following

Proposition 5. *If we consider only "axial cracks", that is cracks of the form* $\Gamma = \gamma \times (0,L)$, $\gamma \subset \bar{S}$, *then the unique evolution which is the limit of an evolution satisfying Law 1 when the step of discretization goes to 0 is the following:*

The cylinder remains crack-free as long as

$$t < t_i = t_f \equiv \sqrt{\frac{2kR_i}{\mu}\log(\frac{R_o}{R_i})}, \tag{31}$$

while there is total debonding of the reinforcement from the inner surface of the cylinder as soon as $t > t_f$.

Proof. The proof is very similar to that of Proposition 4; it is sketched below. We must note that, by virtue of the assumptions on the axial form of the cracks and of the assumed boundary conditions, the problem becomes an antiplane problem.

The crack-free equilibrium fields are

$$u(\emptyset,t) = \frac{\log(r/R_o)}{\log(R_i/R_o)} te_3, \tag{32}$$

$$\sigma(\emptyset,t) = \frac{t\mu}{2r\log(R_i/R_o)}(e_r \otimes e_3 + e_3 \otimes e_r), \tag{33}$$

with corresponding energies

$$E(\emptyset, t) = E_d(\emptyset, t) = \pi\mu L \frac{t^2}{\log(R_o/R_i)}. \tag{34}$$

Let $\Gamma = \gamma \times (0, L)$ be an arbitrary axial cracked configuration. A lower bound on the energy $E(\Gamma, t)$ is first established by projecting Γ onto the cylinder with radius R_i.

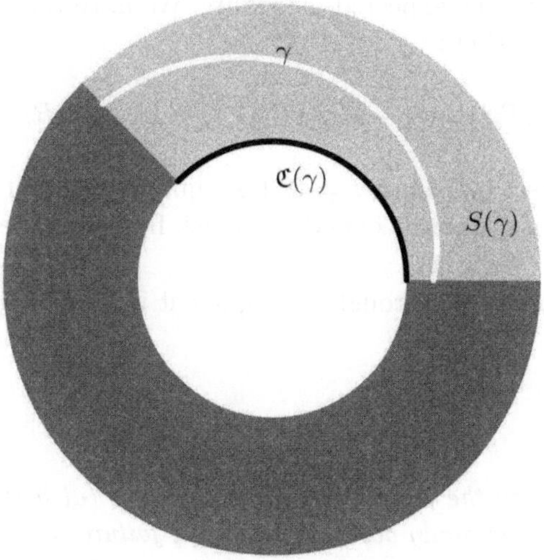

Figure 4. The trace of the axial crack on the cross section and its projection.

Let $\mathfrak{C}(\gamma)$ be the radial projection of γ onto the inner circle $\mathfrak{C}_i = \{[R_i, \theta] \ : \ 0 \leq \theta \leq 2\pi\}$[1], *i.e.*, see Figure 4,

$$\mathfrak{C}(\gamma) = \{[R_i, \theta] \ : \ \exists r \ \textit{such that} \ [r, \theta] \in \gamma\} \subset \mathfrak{C}_i, \tag{35}$$

$\omega(\gamma)$ the length fraction of $\mathfrak{C}(\gamma)$ with respect to $\mathfrak{C}_i$:

$$\omega(\gamma) = \frac{\mathcal{H}^1(\mathfrak{C}(\gamma))}{2\pi \ R_i} \tag{36}$$

and $S(\gamma)$ the angular sector of the cross section which contains the crack, see Figure 4 :

$$S(\gamma) = \left\{[r, \theta] \in \overline{S} \ : \ [R_i, \theta] \in \mathfrak{C}(\gamma)\right\}. \tag{37}$$

We first establish the following inequalities :

$$E_s(\Gamma, t) \geq \omega(\gamma) k 2\pi R_i L, \qquad E_d(\Gamma, t) \geq (1 - \omega(\gamma))\, E_d(\emptyset, t). \tag{38}$$

[1] The square brackets denote polar coordinates of a point of the cross section, e.g. $[r, \theta] \in S$.

The first inequality is a direct consequence of the definition of $\mathfrak{C}(\gamma)$, projection of γ onto $\mathfrak{C}_i$. Moreover the equality holds if and only if $\gamma = \mathfrak{C}(\gamma) \subset \mathfrak{C}_i$. To obtain the second inequality, we use, like in the case of the traction test, dual estimates by choosing as the statically stress field τ :

$$\tau(x) = \begin{cases} 0 & if\ x \in S(\gamma) \times (0, L), \\ \sigma(\emptyset, t) & otherwise. \end{cases} \tag{39}$$

A simple calculation gives the expected inequality. We have then obtained the following estimate of the energy $E(\Gamma, t)$:

$$E(\Gamma, t) \geq (1 - \omega(\gamma))\, E(\emptyset, t) + \omega(\gamma) k 2\pi R_i L. \tag{40}$$

Furthermore, we deduce that, when $t \neq 0$, the equality holds if and only if $\gamma = \emptyset$ or $\gamma = \mathfrak{C}_i$, *i.e.* if and only if either the cylinder is crack free or the reinforcement is entirely debonded.

The result is then obtained through an argument identical to that developed in the proof of Proposition 4. □

Remark 5.

1. *The solutions given by the least energy principle are (at least qualitatively) in good agreement with the commonly observed modes of failure for these two basic tests.*
2. *In both cases, since the body is initially crack free, Griffith law is unable to assess the onset of cracking.*
3. *By using the three items of the Law 2, the solution is unique for the latter test. (The non uniqueness for the traction test is of course expected.)*
4. *In both cases the cracking is brutal.*

2 General Properties

This section investigates various characteristics of the crack evolution as predicted by the least energy principle. We compare them with those foreseen by Griffith's theory of brittle fracture in which the crack evolution is based on the criterion of the critical energy release rate, see Proposition 8 where a precise statement of Griffith's law is given (in a particular case). Our analysis is, as in the previous section, restricted to the no-contact case between the lips of the cracks, to linear elastic materials and to increasing boundary displacement loadings of the form

$$U(x, t) = \begin{cases} tU_0(x)\ , t \geq 0, \\ 0 \qquad\ \ , t < 0. \end{cases}$$

2.1 Crack Initiation and Failure of the Sample

We propose to demonstrate in this subsection that the model predicts crack initiation in a crack-free environment, in striking contrast with Griffith's theory; we also evidence eventual mechanical failure under monotonic loadings. Let us denote by t_i the initiation load, *i.e.*, the load when the cracked state first evolves (recall that Γ_0 can be empty),

$$t_i = \sup\{t \; : \; \Gamma(t) = \Gamma_0\}. \tag{41}$$

By using the infimum $E_d^{min}(U_0)$ of the bulk energy introduced in (8), we immediately obtain:

Proposition 6.

$$t_i = \infty \text{ if and only if } E_d(\Gamma_0, U_0) = E_d^{min}(U_0).$$

Proof. Since $\Gamma(t) \supset \Gamma_0$ by *(i)* of Law 2, $E_s(\Gamma(t)) \geq E_s(\Gamma_0)$. If $E_d(\Gamma_0, U_0) = E_d^{min}(U_0)$, then, because of the monotonicity and the positivity of the bulk energy, $E_d(\Gamma(t), U_0) = E_d^{min}(U_0)$ for all $t > 0$. According to *(iii)* in Law 2, $E_s(\Gamma(t)) \leq E_s(\Gamma_0)$. Then the surface energies are equal and, recalling P1, $\Gamma(t) = \Gamma_0$.

Conversely, if $t_i = \infty$, then $\Gamma(t) = \Gamma_0, \forall t > 0$, and by virtue of (ii) in Law 2,

$$t^2 E_d(\Gamma_0, U_0) + E_s(\Gamma_0) \leq t^2 E_d(\Gamma, U_0) + E_s(\Gamma), \text{ for all } \Gamma \supset \Gamma_0.$$

Let Γ_n' be a minimizing sequence for the bulk energy, *i.e.* a sequence such that

$$\lim_{n \to \infty} E_d(\Gamma_n', U_0) = E_d^{min}(U_0).$$

Take $\Gamma_n = \Gamma_0 \cup \Gamma_n'$, then Γ_n is also a minimizing sequence because $E_d(\Gamma_n, U_0) \leq E_d(\Gamma_n', U_0)$. We first obtain

$$t^2 E_d(\Gamma_0, U_0) + E_s(\Gamma_0) \leq t^2 E_d(\Gamma_n, U_0) + E_s(\Gamma_n), \qquad \forall t > 0,$$

then dividing by t^2 and passing to the limit when t goes to ∞ give

$$E_d(\Gamma_0, U_0) \leq E_d(\Gamma_n, U_0)$$

which is only possible if $E_d(\Gamma_0, U_0) = E_d^{min}(U_0)$. □

Remark 6. *An immediate consequence of the above Proposition is that: if $\Gamma_0 = \emptyset$ (no preexisting crack), all materials and interfaces are breakable and U_0 is not compatible with a rigid displacement, then* **a crack will always appear at a finite load**. *(Indeed, in such a case $E_d(\emptyset, U_0) > E_d^{min}(U_0) = 0$.)*

Concerning the complete fracture, we have

Proposition 7.

(i) $\lim_{t \to \infty} E_d(\Gamma(t), U_0) = E_d^{min}(U_0)$,
(ii) If there exists a finite load t_r such that $E_d(\Gamma(t_r), U_0) = E_d^{min}(U_0)$, then $\Gamma(t) = \Gamma(t_r), \forall t \geq t_r$.

The greatest lower bound of such loads t_r is referred to as the failure load for the sample and denoted by t_f.

Proof. Let Γ'_n be a minimizing sequence for the bulk energy. For a given n, let $\Gamma = \Gamma'_n \cup \Gamma(t)$. Therefore, (ii) in Law 2 implies that

$$\begin{aligned} t^2 E_d(\Gamma(t), U_0) + E_s(\Gamma(t)) &\leq t^2 E_d(\Gamma, U_0) + E_s(\Gamma) \\ &\leq t^2 E_d(\Gamma'_n, U_0) + E_s(\Gamma'_n) + E_s(\Gamma(t)). \end{aligned}$$

Dividing by t^2 in the above inequality and letting t tend to ∞ yields

$$\lim_{t\to\infty} E_d(\Gamma(t), U_0) \leq E_d(\Gamma'_n, U_0),$$

then passing to the limit in n gives *(i)* .

If $E_d(\Gamma(t_r), U_0) = E_d^{min}(U_0)$ for some t_r, then *(iii)* in Law 2 implies that $E_s(\Gamma(t)) \leq E_s(\Gamma(t_r))$ for $t \geq t_r$, and, since $\Gamma(t) \supset \Gamma(t_r)$, (ii) is established. □

Remark 7. *This result demonstrates that, under monotonic loadings, the crack will expand until a configuration of minimum elastic energy. This can be interpreted as a configuration of mechanical failure. This, together with Remark 6, provides a rather complete picture of crack evolution in our main setting, namely that of no-contact between the lips of the cracks and linear elasticity:* **A crack will appear at a finite load and will expand until mechanical failure at a finite or infinite load.**

Remark 8. *Whenever $0 < t_i = t_f < \infty$ the crack evolution is as follows:*

$$\Gamma(t) = \begin{cases} \Gamma_0 & \text{if} \quad t < t_i = t_f, \\ \Gamma_f & \text{if} \quad t > t_i = t_f, \end{cases} \tag{42}$$

where Γ_f is such that $E_d(\Gamma_f, U_0) = E_d^{min}(U_0)$. Such is the case in the three examples developed in Section 1.4.

2.2 Griffith versus Energy Minimization

This subsection is devoted to a comparison between Griffith's theory and the least energy principle. The framework is that of two-dimensional linear elasticity. The following is assumed throughout this subsection:

Hypothesis (Predefined crack path). *The crack path is a rectifiable curve Γ of finite length L with origin x_0 and end point x_L. It is parameterized by its arc-length $x(s)$ with $x(0) = x_0$. Define*

$$\Gamma(l) \equiv \Gamma_0 \cup \{x(s); 0 \leq s \leq l\}. \tag{43}$$

Then the bulk and surface energies become functions of l, namely,

$$\mathcal{E}(l) \equiv E_d(\Gamma(l), U_0), \qquad \mathcal{G}(l) \equiv E_s(\Gamma(l)). \tag{44}$$

These functions are respectively monotonically decreasing and increasing with l; we also assume that they are as smooth as needed.

It now remains to investigate the trajectory of the crack along its path, *i.e.*, the function $t \to l(t)$. The first result concerns progressive crack evolution.

Proposition 8. *If $l(t)$ is an absolutely continuous function of t, then the crack evolution follows Griffith's law. In other words,*
(i) $l(0) = 0$,
(ii) $\dot{l}(t) \geq 0$,
(iii) $-t^2\mathcal{E}'(l(t)) \leq k(x(l(t)))$,
(iv) $\dot{l}(t) = 0$ *if the inequality is strict in (iii)* .

Proof. Items *(i)* and *(ii)* are obvious. Item *(iii)* is obtained by applying *(ii)* of Law 2 at load t, with $\Gamma(l(t)+h), h \geq 0$ as test crack:

$$t^2\mathcal{E}(l(t)) + \int_0^{l(t)} k(x(s))ds \leq t^2\mathcal{E}(l(t)+h) + \int_0^{l(t)+h} k(x(s))ds. \tag{45}$$

Dividing by h and passing to the limit in h give *(iii)* .

Item *(iv)* is obtained by applying *(iii)* of Law 2 at load t with $\Gamma(l(t-h))$ as test crack:

$$t^2\mathcal{E}(l(t)) + \int_0^{l(t)} k(x(s))ds \leq t^2\mathcal{E}(l(t-h)) + \int_0^{l(t-h)} k(x(s))ds. \tag{46}$$

Dividing by h and passing to the limit in h give $\Big(t^2\mathcal{E}'(l(t)) + k\big(x(l(t))\big)\Big)\dot{l}(t) \leq 0$. By using items *(ii)* and *(iii)* of Griffith's law the result follows. □

Remark 9. *Item (iii) in the above Proposition is precisely Griffith's criterion, since $-t^2\mathcal{E}'(l(t))$ represents the energy release rate. Item (iv) is the classical statement that propagation does not take place unless the energy release rate becomes critical.*

If now $l(t)$ experiences a jump at load t_0, the following Proposition holds:

Proposition 9 (The jump condition). *If at load t_0, $l^-(t_0) \neq l^+(t_0)$, then,*

$$-t_0^2\Big(\mathcal{E}(l^+(t_0)) - \mathcal{E}(l^-(t_0))\Big) = \mathcal{G}(l^+(t_0)) - \mathcal{G}(l^-(t_0)). \tag{47}$$

Proof. Apply *(ii)* of Law 2 to load $t_0 - h$ with $\Gamma(l(t_0+h))$ as test crack; then,

$$(t_0-h)^2\mathcal{E}(l(t_0-h)) + \mathcal{G}(l(t_0-h)) \leq (t_0-h)^2\mathcal{E}(l(t_0+h)) + \mathcal{G}(l(t_0+h)), \tag{48}$$

and passing to the limit in h yields one inequality. The reverse inequality is obtained by applying *(iii)* of Law 2 to $t = t_0 + h$, taking $\Gamma(l(t_0-h))$ as test crack, so that

$$(t_0+h)^2\mathcal{E}(l(t_0+h)) + \mathcal{G}(l(t_0+h)) \leq (t_0+h)^2\mathcal{E}(l(t_0-h)) + \mathcal{G}(l(t_0-h)) \tag{49}$$

which yields the result upon passing to the limit in h. □

Remark 10. *The above condition ensures nothing but the continuity of the total energy with respect to the load parameter : even when the crack evolves brutally, the finite release of strain energy is just used to produce the needed finite surface energy. The reader must note that this "energy conservation" during a brutal evolution of a crack is deduced from the least energy principle, it is thus an optimality condition and not an extra condition like in* Hashin *(1996). In a homogeneous medium it can be reinterpreted as a propagation criterion for the mean energy release rate, i.e.,*

$$\frac{-t_0^2}{l^+(t_0)-l^-(t_0)}\int_{l^-(t_0)}^{l^+(t_0)}\frac{d\mathcal{E}}{dl}(l)dl=k. \tag{50}$$

2.3 Progressive Evolution and Brutal Evolution

In this subsection, we investigate the circumstances that preside over the brutal versus progressive character of crack growth. We always adopt the assumption of "predefined crack path", as detailed in the previous subsection, and we reparameterize the crack path in terms of its surface energy by setting

$$\lambda=\mathcal{G}(l),\qquad \hat{\mathcal{E}}(\lambda)=\mathcal{E}(\mathcal{G}^{-1}(\lambda)); \tag{51}$$

such a change is licit since, in view of Remark 1, $\mathcal{G}$ is a continuous strictly increasing function of l. Note that $\hat{\mathcal{E}}$ is a monotonically decreasing function of λ in view of (44).

Remark 11. *In the present context the evolution law described in Law 2 requires that the crack trajectory $t\to\lambda(t)$ must be such that*
(i) $\lambda(t)$ *is a monotonically increasing function of* t,
(ii) $t^2\hat{\mathcal{E}}(\lambda(t))+\lambda(t)\le t^2\hat{\mathcal{E}}(\lambda)+\lambda,\quad \forall\lambda\ge\lambda^-(t),$
(iii) $t^2\hat{\mathcal{E}}(\lambda(t))+\lambda(t)\le t^2\hat{\mathcal{E}}(\lambda)+\lambda,\quad \forall\lambda\le\lambda(t).$

We immediately obtain the following

Proposition 10. *If $\hat{\mathcal{E}}$ is continuous, then $\lambda(t)$ lives, for each t, among the minimizers of $t^2\hat{\mathcal{E}}(\lambda)+\lambda$ on $[0,\mathcal{G}(L)]$.*

Proof. A minimizer exists for each t because the energy function is continuous on a compact set. Let $\lambda(t)$ be (one of) the minimizer(s) of $t^2\hat{\mathcal{E}}(\lambda)+\lambda$ on $[0,\mathcal{G}(L)]$. Of course, it satisfies *(ii)* and *(iii)* . Moreover by using the optimality of $\lambda(t_1)$ and $\lambda(t_2)$ at t_1 and t_2 respectively, we get

$$t_1^2\hat{\mathcal{E}}(\lambda(t_1))+\lambda(t_1)\le t_1^2\hat{\mathcal{E}}(\lambda(t_2))+\lambda(t_2),$$
$$t_2^2\hat{\mathcal{E}}(\lambda(t_2))+\lambda(t_2)\le t_2^2\hat{\mathcal{E}}(\lambda(t_1))+\lambda(t_1).$$

Adding the two inequalities yields $(t_1^2-t_2^2)\big(\hat{\mathcal{E}}(\lambda(t_1))-\hat{\mathcal{E}}(\lambda(t_2))\big)\le 0$ and the required growth *(i)* of $\lambda(t)$ follows from the monotonicity of $\hat{\mathcal{E}}$.

Conversely, if $\lambda(t)$ verifies *(i)* –*(iii)* , then it is a minimizer (note that $\lambda^-(t)\le\lambda(t)$ by *(i)*). □

The brutal versus progressive character of crack growth is then intimately linked to the convexity properties of $\hat{\mathcal{E}}$. Specifically, we obtain the following

Proposition 11.

(i) If $\hat{\mathcal{E}}$ is C^1 and strictly convex, then $\lambda(t)$ is given by

$$\lambda(t) = \begin{cases} 0 & if\ 0 \le t \le t_i, \\ \hat{\mathcal{E}}'^{-1}(-\frac{1}{t^2}) & if\ t_i \le t \le t_L, \\ \mathcal{G}(L) & if\ t_L \le t, \end{cases}, \tag{52}$$

with

$$\begin{cases} t_i = \sqrt{1/(-\hat{\mathcal{E}}'(0))}, \\ t_L = \sqrt{1/(-\hat{\mathcal{E}}'(\mathcal{G}(L)))}; \end{cases} \tag{53}$$

in other words, the crack growth is progressive,

(ii) If $\hat{\mathcal{E}}$ is concave, then $\lambda(t)$ is given by

$$\lambda(t) = \begin{cases} 0 & if\ 0 \le t < t_i = t_L, \\ \mathcal{G}(L) & if\ t_i = t_L < t, \end{cases}, \tag{54}$$

with

$$t_L \equiv \sqrt{\mathcal{G}(L)/(\hat{\mathcal{E}}(0) - \hat{\mathcal{E}}(\mathcal{G}(L)))}; \tag{55}$$

in other words, the crack growth is brutal.

Proof. If $\hat{\mathcal{E}}$ is convex and C^1, $t^2\hat{\mathcal{E}}(\lambda) + \lambda$ attains its minimum at $\lambda = 0$ if $t^2\hat{\mathcal{E}}'(0) + 1 \ge 0$, at $\mathcal{G}(L)$ if $t^2\hat{\mathcal{E}}'(\mathcal{G}(L)) + 1 \le 0$, and at λ such that $t^2\hat{\mathcal{E}}'(\lambda) + 1 = 0$ otherwise ; application of Proposition 10 then yields *(i)* . If however $\hat{\mathcal{E}}$ is concave, $t^2\hat{\mathcal{E}}(\lambda) + \lambda$ can only attain its minimum at $\lambda = 0$ if $t < t_L$, and at $\lambda = \mathcal{G}(L)$ if $t > t_L$; application of Proposition 10 then yields *(ii)* . □

Remark 12. *Whenever $\hat{\mathcal{E}}$ is not convex, the minima of $t^2\hat{\mathcal{E}}(\lambda) + \lambda$ will be those of $t^2\hat{\mathcal{E}}^{**}(\lambda) + \lambda$, where $\hat{\mathcal{E}}^{**}$ denotes the convexification of $\hat{\mathcal{E}}$. Thus, the crack evolution will be progressive on the strictly convex parts of the graph of $\hat{\mathcal{E}}^{**}$ and brutal on the linear parts of that graph.*

When comparing the above results to those yielded by Griffith's theory, it is easily seen that the results are identical in the strictly convex case while in the concave case Griffith's criterion will actually predict initiation at load $t_G = \sqrt{1/(-\hat{\mathcal{E}}'(0))}$ (cf. (iv) of Proposition 8), but then, the criterion will not be met for $t > t_G$ by any λ. The crack is then baptized "unstable" and henceforth intractable. Note also that the predicted initiation load t_G is greater than that, namely t_L, obtained through our formulation. If, in general, $\hat{\mathcal{E}}'(0) = 0$, Griffith's theory will not permit crack initiation (cf. (iv) of Proposition 8), whereas ours will generate brutal growth, as is immediately deduced from the preceding results.

3 An Application : The Pull-Out Problem

The fiber pull-out problem is a particularly interesting case so as to illustrate the properties presented in the previous section. However, we treat it in a very simplified context by neglecting internal stresses and friction. The reader interested by a more realistic approach should refer to Hutchinson and Jensen (1990) and Kerans and Parthasarathy (1991).

3.1 Statement of the Problem

We consider a straight cylindrical beam of length L with cross circular section of radius R reinforced by a centered fiber with circular cross section of radius R_f (see Figure 5). We denote by $S = \{(x_1, x_2) \ : \ r < R\}$ and $\Omega = S \times (0, L)$ the section and the domain,

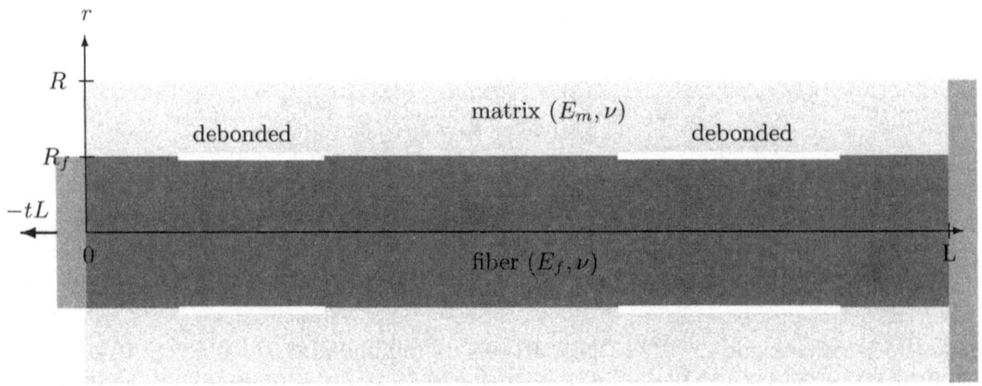

Figure 5. *Geometry and loading of the pull-out problem.*

as well as S_f and $V_f = R_f^2/R^2$ (respectively, S_m et V_m) the corresponding section and the volume fraction of the fiber (respectively, of the matrix). Both fiber and matrix are assumed to be perfectly bonded before loading and made of linearly elastic, unbreakable and isotropic materials with respective Young moduli E_f and E_m, while Poisson ratios are equal $\nu_f = \nu_m = \nu$. Their toughness k_f and k_m is then infinite, only the interface is breakable, its toughness being k. The beam is clamped on its section $S \times \{L\}$ while a longitudinal displacement $-tL$ is prescribed on $S_f \times \{0\}$, t increasing from 0 and playing the role of a loading parameter. The system is free of body forces and the lateral surface and the extremity $S_m \times \{0\}$ are not submitted to surface forces. Moreover we assume that the lips of the debonded surfaces are never in contact. Due to the symmetry of the problem we only consider *axisymmetric* debonding along the interface I. Therefore the debonding state is defined by a characteristic function χ defined on $[0, L]$ which takes the value 0 at sections z where fiber and matrix are bonded and the value 1 elsewhere.

We denote by

$$I_\chi = \{x = (x_1, x_2, x_3) \ : \ x_1^2 + x_2^2 = R_f^2, \quad \chi(x_3) = 1\} \tag{56}$$

the debonded part of the interface I and Ω_χ refers to the beam Ω in "the cracked configuration χ". The set of the kinematically admissible displacements, denoted $\mathcal{C}(\chi)$, is :

$$\mathcal{C}(\chi) = \{\mathbf{v} \in W^{1,2}(\Omega_\chi; \mathbb{R}^3) : \ v_z = -1 \text{ on } S_f \times \{0\}, \ \mathbf{v} = 0 \text{ on } S \times \{L\}\}. \tag{57}$$

The pull-out problem consists in finding the load-dependent characteristic function $\chi(t)$ of the debonded zone which has to satisfy

Law 3. Debonding evolution:

(i) $\chi(t)$ *is increasing with* t, $\quad \chi(0) = 0$,
(ii) $t^2 E_d(\chi(t)) + E_s(\chi(t)) \le t^2 E_d(\chi) + E_s(\chi), \quad \forall \chi \ge \chi(t-)$,
(iii) $t^2 E_d(\chi(t)) + E_s(\chi(t)) \le t^2 E_d(\chi) + E_s(\chi), \quad \forall \chi \le \chi(t)$,

where $E_d(\chi)$ is the elastic energy stored in the χ-damaged beam at equilibrium under a *unit* load and $E_s(\chi)$ is the Griffith surface energy

$$E_d(\chi) = \min_{\mathbf{v} \in \mathcal{C}(\chi)} \int_{\Omega_\chi} \tfrac{1}{2} A\varepsilon(v) \cdot \varepsilon(v) dx, \tag{58}$$

$$E_s(\chi) = \kappa \int_0^L \chi(z) dz, \qquad \kappa = 2\pi R_f k. \tag{59}$$

3.2 First Properties

Since the fiber and the matrix are unbreakable, the greatest crack with finite surface energy corresponds to the entire debonding of the interface, hence the minimum of the bulk energy is reached when $\chi(z) = 1$, $\forall z$. Let us show that the bulk energy of the beam when the interface is perfectly bonded is greater than this lower bound :

$$E_d(0) > E_d(1) = E_d^{min}. \tag{60}$$

The inequality $E_d(0) \ge E_d(1)$ holds because of the monotonicity of E_d. Let $u(0)$ and $u(1)$ be the displacement fields which achieve the minimum in (58) when $\chi = 0$ and $\chi = 1$, respectively. Assume that $E_d(0) = E_d(1)$, then, since $u(0) \in \mathcal{C}(1)$ and since the minimizer of $E_d(1)$ is unique, we should have $u(0) = u(1)$. Let $u_f(1)$ and $u_m(1)$ be the restrictions of $u(1)$ to the fiber and matrix, respectively. Since these two fields are linked by no relation and since $u_m(1)$ is only submitted to homogeneous boundary condition, then $u_m(1) = 0$. On the other hand, $u_m(0)$ and $u_f(0)$ are equal along the interface. Since $u_f(0)_z = -1$ at $z = 0$, $u_m(0)$ cannot be identically zero. It is a contradiction, thus $E_d(0) \ne E_d(1)$.

We deduce from Proposition 6 that debonding will appear at the interface at a finite load t_i. Moreover Proposition 7 says that the debonding will continue until the entire surface is debonded. If the load of total debonding t_f is finite, then the fiber will behave elastically for $t > t_f$ (whereas the matrix is unstrained).

The main issues are to find accurate estimates for t_i, t_f and for the debonding process in the interval (t_i, t_f). The next subsections are devoted to this task.

3.3 A First Approximation for Slender Beams

We now consider slender beams and we introduce the small parameter ϵ :

$$\epsilon = R/L. \tag{61}$$

The asymptotic bulk energy. Let χ be a given debonding state. Following the asymptotic analysis of Geymonat et al. (1987) for slender structures made of an unidirectional fibrous composite, it is shown that the displacement field $\mathbf{u}^\epsilon$ which achieves the minimum in (58) converges (when ϵ goes to 0) to a displacement field $\mathbf{u}^0 = u^0\,\mathbf{e}_3$ which satisfies the classical Navier-Bernoulli's kinematical assumptions ($\varepsilon_{rr}(\mathbf{u}^0) = \varepsilon_{\theta\theta}(\mathbf{u}^0) = \varepsilon_{rz}(\mathbf{u}^0) = 0$) and takes the form

$$u^0(r,z) = \begin{cases} U_m(z) \text{ when } & r > R_f, \\ U_f(z) \text{ when } & r < R_f, \end{cases} \tag{62}$$

U_m and U_f representing the axial displacement of the matrix and the fiber, respectively. Moreover, these displacements have to be equal in the bonded zone and thus they satisfy

$$(1-\chi(z))(U_f(z) - U_m(z)) = 0, \quad \forall z \in [0, L]. \tag{63}$$

Furthermore the elastic energy $E_d^\epsilon(\chi)$ converges to $E_d^0(\chi)$ which is obtained by solving the following minimization problem, the minimizer of which is just (U_f, U_m)

$$E_d^0(\chi) = \min_{(V_f,V_m)\in\mathcal{C}^0(\chi)} \frac{1}{2}\int_0^L \left(A_f V_f'^2 + A_m V_m'^2\right) dz. \tag{64}$$

Here $\mathcal{C}^0(\chi)$ denotes the set of associated admissible Navier-Bernoulli displacements,

$$\mathcal{C}^0(\chi) = \Big\{(V_f, V_m) \in W^{1,2}(0,L)^2 \ : \ V_f(0) = -1, V_f(L) = V_m(L) = 0, \\ (1-\chi(z))(V_f(z) - V_m(z)) = 0 \ a.e.\Big\}, \tag{65}$$

whereas A_f and A_m represent the stiffness of the fiber and matrix cross section, respectively, that is :

$$A_f = \pi R_f^2 E_f, \qquad A_m = \pi(R^2 - R_f^2)E_m. \tag{66}$$

Let us denote by $\ell(\chi)$ the length of the debonded zone : $\ell(\chi) = \int_0^L \chi(z)dz$. So as to solve

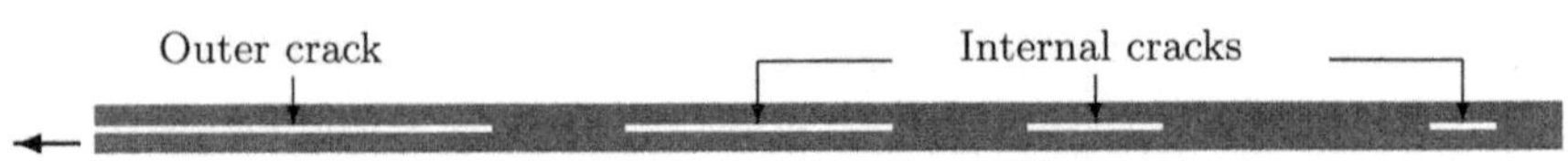

Figure 6. *Outer crack, internal cracks and bonded zone.*

(64) and to obtain the minimizer (U_f, U_m), we divide the interval $(0, L)$ into three parts :

$I_d^1(\chi) = (0, \ell_1(\chi))$, $I_d^2(\chi)$ and $I_c(\chi)$. The first corresponds to the debonded zone in the neighborhood of $z = 0$ (it is called *the outer crack*); the second, empty if $\ell_1(\chi) = \ell(\chi)$, corresponds to the remainder of the debonded zone, made of *the internal cracks*; the third, empty if $\ell(\chi) = L$, corresponds to the bonded zone. The Euler equations corresponding to the equilibrium equations read as :

$$In\ I_c(\chi) : (A_f + A_m)U_f'(z) = F, \quad U_f(z) = U_m(z) \tag{67}$$

$$In\ I_d^1(\chi) : A_f U_f'(z) = F, \quad A_m U_m'(z) = 0 \tag{68}$$

$$In\ I_d^2(\chi) : A_f U_f'(z) = G, \quad A_m U_m'(z) = F - G, \tag{69}$$

where F and G are two constants. From (69) we deduce that $U_f' - U_m'$ is constant on the internal cracks, but, since they are equal at the ends of these inner cracks (by continuity), we get $U_f = U_m$ on $I_d^2(\chi)$, *i.e.* the fiber does not slip along the matrix along the internal cracks. The system becomes

$$A_f U_f'(z) = F, \quad A_m U_m'(z) = 0, \quad \text{when} \quad 0 < z < \ell_1(\chi), \tag{70}$$

$$(A_f + A_m)U_f'(z) = F, \quad U_f(z) = U_m(z), \quad \text{when} \quad \ell_1(\chi) < z < L, \tag{71}$$

with the boundary conditions : $U_f(0) = -1$, $U_m(L) = U_f(L) = 0$, and the continuity at $z = \ell_1(\chi) : U_m(\ell_1(\chi)) = U_f(\ell_1(\chi))$. The determination of F is now straightforward and we finally get

$$E_d^0(\chi) = \frac{F}{2} = \frac{1}{2} \frac{A_f(A_f + A_m)}{A_f L + A_m \ell_1(\chi)}. \tag{72}$$

Remark 13. *The simplicity of the expression of the stiffness coefficients and of the bulk energy is essentially due to the assumption that the Poisson ratio is constant and that the lips of the crack are free. The same coefficients can be obtained in a closed (but intricate) form in a more general context and the interested reader is referred to* Bilteryst *and* Marigo *(2002).*

The asymptotic debonding problem. By approximating the true elastic energy $E_d^\epsilon(\chi)$ of the beam in a debonding state χ by its asymptotic limit $E_d^0(\chi)$, the "limit" debonding evolution $t \mapsto \chi^0(t)$ must satisfy :

Law 4. Asymptotic debonding evolution:

(i) $\chi^0(t)$ *is increasing with* t, $\quad \chi^0(0) = 0$,

(ii) $t^2 E_d^0(\chi^0(t)) + E_s(\chi^0(t)) \leq t^2 E_d^0(\chi) + E_s(\chi), \quad \forall \chi \geq \chi^0(t-)$,

(iii) $t^2 E_d^0(\chi^0(t)) + E_s(\chi^0(t)) \leq t^2 E_d^0(\chi) + E_s(\chi), \quad \forall \chi \leq \chi^0(t)$.

The asymptotic solution. Let $\ell \in [0, L]$ and let χ be a characteristic function such that

$$\int_0^L \chi(z)dz = \ell. \tag{73}$$

The characteristic function of the interval $[0, \ell]$ is noted $\chi_{[0,\ell]}$ ($\chi_{[0,\ell]}(z)$ equals 1 when $0 \leq z \leq \ell$ and 0 elsewhere). Since only the length of the outer crack appears in (72), we have

$$E_d^0(\chi) = E_d^0(\chi_{[0,\ell_1(\chi)]}) \geq E_d^0(\chi_{[0,\ell]}) \equiv \mathcal{E}_0(\ell) \tag{74}$$

and equality holds only if $\chi = \chi_{[0,\ell]}$.

Let us prove that, if there exists a solution, it corresponds to an outer crack, *i.e.*

$$\chi^0(t) = \chi_{[0,\ell^0(t)]}. \tag{75}$$

Indeed, let $\chi^0(t)$ be a solution and let $\ell^0(t) = \ell_1(\chi^0(t))$ the length of its outer crack. Using $\chi_{[0,\ell^0(t)]}$ as a test function in *(iii)* of Law 4 and referring to (74) yields

$$\begin{aligned} t^2 E_d^0(\chi^0(t)) + E_s(\chi^0(t)) &\leq t^2 E_d^0(\chi_{[0,\ell^0(t)]}) + E_s(\chi_{[0,\ell^0(t)]}) \\ &= t^2 E_d^0(\chi^0(t)) + E_s(\chi_{[0,\ell^0(t)]}). \end{aligned} \tag{76}$$

Since, on the other hand, $\ell(\chi^0(t)) \geq \ell^0(t)$, the equality holds and since $\chi^0(t) \geq \chi_{[0,\ell^0(t)]}$, Eq. (75) follows.

Thus we are exactly arrived at the situation of a predefined crack path like in Section 2.3. Since $\mathcal{E}_0$ is a strictly convex function of ℓ, Propositions 10 and 11 yield $\ell^0(t)$ as the unique minimizer of $t^2\mathcal{E}_0(\ell) + \kappa\ell$. The minimum is at 0 as long as $t^2\mathcal{E}_0'(0) + \kappa \geq 0$, at L as soon as $t^2\mathcal{E}_0'(1) + \kappa \leq 0$, and at ℓ such that $t^2\mathcal{E}_0'(\ell) + \kappa = 0$ otherwise. Thus the debonding length is given by :

$$\ell^0(t) = \begin{cases} 0 & if \quad 0 \leq t \leq t_i^0 \\ \dfrac{t - t_i^0}{t_r^0 - t_i^0} & if \quad t_i^0 \leq t \leq t_f^0 \\ 1 & if \quad t \geq t_f^0 \end{cases} \tag{77}$$

where t_i^0 and t_f^0 are given by

$$t_i^0 = \sqrt{\frac{4\pi R_f k L^2 A_f}{(A_m + A_f)A_m}}, \qquad t_f^0 = \sqrt{\frac{4\pi R_f k L^2 (A_f + A_m)}{A_m A_f}}. \tag{78}$$

It remains to verify that this candidate satisfies all the items of Law 4. The growth of $\ell^0(t)$ is verified.

Let us check that *(ii)* of Law 4 is satisfied. Remark that ℓ^0 is a continuous function of t and let $\chi \geq \chi_{[0,\ell^0(t)]}$. Since $\ell^0(t)$ is a minimizer of $t^2\mathcal{E}_0(\ell) + \kappa\ell$, we have

$$\begin{aligned} t^2 E_d^0(\chi_{[0,\ell^0(t)]}) + E_s(\chi_{[0,\ell^0(t)]}) &= t^2\mathcal{E}_0(\ell^0(t)) + \kappa\ell^0(t) \\ &\leq t^2\mathcal{E}_0(\ell_1(\chi)) + \kappa\ell_1(\chi) \\ &= t^2 E_d^0(\chi) + \kappa\ell_1(\chi) \\ &\leq t^2 E_d^0(\chi) + E_s(\chi), \end{aligned}$$

which is the required inequality.

Let us check that *(iii)* of Law 4 is satisfied. Let $\chi \leq \chi_{[0,\ell^0(t)]}$. By using the optimality of $\ell^0(t)$ and (74), we obtain

$$\begin{aligned} t^2 E_d^0(\chi_{[0,\ell^0(t)]}) + E_s(\chi_{[0,\ell^0(t)]}) &= t^2\mathcal{E}_0(\ell^0(t)) + \kappa\ell^0(t) \\ &\leq t^2\mathcal{E}_0(\ell(\chi)) + \kappa\ell(\chi) \\ &= t^2 E_d^0(\chi_{[0,\ell(\chi)]}) + E_s(\chi) \\ &\leq t^2 E_d^0(\chi) + E_s(\chi), \end{aligned}$$

which is the required inequality.

We have established the

Proposition 12 (Asymptotic solution of the debonding). *The asymptotic pull-out problem (see Law 4), admits a unique solution*

$$\chi^0(t) = \chi_{[0,\ell^0(t)]}$$

with $\ell^0(t)$ *given by* (77) *and* (78) *which is interpreted as follows :*

When the load reaches the critical value t_i^0 *defined by* $-t_i^{0\,2}\mathcal{E}_0'(0) = \kappa$, *then Griffith's condition of propagation is satisfied and a crack appears at* $z = 0$. *Then its length* $\ell^0(t)$ *grows with the loading* t *in such a manner that Griffith's condition of propagation is always fulfilled,* i.e. *the equality* $-t^2\mathcal{E}_0'(\ell^0(t)) = \kappa$ *holds. It reaches the end* $z = 1$ *when the load attains the critical value* t_f^0 *given by* $-t_f^{0\,2}\mathcal{E}_0'(1) = \kappa$, *after which the fiber behaves elastically since the end* $z = 1$ *is assumed to be clamped.*

3.4 Possible Scenarios

We return now to the 3D problem and, since we do not know enough about $\mathcal{E}(\chi)$ for an arbitrary characteristic function, we will only consider outer cracks by assuming that $\chi(t) = \chi_{[0,\ell(t)]}$. According to the previous asymptotic result, this is not too restrictive an assumption. The bulk energy of a such debonding configuration is denoted by $\mathcal{E}(\ell)$:

$$\mathcal{E}(\ell) \equiv E_d(\chi_{[0,\ell]}). \tag{79}$$

By virtue of Proposition 10, the problem is then reduced to the determination of $\ell(t)$ such that

$$t^2\mathcal{E}(\ell(t)) + \kappa\ell(t) \leq t^2\mathcal{E}(\ell) + \kappa\ell \quad \forall\ell \in [0, L]. \tag{80}$$

For small values of ℓ, it is shown in Leguillon (1989) that the presence of a singularity at the interface on $S\times\{0\}$ leads to the following expansion of the elastic energy:

$$\mathcal{E}(\ell) = \mathcal{E}(0) - C\ell^{2\alpha} + o(\ell^{2\alpha}), \tag{81}$$

where α is the real part of the power of the singularity of the displacement field and C is a positive factor depending on the applied load and on the shape of the defect, but not on α.

In our case it turns out that $\alpha > 0.5$ when the fiber is stiffer than the matrix ($E_f > E_m$) while $\alpha < 0.5$ when the matrix is stiffer than the fiber ($E_f < E_m$), see Leguillon et al. (2000). Therefore, when the fiber is stiffer than the matrix the energy release rate is of the order of $\ell^{2\alpha-1}$ and tends to 0 with ℓ, whereas when the matrix is stiffer than the fiber the energy release rate tends to $-\infty$ when $\ell \to 0$:

$$\mathcal{E}'(0) = 0 \quad if \quad E_f > E_m, \qquad \mathcal{E}'(0) = -\infty \quad if \quad E_f < E_m. \tag{82}$$

The main features. We recall or establish now a series of properties (several of them are already contained in Section 2) concerning this problem; these do not require a precise expression for $\mathcal{E}(\ell)$. We simply use the fact that $\mathcal{E}$ is a continuously differentiable and (strictly) decreasing function of ℓ.

(i): ***A solution of*** (80) ***exists and*** $\ell(t)$ ***increases with*** t. The existence is ensured because one has to minimize a continuous function on a compact set. However uniqueness is not ensured.

(ii): ***The solution of*** (80) ***is not trivial, a debonded zone will appear at a load*** $t_i, 0 \leq t_i < \infty$.

(iii): ***The debonding process continues until the interface is entirely debonded.***

(iv): ***At each load, Griffith's criterion holds*** :$-t^2\mathcal{E}'(\ell(t)) \leq \kappa$. It suffices to apply (80) at t with $\ell = \ell(t) + h$ and $h > 0$, to divide by h, and to pass to the limit when h tends to 0.

(v): ***At each load*** t ***where the debonding evolves continuously, Griffith's condition of propagation holds*** : $-t^2\mathcal{E}'(\ell(t)) = \kappa$. See Proposition 8.

(vi): ***The total energy is a continuous function of*** t. See Proposition 9.

The above properties deserve several remarks. Thus, (i)–(iii) demonstrate the advantage of the minimization principle in investigating the debonding process. However, the principle is intimately related to Griffith's law as it is showed by (iv) and (v). The fundamental difference lies in the ability of the minimization principle to take into account discontinuous debonding for which a new criterion emerges from energy continuity (vi).

(vii): ***If the fiber is stiffer than the matrix, then the onset of debonding starts at*** $t_i > 0$ ***and it is brutal, whereas, if the matrix is stiffer than the fiber, then the debonding starts at*** $t = 0$ ***and grows continuously at first.***

Proof. Assume that $E_f > E_m$. We know by (ii) that a debonding will appear at t_i, but it cannot appear and grow smoothly from $\ell = 0$, because it should satisfy Griffith's condition of propagation by (v), which is impossible since $\mathcal{E}'(0) = 0$. Thus at t_i, a crack of finite length, say ℓ_i, appears. Formally : $\ell(t_i-) = 0$, $\ell(t_i+) = \ell_i > 0$. By (vi), t_i and ℓ_i must satisfy $t_i^2\mathcal{E}(\ell_i) + \kappa\ell_i = t_i^2\mathcal{E}(0)$ which forces $t_i > 0$.

Assume now that $E_m > E_f$. Since $\mathcal{E}'(0) = -\infty$, (iv) yields $t_i = 0$. Moreover, if the onset was brutal at $t = 0$, we should have by (vi) $\ell(0+) = 0$ which is a contradiction. □

Another important property is the stability of the minimization principle with respect to the limit process when ϵ goes to 0. By reintroducing the parameter ϵ, the elastic energy and the debonding length are denoted by $\mathcal{E}_\epsilon$ and ℓ^ϵ. Since $\mathcal{E}_\epsilon$ converges uniformly to $\mathcal{E}_0$, then the following convergence result holds :

(viii): $\ell^\epsilon(t)$ ***converges to*** $\ell^0(t)$ ***when*** $\epsilon \to 0$. For a given t, since $\ell^\epsilon(t)$ is bounded, it converges to a limit, say $\ell^*(t)$ (*a priori* up to the possible extraction of subsequences, but, since the limit is unique, the whole sequence really converges). We can pass to the limit in each term of (80) and immediately see that $\ell^*(t)$ is necessarily the minimizer of $t^2\mathcal{E}_0(\ell) + \kappa\ell$, namely $\ell^0(t)$.

The case of a stiffer fiber. We consider the case when the fiber is stiffer than the matrix. We cannot continue the analysis of the debonding process without additional information concerning the detailed behavior of the elastic energy. Since there are many parameters in the problem, it is not possible to investigate all possibilities. We first remark that $\mathcal{E}_\epsilon$ is necessarily concave in a neighborhood of $\ell = 0$ since it is decreasing and its derivative at 0 vanishes. On the other hand, since $\mathcal{E}_\epsilon$ converges (when ϵ tends to 0) to $\mathcal{E}_0$ which is strictly convex, we can expect that the concavity of $\mathcal{E}_\epsilon$ should change for sufficiently large value of ℓ when ϵ is small enough. As a consequence, we need to consider only two potential situations : the first, where the graph of $\mathcal{E}_\epsilon$ lies always above the chord $A_\epsilon C_\epsilon$ joining the point $A_\epsilon = (0, \mathcal{E}_\epsilon(0))$ and $C_\epsilon = (1, \mathcal{E}_\epsilon(1))$, see Figure 7; the second, where the graph of $\mathcal{E}_\epsilon$ intersects the chord $A_\epsilon C_\epsilon$ and is at first concave, then convex, see Figure 8.

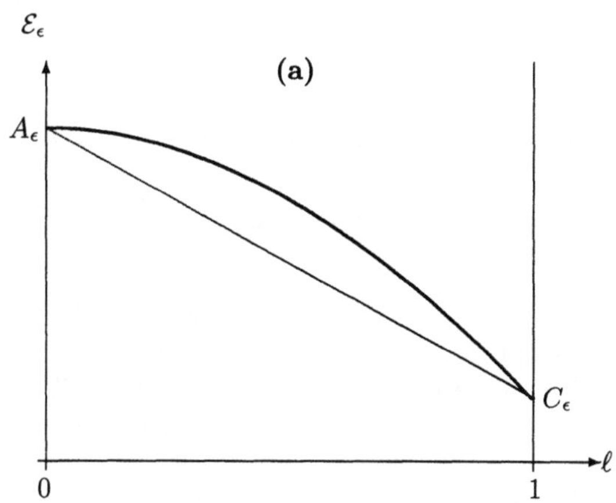

Figure 7. *Graphs of the elastic energy versus the debond length for short beams.*

(a): ***The graph of $\mathcal{E}_\epsilon$ is above its chord***. We can expect that it is the case for short beams, that is, when ϵ is large, but the rigorous proof of such a result requires a careful analysis of the asymptotic behavior of $\mathcal{E}_\epsilon$ when ϵ becomes infinite, which is out of the scope of the present study. In any case, the minimum of $t^2\mathcal{E}_\epsilon(\ell) + \kappa\ell$ is then reached at $\ell = 0$ or at $\ell = 1$. Specifically, we obtain

$$\ell^\epsilon(t) = \begin{cases} 0 & if \quad t < t_i^\epsilon \\ 1 & if \quad t > t_i^\epsilon \end{cases} \qquad with \quad t_i^\epsilon = \sqrt{\frac{\kappa}{\mathcal{E}_\epsilon(0) - \mathcal{E}_\epsilon(1)}}. \tag{83}$$

In other words, when the load attains the critical value t_i^ϵ, the fiber is suddenly entirely debonded.

(b): ***The graph of $\mathcal{E}_\epsilon$ intersects its chord, and is at first concave, then strictly convex***. By virtue of (vii) a crack of length ℓ_i^ϵ appears at $t = t_i^\epsilon$ and by virtue of

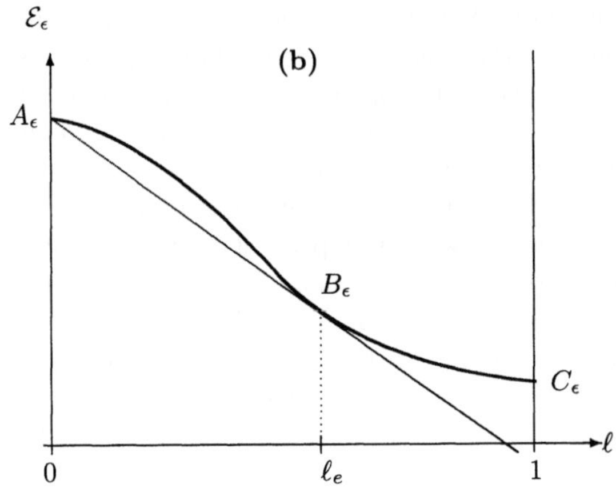

Figure 8. *Graph of the elastic energy versus the debond length for slender beams.*

the continuity property (vi) and of the minimum property (80), ℓ_i^ϵ and t_i^ϵ have to satisfy

$$\mathcal{E}_\epsilon(0) - \mathcal{E}_\epsilon(\ell_i^\epsilon) = \frac{\kappa}{t_i^{\epsilon 2}}\ell_i^\epsilon, \qquad \mathcal{E}_\epsilon(0) - \mathcal{E}_\epsilon(\ell) \leq \frac{\kappa}{t_i^{\epsilon 2}}\ell, \quad \forall \ell \in [0,1]. \tag{84}$$

Hence the graph of $\mathcal{E}_\epsilon$ must be above the chord passing through A_ϵ and $B_\epsilon = (\ell_i^\epsilon, \mathcal{E}_\epsilon(\ell_i^\epsilon))$. Since by hypothesis, the graph of $\mathcal{E}_\epsilon$ intersects the chord $(A_\epsilon C_\epsilon)$, B_ϵ is not at C_ϵ and consequently the chord $A_\epsilon B_\epsilon$ is tangent at B_ϵ to the graph of $\mathcal{E}_\epsilon$, cf Figure 8. This last property gives us a second relation between ℓ_i^ϵ and t_i^ϵ which have thus to satisfy

$$-t_i^{\epsilon 2}\mathcal{E}_\epsilon'(\ell_i^\epsilon) = \kappa, \qquad t_i^{\epsilon 2}\Big(\mathcal{E}_\epsilon(0) - \mathcal{E}_\epsilon(\ell_i^\epsilon)\Big) = \kappa \ell_i^\epsilon. \tag{85}$$

The assumed strict convexity of $\mathcal{E}_\epsilon$ after its inflection point implies the existence of a unique pair $(\ell_i^\epsilon, t_i^\epsilon)$ solution of (85). Moreover, by inserting the expression of t_i^ϵ given by (85a) into (85b) we obtain the following equation for ℓ_i^ϵ

$$\mathcal{E}_\epsilon(0) - \mathcal{E}_\epsilon(\ell_i^\epsilon) + \mathcal{E}_\epsilon'(\ell_i^\epsilon)\ell_i^\epsilon = 0 \tag{86}$$

in which the toughness κ does not appear. The first debonding length is independent of the toughness but depends only on the elastic coefficients and on the geometric parameters. Of course, the critical load t_i^ϵ depends on the toughness.

After this brutal initial debonding, the length of the crack grows continuously with the load until it reaches 1; the evolution is then governed by Griffith law. The crack arrives at 1 when t attains t_f^ϵ :

$$t_f^\epsilon = \sqrt{\frac{\kappa}{-\mathcal{E}_\epsilon'(1)}}. \tag{87}$$

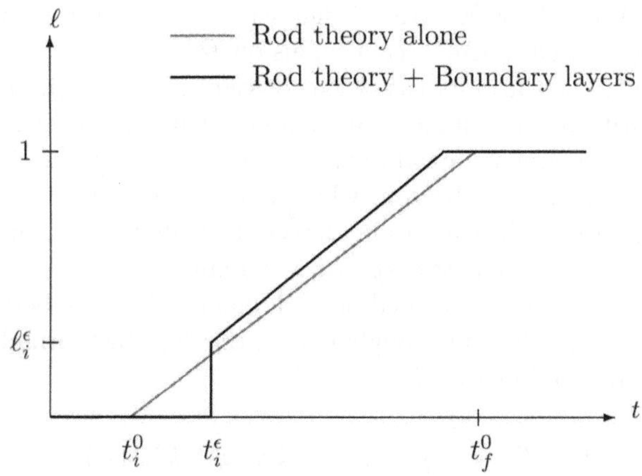

Figure 9. *The debonding process for slender beams and stiffer fiber.*

Remark 14. *In Figure 9 we compare the debonding evolution obtained from the asymptotic expression of the bulk energy with that given by the true 3D problem. The difference essentially lies into the onset of the debonding. In the former case, the onset is smooth and follows Griffith law, while it is brutal in the latter case and cannot be deduced from Griffith's law. There is no contradiction, as is proved in* Bilteryst *and* Marigo *(2002) by using matched asymptotic expansions like in* Abdelmoula *and* Marigo *(1991), because the finite initial debond length ℓ_i^ϵ tends to 0 as $\sqrt{\epsilon}$. On the other hand, it is necessary to use the least energy principle instead of Griffith law in the full 3D context, otherwise debonding never starts.*

4 Numerical Treatment

The numerical implementation of the revisited theory of brittle fracture developed in section 1 is presented. A computational method based on a variational approximation of the original functional is proposed. It is tested on two examples where we consider the discrete evolution Law 1 which consists, given Γ_0, of finding Γ_i, for $i \geq 1$ such that

(i) $\Gamma_i \supset \Gamma_{i-1}$,

(ii) $E(\Gamma_i, t_i) \leq E(\Gamma, t_i)$, for all $\Gamma \supset \Gamma_i$.

4.1 Numerical Implementation

Antiplane problems. Our law of brittle fracture is close to a model of image segmentation, namely that obtained through the minimization of the Mumford-Shah functional (Mumford and Shah, 1989). For a given grey level image, (*i.e.*, a real valued function g, defined on a bounded open domain Ω), the goal is to minimize the following "energy"

$$\int_{\Omega \setminus \Gamma} |\nabla u|^2 + |u - g|^2 \, dx + \mathcal{H}^{n-1}(\Gamma), \tag{88}$$

over all compact subsets Γ of $\mathbb{R}^n$ (the "edge set" of the image) and all real valued functions u (the "segmented image"), continuous on $\Omega \setminus \Gamma$.

The antiplane isotropic elasticity case with constant elasticity and fracture toughness is the closest in spirit to the Mumford-Shah functional. In the former however, non-homogeneous Dirichlet boundary conditions will replace the $\int_\Omega |u-g|^2\,dx$ term while the set of admissible cracks will be somewhat different from the set of edges. Indeed, while in the image context, edges are to be detected *inside* the domain, cracks are to be accounted for *inside* and *at the boundary* of the domain.

In what follows, Ω denotes a bounded open domain of $\mathbb{R}^2$. The body is homogeneous and isotropic with constant fracture toughness. In the "strong" formulation, the energy of the body is given by the functional

$$\mathcal{E}(u,\Gamma) = \frac{1}{2}\int_{\Omega\setminus\Gamma} \mu \nabla u \cdot \nabla u\, dx + k\mathcal{H}^1(\Gamma), \tag{89}$$

for each admissible crack set $\Gamma \subset \overline{\Omega}$ and each admissible function, $u \in W^{1,2}(\Omega \setminus \Gamma)$ such that $u = U$ on $\partial_d\Omega$. The complementary boundary part $\partial_f\Omega = \partial\Omega \setminus \partial_d\Omega$ will then be traction free. Since jumps of u on $\partial\Omega$ are allowed, the functional $\mathcal{E}$ is redefined on a "large enough" open set $\widetilde{\Omega}$, containing $\overline{\Omega}$ (an estimate of "how large" $\widetilde{\Omega}$ should be will be given later) and the boundary condition is automatically satisfied by imposing $u = U$ in $\widetilde{\Omega} \setminus \Omega$. The "weak" formulation (that is not detailed here, see Bourdin et al. (2000)) consists of introducing a larger space of displacement fields (namely, the space of special functions of bounded variation) and of taking as the cracked state Γ the set of points S_u where u is discontinuous, see Ambrosio and Tortorelli (1990) or Braides (1998).

Approximation by an elliptic functional. From a numerical standpoint, the main difficulty in solving the minimization problem in its original form is that we have to investigate a family of free discontinuity surfaces. An alternative solution is to use a regularized formulation where the displacements are not discontinuous but can undergo large gradients. It is the method used in Ambrosio and Tortorelli (1990) for the Mumford-Shah functional; we will adapt it in our context. The main idea in this kind of approximation is to introduce an auxiliary variable (subsequently denoted by v) taking its values in the interval [0,1] and approaching (in some sense) the crack. We introduce the following regularized functional, for each $u \in W^{1,2}(\widetilde{\Omega})$ and $v \in W^{1,2}(\widetilde{\Omega};[0,1])$,

$$\mathcal{E}_c(u,v) = \frac{1}{2}\int_{\widetilde{\Omega}} (v^2 + k_c)\mu\nabla u \cdot \nabla u\, dx + \int_{\widetilde{\Omega}} k\Big(c|\nabla v|^2 + \frac{(1-v)^2}{4c}\Big)\, dx, \tag{90}$$

where c and k_c are positive constants. Then, if $k_c \ll c$ when $c \to 0$, it can be proved by using $\Gamma-$*convergence* arguments that the minimum of $\mathcal{E}_c$ converges to the minimum of $\mathcal{E}$, the sequence of the (smooth) minimizers u_c converge to the non smooth field u, the set where u is discontinuous, S_u playing the role of the crack in the "weak" formulation. (The interested reader is referred to Ambrosio and Tortorelli (1990) or Bourdin et al. (2000) for precise statements and proofs.) Thus, in the limit, the minimization of $\mathcal{E}$ and that of $\mathcal{E}_c$ are equivalent. The proofs of the Γ–limit estimates further demonstrate that the first part of the regularized energy $\int_{\widetilde{\Omega}}(v_c^2+k_c)|\nabla u_c|^2\,dx$ converges to the bulk energy $\int_{\widetilde{\Omega}}|\nabla u|^2\,dx$ while the second one converges to the surface energy $\mathcal{H}^1(S_u)$. Finally the auxiliary function v in $\mathcal{E}_c$ converges pointwise to 1 on $\widetilde{\Omega}\setminus S_u$ and to 0 on S_u.

The same Γ–convergence result holds for the discrete functional $\mathcal{E}_{c,h}$, defined by the projection of $\mathcal{E}_c$ over a piecewise affine finite element space, provided that the characteristic length of the mesh h tends to zero faster than the infinitesimal k_c and c, see Bellettini and Coscia (1994). In the following subheading, we discuss the numerical solving of the discrete minimization problem.

Numerical solving of the discrete problem. We propose to minimize the regularized functional for small c's. At this stage, quite a few technical issues are to be addressed. Firstly, because of the presence of the term $v^2|\nabla u|^2$, the functional $\mathcal{E}_{c,h}$ is not convex in (u,v). However, it is convex and coercive in *each* variable, so that, once one of the fields is fixed, the minimization with respect to the other variable is easy. The idea is then to iterate minimizations in each variable until the successive minimizers are close enough to one another. We are unfortunately unable to prove the convergence of this algorithm; the sequence of optimal energies does converge and, up to a subsequence, the alternate minimizers converge to some critical point of $\mathcal{E}_{c,h}$. Both minimization problems are implemented by a standard finite element method, with triangular first order Lagrange elements. To enforce irreversibility on the crack set evolution, one should disconnect, at the end of each loading step, the nodes where a crack is detected, or, equivalently, add restrictions on the admissible set for the crack field by imposing some homogeneous Dirichlet conditions on the detected crack field v. The second solution is that which has been implemented.

As far as the choice of the different numerical parameters is concerned, we simply make the following comments (the interested reader is referred to Bourdin et al. (2000) for a more detailed discussion):

- c must be chosen small enough to prevent a softening effect that causes the bulk energy to be underestimated and large enough, compared to the discretization step h, close to the cracks to avoid overestimating the surface energy.
- k_c must be large enough to prevent the divergence of the numerical scheme but small enough so that no rigidity remains on the cracks.
- The logical domain $\widetilde{\Omega}$ must be chosen according to c so that the energy of cracks near the boundary of the physical domain Ω is well represented.

Remarks on the plane elasticity case. The study of the plane elasticity problem is still in its infancy. The equivalence between strong and weak formulation is not established at this time, nor is the weak formulation so clear. Furthermore, non-interpenetration of the crack lips should be imposed . The numerical adaptation of the algorithm to a linear isotropic elasticity problem in the absence of unilateral conditions is similar to that of the antiplane case. The regularized functional that has to be minimized is

$$\int_{\widetilde{\Omega}}(v^2+k_c)W(\varepsilon(u))\,dx+\int_{\widetilde{\Omega}}k\Big(c|\nabla u|^2+\frac{(1-v)^2}{4c}\Big)\,dx, \tag{91}$$

where $2W(\xi) = \lambda tr(\xi)^2 + 2\mu\, \xi \cdot \xi$. This is the method we used for the computations of the traction on a reinforced matrix, Section 4.2. Implementing unilateral conditions is however an open problem as of yet, for want of the proper regularized functional in place of (91).

4.2 Numerical experiments

In this section two numerical experiments, the first in antiplane elasticity, the second in plane elasticity, are described and compared with the theoretical predictions in Sections 1 and 2. The figures presented below represent the "damage" v-field. The crack site is included within the set of points where v is near zero (black or red color-coded in the figures).

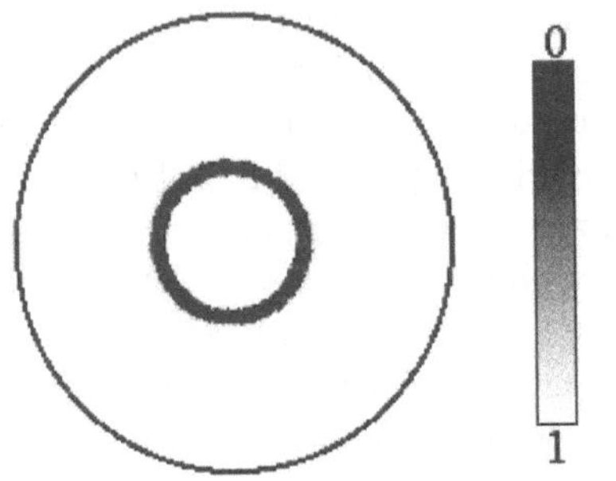

Figure 10. The computed damage field at $t = 1.5$.

Tearing of a reinforcement. The tearing of a rigid fiber out of a three-dimensional axisymmetric cylinder is solved in a closed form in Section 1.4. It was obtained there that the response is elastic without any crack as long as the axial displacement t prescribed to the fiber is smaller than a critical value t_f given by (31) while there is total debonding of the reinforcement from the inner surface of the cylinder (where the fiber is initially glued) as soon as $t > t_f$. Here, we take the following values of the material and geometrical parameters :

$$\mu = 1, \quad k = 1, \quad R_i = 1, \quad R_o = 3.$$

The failure load t_f, the bulk energy and the surface energy are given by

$$t_f = \sqrt{2 \ln 3} \sim 1.48, \tag{92}$$

$$E_d(t) \sim \begin{cases} 2.86\, t^2 & if\ t < 1.48 \\ 0 & if\ t > 1.48 \end{cases}, \quad E_s(t) \sim \begin{cases} 0 & if\ t < 1.48 \\ 6.28 & if\ t > 1.48 \end{cases} \tag{93}$$

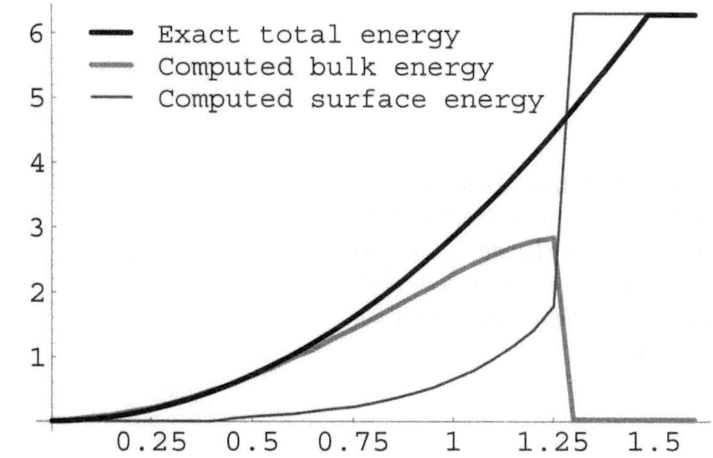

Figure 11. Evolution of the computed or exact energies with t.

Our goal is to test the numerical scheme on this example. Since the computer code has not been developed in the full $3D$ context, we will test the antiplane version. So we assume that the problem is independent of the axial direction and we try to describe the crack evolution on a cross section. We make no assumption concerning the axisymmetry of the solution. The parameters are $h = 10^{-3}$, $c = 10^{-1}$, $k_c = 10^{-4}$. In Figure 11, we plot the computed bulk energy, the computed surface energy and the exact total energy versus the load t, whereas in Figure 10 we plot the computed "damage" field v at $t = 1.5$.

Figure 11 shows that, at $t \sim 1.30$, brutal cracking appears in the section because the surface energy jumps from 1.78 to 6.30, while, at the same time, the computed bulk energy jumps from 2.83 to nearly 0. Then, the energy of the body does not evolve. Figure 10 indicates what happens at $t \sim 1.30$, namely the total debonding of the fiber. Therefore the computations are qualitatively in perfect agreement with the theoretical predictions. From a quantitative point of view, one sees that the computed critical load is underestimated (1.30 instead of 1.48). Note also that the surface energy is not 0 before the critical load, even if there is no localization of the damage at this stage. The reason is probably due to an insufficiently small value of the parameter c (and of the mesh parameter). On the other hand, the surface energy is correct within 1%, once the fiber is debonded.

Traction experiment on a fiber reinforced matrix. A square elastic matrix is reinforced by a rigid circular fiber as shown in the figure below. The fiber remains fixed, while a uniform displacement field te_2 is imposed on the upper side of the square; the remaining sides are traction-free.

The computed crack evolution. The following evolution is observed as t grows:

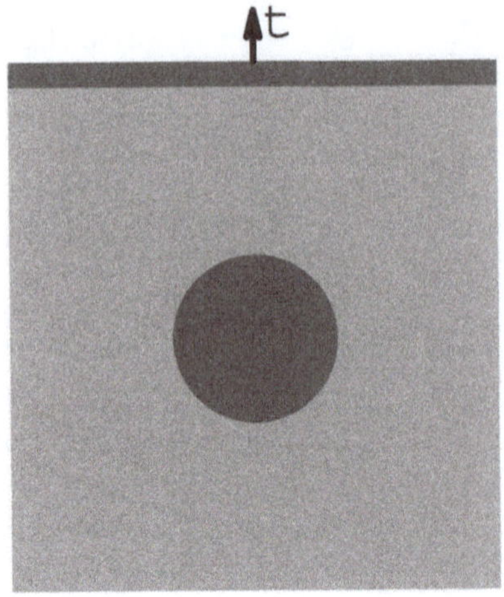

Phase 1 : Elastic response. As long as $t < 0.2$, the matrix remains purely elastic. Note however on Figure 12, that the surface energy is not exactly 0 during this phase, probably because the chosen value of the parameter c is not sufficiently small.

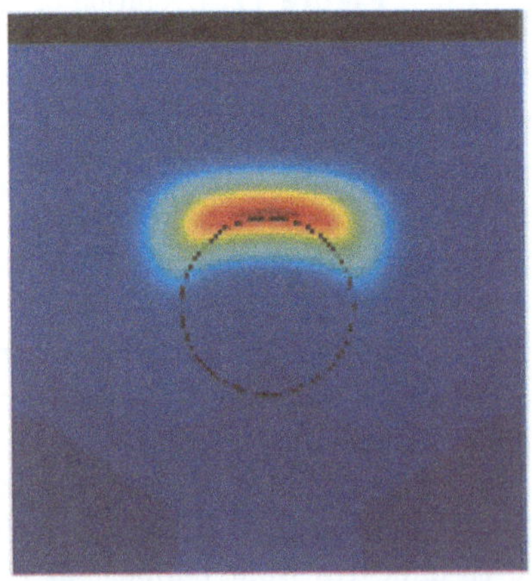

Phase 2 : Brutal onset. At $t \sim 0.2$, a crack of finite length brutally appears near the north pole of the inclusion. Note that this first crack is symmetric with respect to the 2 axis, but that is not straight. (It is not a purely numerical effect, because this path does not follow the mesh.) This brutal onset is confirmed in Figure 12 where, at this load, the surface energy increases while the bulk energy decreases.

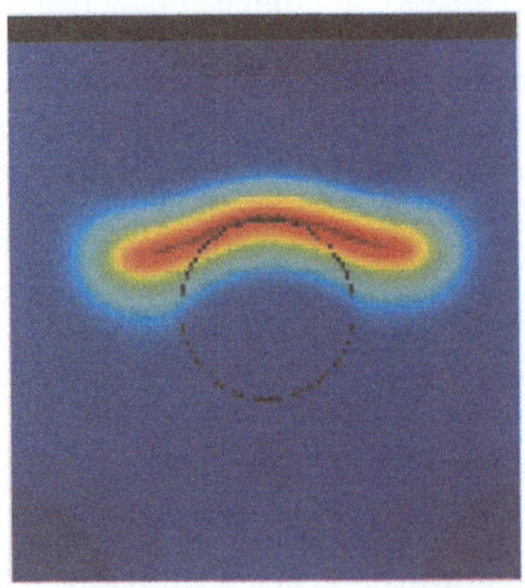

Phase 3 : Progressive and symmetric evolution of the cracking. When t varies between 0.2 and 0.32, the crack progressively grows in the matrix. The evolution is smooth as is shown in Figure 12 where the surface energy increases smoothly, while the bulk energy is nearly constant. Note that the propagation is symmetric but not straight.

Phase 4 : Rupture of the right ligament. At $t \sim 0.32$, the right hand-side of the matrix is brutally cut. The brutal character of the phenomenon can be verified on Figure 12 with jump discontinuity of both the surface and bulk energies. Note that the cracked state is no longer symmetric.

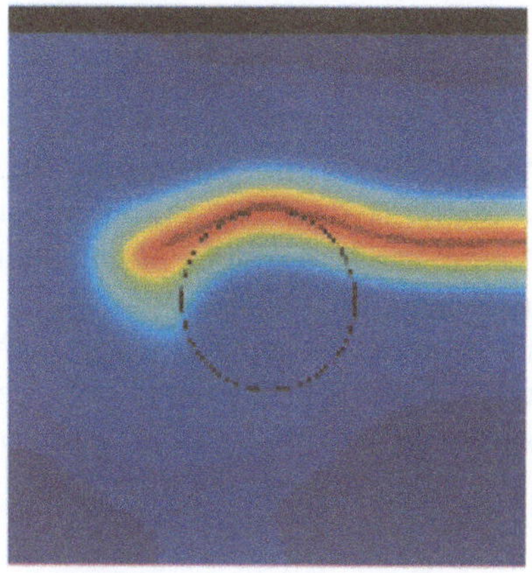

Phase 5 : Progressive propagation of the left branch. When t varies between 0.32 and 0.37, the left part of the crack progressively grows, the surface energy slowly increases, whereas the bulk energy is nearly constant. Note also that the crack is trying to recover symmetry.

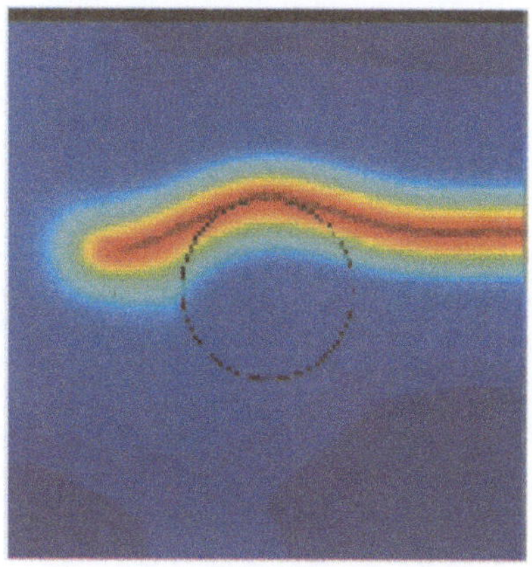

Phase 6 : Rupture of the left ligament. At $t \sim 0.37$, the crack brutally severs the remaining filament of uncracked material, see also Figure 12 where the jump of the energies is clearly evidenced. Note also that, in the final stage the right and left parts of the crack are nearly symmetric.

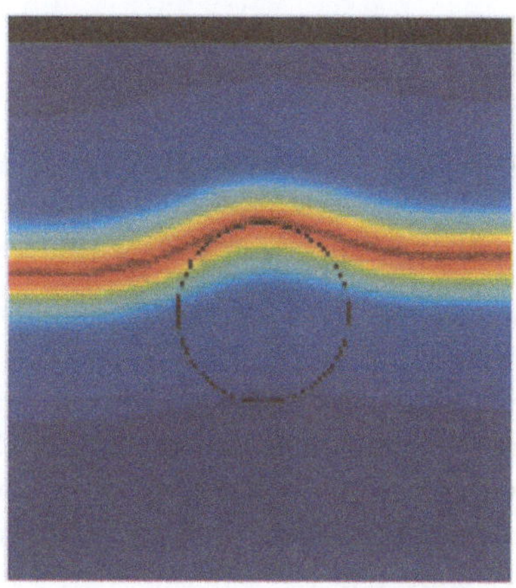

Phase 7 : The sample is broken. For $t > 0.37$, the sample is split into two parts. Theoretically, the lower piece should be now in its natural configuration, while the upper one should be simply translated. Note however in figure 12 that the elastic energy is not exactly equal to 0, because of the presence of the coefficient k_c in the functional.

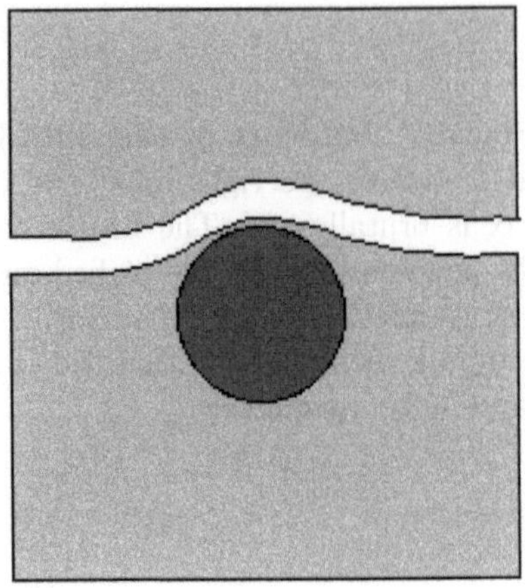

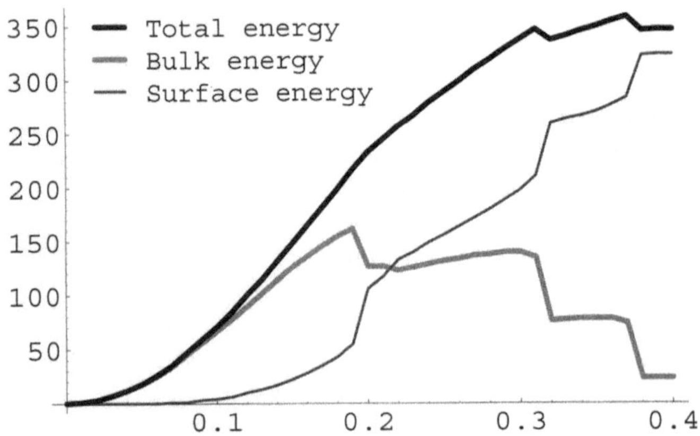

Figure 12. *Evolution of the energies with the load.*

Comparisons with the theoretical properties. Of course, an exact solution is not available for this example. We can however compare some features of the computations with the theoretical properties established in Section 2 or in Francfort and Marigo (1998).

1. The onset of cracking is brutal. This is in agreement with item 4 of Proposition 4.19 in Francfort and Marigo (1998) and Proposition 4.2 in Section 2. Indeed, if x_0 is a non singular point for the purely elastic solution, the bulk energy release due to a crack of small length ℓ near x_0 is of the order of ℓ^2. Therefore its derivative tends to 0 with ℓ and the bulk energy is a concave function of the crack length near $\ell = 0$.
2. The energy is (nearly) conserved during phases 2, 4, 6 of brutal growth, as theoretically expected.

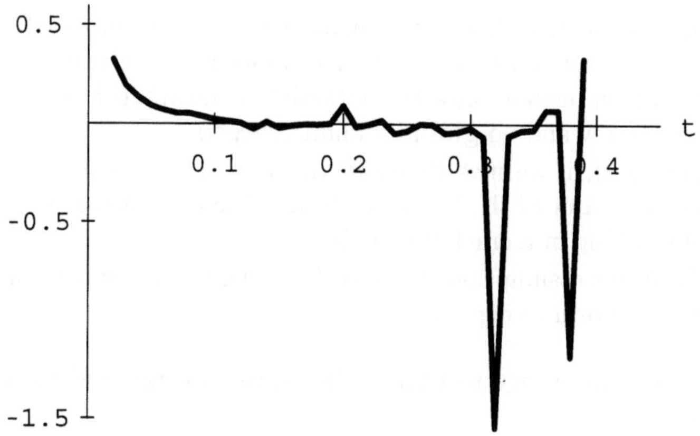

Figure 13. *Verification of the Griffith law.*

3. Griffith law is satisfied during the phases 3 and 5 of progressive growth of the crack. Figure 13 is an attempt at checking this. Denoting by $\ell(t)$, $E(t)$ and $E_d(t) = t^2\mathcal{E}_d(\ell(t))$ the length of the crack, the total energy and the bulk energy, respectively, when the load is t we have

$$E(t) = E_d(t) + k\ell(t), \tag{94}$$

while Griffith law reads as

$$t^2 \frac{d\mathcal{E}_d}{d\ell}(\ell(t)) + k = 0. \tag{95}$$

Differentiating (94) with respect to t and using (95) yields

$$E'(t) - \frac{2E_d(t)}{t} = 0.$$

In Figure 13, $E' - 2E_d(t)/t$ is plotted as a function of t, and we see that the criterion is indeed met during the progressive phases of the evolution, while it is not satisfied at $t = 0.32$ and at $t = 0.37$, *i.e.* when the right and left ligaments break. Note however that it is nearly satisfied at $t = 0.2$ which is in agreement with the results obtained in Section 3 for the pull-out test.

Remark 15. *This numerical experiment is exemplary because of the wide range of associated crack behaviors throughout the evolution: brutal versus progressive fracture, edge versus bulk crack, symmetric versus asymmetric path, curvilinear crack, ...*

5 Obstacles and Perspectives

Before illustrate them in the one dimensional context let us briefly present the main limitations of our formulation.

1. As mentioned in Section 1, our formulation is severely impaired when surface or body force-loads are applied to the sample. Indeed, as illustrated in the sequel, the total energy (which includes now the potential of the applied forces) is generally not bounded from below and no **global** minimizer exists.
2. On the other hand, if we investigate **local** minimizers, since the purely elastic response is, for any value of the loading, a local minimizer, it is not possible to predict the onset of cracking in a crack free body.
3. Moreover, Griffith's assumption on the surface energy leads in general to **size effects** which are not conform to experiments.

Accordingly, it is necessary to change the surface energy and to weaken the least energy principle.

5.1 Limits of the Current Formulation.

No global minimum when forces are prescribed. We consider a homogeneous bar of natural length L, with cross-sectional area S, constituted of an elastic breakable material with Young modulus E and toughness k. Before any loading, the bar is assumed to be crack free. The end $x = 0$ is fixed, a tensile force tF $(F > 0)$ is applied at $x = L$ and the bar is submitted to a uniform distribution of tensile body forces tg, $g > 0$, t being a loading parameter increasing from 0. If we adopt Griffith's surface energy, the total energy of the bar is the following functional on admissible displacement fields v :

$$\mathcal{E}_t(v) = \frac{ES}{2}\int_0^L v'(x)^2 dx + kS\, card(\Gamma(v)) - tg\int_0^L v(x)dx - tFv^+(L), \qquad (96)$$

where $\Gamma(v)$ denotes the set of points where v is discontinuous and will play the role of the cracked state. The set $\mathcal{C}$ of admissible displacements is the space of functions v of bounded variation such that $v^-(0) = 0$[2] and with positive jumps ($[\![v]\!] = v^+ - v^- \geq 0$) :

$$\mathcal{C} = \Big\{ v \in BV(\mathbb{R})\ :\ v = 0\ on\ (-\infty, 0),\ v = const\ on\ (L, \infty),\ [\![v]\!] \geq 0\ on\ \Gamma(v) \Big\}. \qquad (97)$$

We immediately obtain the

Proposition 13. *When $t > 0$, the energy $\mathcal{E}_t$ is not bounded from below. Thus a global minimizer does not exist.*

Proof. Let v be the field such that $v(x) = 0$ for $x < L/2$ and $v(x) = U > 0$ for $x > L/2$. This field is admissible, its elastic energy vanishes, its surface energy is equal to kS and then its total energy is $kS - t(F + gL/2)U$. Since U can be taken arbitrarily large, the result follows. □

[2] Since v can be discontinuous at 0 or at L, v is defined on the whole real line by setting $v(x) = 0$ when $x < 0$ and $v(x) = constant$ when $x > L$.

The elastic response is the unique local minimum. We say that an admissible field u is a *local minimizer* of $\mathcal{E}_t$ if there exists a ball (in the sense of the natural norm of functions of bounded variation, see Braides (1998)) $\mathcal{B}(u,r)$ of center u and of radius $r>0$ such that

$$\mathcal{E}_t(u) \leq \mathcal{E}_t(v) \quad \forall v \in \mathcal{B}(u,r) \cap \mathcal{C}. \tag{98}$$

Denoting by u_t the purely elastic response of the bar for the loading parameter t, *i.e.*

$$u_t(x) = \frac{tg}{2ES}\, x(2L-x) + \frac{tF}{ES}\, x, \tag{99}$$

the following Proposition holds true :

Proposition 14. *For any value of the loading parameter t, the elastic response u_t is a local minimizer.*

Proof. Let $r>0$ and $\phi \in \mathcal{C}$ with $\|\phi\| = 1$ and $card(\Gamma(\phi)) \geq 1$. Direct calculations give

$$\begin{aligned}
\mathcal{E}_t(u_t + r\phi) - \mathcal{E}_t(u_t) &= r\left(\int_0^L \Big(ESu_t'(x)\phi'(x) - tg\phi(x)\Big)dx - tF\phi^+(L)\right) \\
&\quad + kS\; card(\Gamma(\phi)) + \frac{r^2}{2}\int_0^L \phi'(x)^2 dx \\
&\geq rt\int_0^L \Big(\big(g(L-x)+F\big)\phi'(x) - g\phi(x)\Big)dx - rtF\phi^+(L) + kS \\
&= -rt\sum_{x\in\Gamma(\phi)} \Big(g(L-x)+F\Big)[\![\phi]\!](x) + kS \\
&\geq -rt(gL+F) + kS,
\end{aligned}$$

where the last inequality follows from $\|\phi\| = 1 \geq \sum_{x\in\Gamma(\phi)} |[\![\phi]\!](x)|$. Thus, if we choose r sufficiently small we obtain

$$\mathcal{E}_t(v) \geq \mathcal{E}_t(u_t), \qquad \forall v \in \mathcal{B}(u_t, r) \cap \{v \in \mathcal{C} \;:\; card(\Gamma(v)) \geq 1\}. \tag{100}$$

If we now consider $v \in \mathcal{C}$ such that $card(\Gamma(v)) = 0$, *i.e.* smooth v, we classically obtain

$$\mathcal{E}_t(v) - \mathcal{E}_t(u_t) = \frac{1}{2}\int_0^L \left(v'(x) - u_t'(x)\right)^2 dx \geq 0, \tag{101}$$

and the result follows. □

Remark 16. *This result means that the elastic response is never unstable if we consider Griffith's surface energy. It remains true in the general context of 3D Fracture Mechanics, provided that the elastic field u_t does not contain strong singularities, see* Francfort *and* Marigo *(1998).*

It is possible to obtain a stronger result by proving that, in our $1D$ context of prescribed forces, the elastic response $u_t, t > 0$, is the unique local minimizer, see Charlotte et al. (2000). Let us simply verify here that no discontinuous field v is a local minimizer. Let $v \in \mathcal{C}$ with $card(\Gamma(v)) \geq 1$, let $x_1 \in \Gamma(v)$ and let v_n be the following sequence of admissible fields converging to v when n goes to infinity :

$$v_n(x) = \begin{cases} v(x) & if\ x < x_1 \\ v(x) + \dfrac{1}{n} & if\ x > x_1 \end{cases}. \tag{102}$$

Noting that $card(\Gamma(v_n)) = card(\Gamma(v))$, a direct calculation gives then

$$\mathcal{E}_t(v_n) - \mathcal{E}_t(v) = -\frac{t}{n}\Big(g(L - x_1) + F\Big) < 0. \tag{103}$$

Spurious size effects. Let us consider the case of applied displacements with Griffith's model. We have found in Section 4.2 that the cylinder breaks when the prescribed displacement U reaches the critical value

$$U_c = \sqrt{\frac{2kL}{E}}$$

corresponding to the critical strain $\sqrt{\frac{2k}{EL}}$ and to the critical stress $\sqrt{\frac{2kE}{L}}$. Thus, the longer is the bar, the smaller is the rupture stress. Moreover, when the bar length tends to infinity, the rupture stress tends to 0. This size effect is general (see also the tearing of the reinforcement in Section 1.4 and the pull-out test in Section 3) and is due to Griffith assumption on the surface energy which is based on the comparison of surface energy with volume energy and brings necessarily into the theory an internal length scale.

5.2 More Realistic Models.

Barenblatt's model As outlined by Griffith himself, the surface energy is proportional to the area of the surface of discontinuity only when the distance between the lips of the crack are large with respect to the characteristic atomic length. In the spirit of what happens at an atomic scale when atomic bonds break, we could assume, following the idea of Barenblatt (1962), that the surface energy depends on the value of the displacement jump, starting from 0 and progressively growing to its effective Griffith value k when the displacement jump becomes large with respect to the characteristic atomic length. It is then more convenient to assume that the surface energy takes the following form :

$$\mathcal{E}_s(u) = \int_{\Gamma(u)} \kappa(x, \nu(x), [\![u(x)]\!]) d\mathcal{H}^{N-1}(x), \tag{104}$$

where $\nu(x)$ denotes the unit normal vector on $\Gamma(u)$ at x. The surface energy is now a function of the displacement field u.

In the present $1D$ context we will assume that the surface energy reads as

$$\mathcal{E}_s(u) = \sum_{\Gamma(u)} S\kappa([\![u]\!](x)), \tag{105}$$

with

$$\kappa(0) = 0, \quad \kappa \text{ twice differentiable and increasing}, \quad \kappa(+\infty) = k. \tag{106}$$

In the analysis of local minima, the derivative of κ at 0 plays a fundamental role. Noticing that κ has the dimension of an energy per unit surface and that its derivative has the dimension of a stress, we put

$$\sigma_c = \kappa'(0). \tag{107}$$

Remark also that κ can be defined only for positive real number, because of the non-penetration condition.

The presence of a critical stress. A quite general study of this model has been achieved in Choksi et al. (1999), Del Piero (1999) or Charlotte et al. (2000), always in the $1D$ context. We will simply show in the present paragraph how Barenblatt's surface energy destabilizes the elastic response when the loading parameter is sufficiently large. The total energy reading now

$$\mathcal{E}_t(v) = \frac{ES}{2}\int_0^L v'(x)^2dx + \sum_{x\in\Gamma(v)} \kappa([\![v]\!](x))S - tg\int_0^L v(x)dx - tFv^+(L), \tag{108}$$

we obtain

Proposition 15. *The elastic response u_t is not a local minimum when $t > t_f$, with*

$$t_f = \frac{\sigma_c S}{gL + F}. \tag{109}$$

Proof. Let v_n be the following sequence of admissible fields converging to u_t when n goes to infinity :

$$v_n(x) = \begin{cases} u_t(x) + \dfrac{1}{n} & if\ x > 0 \\ 0 & if\ x < 0 \end{cases}.$$

Let us note that v_n is discontinuous at $x = 0$, $[\![v_n]\!](0) = 1/n$. We get, for sufficiently large n :

$$\begin{aligned} \mathcal{E}_t(v_n) - \mathcal{E}_t(u_t) &= -\frac{t}{n}(gL + F) + \kappa\left(\frac{1}{n}\right)S \\ &= -\frac{t - t_f}{n}(gL + F) + o(1/n) \\ &< 0, \end{aligned}$$

and the result follows. □

Remark 17. *The value $t(gL+F)$ represents the reaction of the wall at which the bar is fixed. It is the maximal tension that the bar must sustain. Hence, the proposition says that the bar is necessarily broken when the tension at this point is greater than $\sigma_c S = \kappa'(0)S$. Thus σ_c plays the role of a critical stress for the material.*

It is possible to prove the converse result, i.e. *the elastic response remains a local minimum as long as the tension at $x = 0$ is less than σ_c, see* Charlotte *et al. (2000).*

Let us finally remark that this energy based analysis suggests that the bar will break at $x = 0$ as soon as $t > t_f$. But, since the bar is then no more at equilibrium, this "post-mortem" analysis requires to consider dynamical effects.

References

Abdelmoula, R., and Marigo, J.-J. (1991). Etude par la méthode des développements asymptotiques raccordés de l'assouplissement induit par une fissure dans une poutre en traction. *E.D.F. Bulletin de la Direction des Etudes et Recherches* 4: 45–80.

Ambrosio, L., and Tortorelli, V. M. (1990). Approximations of functionals depending on jumps by elliptic functionals via Γ-convergence. *Comm. Pure Applied Math.* XLIII: 999–1036.

Barenblatt, G. I. (1962). The mathematical theory of equilibrium cracks in brittle fracture. *Adv. Appl. Mech.* 7: 55–129.

Bellettini, G., and Coscia, A. (1994). Discrete approximation of a free discontinuity problem. *Numer. Funct. Anal. Optim.* 15: 105–123.

Bilteryst, F., and Marigo, J.-J. (2003). An energy based analysis of the pull-out problem. *Eur. J. Mech., A/Solids* 22: 55–69.

Bourdin, B., Francfort, G. A., and Marigo, J.-J. (2000). Numerical experiments in revisited brittle fracture. *J. Mech. Phys. Solids* 48: 797–826.

Braides, A. (1998). *Approximation of Free-Discontinuity Problems.* n°1694 in Lectures Notes in Mathematics. Springer.

Charlotte, M., Francfort, G. A., Marigo, J.-J., and Truskinovsky, L. (2000). Revisiting brittle fracture as an energy minimization problem : comparison of Griffith and Barenblatt surface energy models. In Benallal, A., ed., *Proceedings of the Symposium on "Continuous Damage and Fracture"*, 7–12. Elsevier, Paris.

Choksi, R., Del Piero, G., Fonseca, I. and Owen, D. R. (1999). Structured deformations as energy minimizers in models of fracture and hysteresis. *Mathematics and Mechanics of Solids* 4: 321–356.

Del Piero, G. (1999). One dimensional ductile-brittle transition, yielding, and structured deformations. In Argoul, P., Fremond, M., and NGuyen, Q. S., eds., *Proceedings of the IUTAM Symposium "Variations de domaines et frontières libres en mécanique"*, 197–202. Paris, 1997.

Francfort, G. A., and Larsen, C. (2002). Existence and convergence for quasistatic evolution in brittle fracture. *(to appear)*

Francfort, G. A., and Marigo, J.-J. (1998). Revisiting brittle fracture as an energy minimization problem. *J. Mech. Phys. Solids* 46(8): 1319–1342.

Geymonat, G., Krasucki, F., and Marigo, J.-J. (1987). Stress distribution in anisotropic elastic composite beams. In Ciarlet, P. G., and Sanchez Palencia, E., eds., *Applications of multiple scalings in Mechanics*, 118–133. Masson, Paris.

Griffith, A. (1920). The phenomena of rupture and flow in solids. *Phil. Trans. Roy. Soc. London* CCXXI-A: 163–198.

Hashin, Z. (1996). Finite thermoelastic fracture criterion with application to laminate cracking analysis. *J. Mech. Phys. Solids* 44(7): 1129–1145.

Hutchinson , J. W., and Jensen, H. M. (1990). Models of fibre debonding and pullout in brittle composites with friction. *Mech. Materials* 9: 139–163.

Kerans, R., and Parthasarathy, T. A. (1991). Theoretical analysis of the fibre pullout and pushout tests. *J. Am. Ceram. Soc.* 74: 1585–1596.

Leguillon, D., Lacroix, C., and Martin, E. (2000). Interface debonding ahead of a primary crack. *J. Mech. Phys. Solids* 48: 2137–2161.

Leguillon, D. (1989). Calcul du taux de restitution d'énergie au voisinage d'une singularité. *C. R. Acad. Sci. Paris Série II b* 309: 945–950.

Mumford, D., and Shah, J. (1989). Optimal approximations by piecewise smooth functions and associated variational problems. *Comm. Pure Applied Math.* XLII: 577–685.

Foundations of the Theory of Structured Deformations

Gianpietro Del Piero
Dipartimento di Ingegneria, Università di Ferrara *

1 Why Structured Deformations?

To introduce the theory of structured deformations, a good starting point is perhaps to illustrate a situation in which such objects arise naturally, although unexpectedly, from a problem in fracture mechanics.[1]

In a one-dimensional setting, consider a bar of length l subject to an axial displacement u. Assume that u is smooth except at a finite set of points $S(u)$ of $(0, l)$, called the *jump set* of u. At each x in $S(u)$, u is assumed to be discontinuous, and the *jump* of u at x is defined as the difference between the right and the left limits of u at x and is denoted by $[u](x)$. The bar is assumed to be made of an elastic material, whose energy density w is a function of the axial deformation, measured by the derivative u' of u. In addition, with every jump of u there is associated an *interfacial energy* θ. In the absence of applied loads, the total energy of the deformed bar is

$$E(u) = \int_0^l w(u'(x))\, dx + \sum_{x \in S(u)} \theta([u](x)) \,. \tag{1.1}$$

If we prescribe the total elongation βl

$$\int_0^l u'(x)\, dx + \sum_{x \in S(u)} [u](x) = \beta l \,, \tag{1.2}$$

*Presently on leave at the Centro Linceo Interdisciplinare "Beniamino Segre", Accademia Nazionale dei Lincei, Rome, Italy. This research was supported by the the Italian Ministry for University and Scientific Research through the Programma Cofinanziato 2002 *Modelli matematici per la scienza dei materiali.* I thank D.R. Owen for his accurate reading of the manuscript and for many valuable suggestions and comments.

[1] In recent years, several authors approached fracture problems within the framework of Continuum Mechanics, by dropping the continuity hypothesis usually made on the deformations of a continuous body. See for example Choksi and Fonseca (1997), Francfort and Marigo (1998), Dal Maso and Toader (2002).

the equilibrium problem for the bar consists in finding the (local and global) minimizers of E in the class of all displacements u with appropriate regularity, which satisfy the *hard device relation* (1.2) and the *non-interpenetration condition*

$$[u](x) > 0 \qquad \forall x \in S(u)\,. \tag{1.3}$$

It is usually assumed that w is convex and that θ is concave. If this is the case, then for all β smaller than a critical value β_c the global energy minimum is attained at the homogeneous unfractured configuration $u(x) = \beta x$, while for $\beta > \beta_c$ it is attained at a fractured configuration with a single fracture.[2]

In what follows I assume that w is strictly convex and that $w(0) = w'(0) = 0$. For the interfacial energy, instead of concavity I assume that θ is increasing, convex in a right neighborhood $(0, \gamma)$ of the origin, and concave outside. Graphs of functions w and θ obeying these assumptions are shown in Fig. 1.

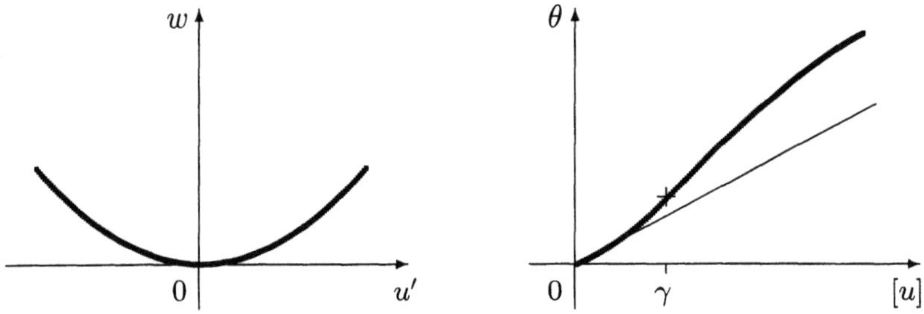

Figure 1.
Graphs of the bulk energy density w and of the interfacial energy θ.

Let u be a displacement field satisfying the conditions (1.2), (1.3), and let η be a smooth function with

$$\int_0^l \eta'(x)\,dx \;=\; 0\,. \tag{1.4}$$

Then the family of perturbed displacements $\lambda \mapsto u + \lambda\eta$ satisfies the above conditions for all λ, and

$$\begin{aligned} E(u+\lambda\eta) - E(u) &= \int_0^l \left(w(u'(x) + \lambda\eta'(x)) - w(u'(x)\right) dx \\ &= \lambda \int_0^l w'(u'(x))\,\eta'(x)\,dx + o(\lambda). \end{aligned}$$

[2]See Choksi et al. (1999), or Del Piero and Truskinovsky (in preparation). In the same papers it is shown that, for fixed w and θ, the post-fracture behavior may be of two types, ductile and brittle, depending on the length l of the bar.

This equality tells us that if u is a minimizer for E then the integral on the right side vanishes. This occurs only if

$$w'(u'(x)) \;=\; constant \;=:\; \sigma\,. \tag{1.5}$$

Therefore, (1.5) is a necessary condition for a (local or global) minimum.

For a strictly convex w the derivative w' is increasing, and (1.5) implies

$$u'(x) \;=\; constant \;=:\; \epsilon. \tag{1.6}$$

The constants σ and ϵ are the *stress* and the *strain* in the bar, respectively. By (1.6), one may restrict the search for minimizers for E to configurations with constant strain. For such configurations, the energy (1.1) and the hard device relation (1.2) take the simplified forms

$$E(u) \;=\; lw(\epsilon) \;+\; \sum_{x\in S(u)} \theta([u](x))\,, \tag{1.7}$$

$$l\epsilon \;+\; \sum_{x\in S(u)} [u](x) \;=\; l\beta\,. \tag{1.8}$$

Further necessary conditions for a minimum are obtained by taking perturbations η with constant derivative η' and with a single jump $[\eta]$ at a point x_0 of $(0,l)$. For such perturbations, the displacement $u+\eta$ satisfies the conditions (1.2), (1.3) if

$$l\eta' \;+\; [\eta] \;=\; 0\,, \qquad [u](x_0)+[\eta] \;>\; 0\,. \tag{1.9}$$

In particular, if x_0 belongs to $S(u)$ then $[u](x_0)>0$, and for all η obeying the conditions (1.9) the family $\lambda\mapsto u+\lambda\eta$ satisfies (1.3) for all λ with $|\lambda|<1$. Then, for all such λ,

$$\begin{aligned} &E(u+\lambda\eta)-E(u)\\ &= lw(\epsilon+\lambda\eta')-lw(\epsilon)+\theta([u](x_0)+\lambda[\eta])-\theta([u](x_0))\\ &= \lambda\left(lw'(\epsilon)\,\eta'+\theta'([u](x_0))\,[\eta]\right)+o(\lambda)\,, \end{aligned}$$

and, by $(1.9)_1$,

$$E(u+\lambda\eta)-E(u) \;=\; \lambda\left(\theta'([u](x_0))-w'(\epsilon)\right)[\eta]+o(\lambda)\,. \tag{1.10}$$

If u is a minimizer for E, the difference on the left-hand side must be non-negative. Then the linear term in λ must vanish, and because x_0 is an arbitrary point in $S(u)$ we conclude that

$$w'(\epsilon) \;=\; \theta'([u](x)) \qquad \text{for all } x \text{ in } S(u)\,. \tag{1.11}$$

If x_0 does not belong to $S(u)$, then $[u](x_0) = 0$ implies $[\eta] > 0$ by $(1.9)_2$. Thus, the family $\lambda \mapsto u + \lambda\eta$ satisfies the non-interpenetration condition (1.3) only if $\lambda > 0$. Then the two sides of (1.10) are non-negative only if

$$w'(\epsilon) \leq \theta'(0+)\,, \tag{1.12}$$

and by the strict monotonicity of w' this implies

$$\epsilon \leq (w')^{-1}(\theta'(0+)) \;=:\; k\,. \tag{1.13}$$

Thus, all minimizers for E must satisfy the conditions (1.6), (1.8), (1.11) and (1.13).

If u is an energy minimizer, then $u'(x) = \epsilon$ by (1.6). If u has no jumps, then $\epsilon = \beta$ by (1.8), and from (1.13) it follows that $\beta \leq k$. Thus, there are no minimizers without jumps for $\beta > k$. On the other hand, if u is a minimizer with jumps then from (1.11) and (1.12) we get

$$\theta'([u](x)) \;=\; w'(\epsilon) \;\leq\; \theta'(0+)\,.$$

If the interfacial energy is of the particular type represented in Fig. 1, it satisfies the condition

$$\theta'([u]) \;>\; \theta'(0+) \qquad \forall [u] > 0\,, \tag{1.14}$$

which contradicts the preceding inequality. Thus, there are no minimizers with jumps for this type of interfacial energy. Because there can be minimizers without jumps only if $\beta \leq k$, we conclude that if (1.14) holds there are no minimizers for $\beta > k$.

The scarce physical plausibility of this conclusion suggests us to broaden the class of functions in which the minimizers are sought. To do this, let us first evaluate a lower bound for the energy. Take a $\beta > k$ and a piecewise smooth displacement field u satisfying the conditions (1.2) and (1.3). Then using Jensen's inequality, the hard device relation (1.2) and the inequality $w(y) > w(x) + (y-x)w'(x)$ valid for all strictly convex differentiable functions, we get

$$\int_0^l w(u'(x))\,dx \;\geq\; l\,w(l^{-1}\int_0^l u'(x)\,dx)$$

$$=\; l\,w(\beta - l^{-1}\sum_{x\in S(u)}[u](x)) \;>\; l\,(w(k) + (\beta - l^{-1}\sum_{x\in S(u)}[u](x) - k)\,w'(k))\,,$$

and recalling that $w'(k) = \theta'(0+)$ by the definition $(1.13)_2$, for the energy of the bar we have

$$E(u) \;>\; l\,(w(k) + (\beta - k)w'(k)) \;+\; \sum_{x\in S(u)}(\theta([u](x)) - \theta'(0+)\,[u](x))\,.$$

The second term on the right-hand side is non-negative, as follows from integration of (1.14) over $(0, [u](x))$. Therefore, the energy of u has the lower bound

$$E(u) > l\left(w(k) + (\beta - k)w'(k)\right), \tag{1.15}$$

depending only on β, l, and on the material constant k.

Now consider a sequence $n \mapsto u_n$, in which each u_n has constant strain $\epsilon = k$ and n jumps of intensity $n^{-1}l(\beta - k)$. Every u_n satisfies the hard device relation (1.8), and its energy is

$$E(u_n) = l\,w(k) + n\,\theta(n^{-1}l(\beta - k)) = l\,w(k) + n\left(\theta'(0+)\,n^{-1}l(\beta - k) + o(n^{-1})\right).$$

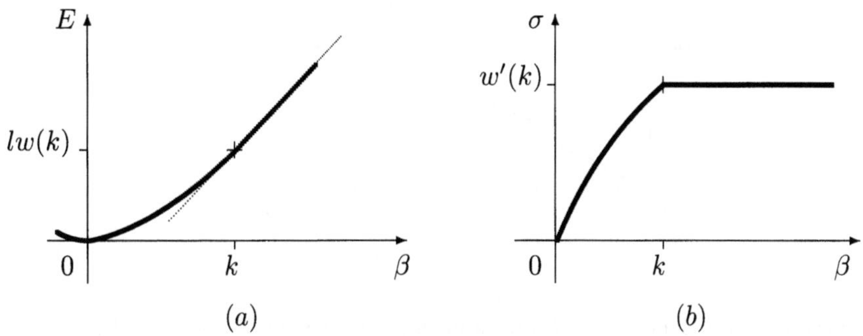

Figure 2.
Global energy minima (a) and the corresponding stresses (b) as functions of β.

When $n \to \infty$, the sequence converges to the right-hand side of (1.15). Therefore, this is the infimum of E for the given β, and $n \mapsto u_n$ is a minimizing sequence. The strict inequality in (1.15) tells us that the infimum cannot be attained at a piecewise smooth u.

Thus, we come to the following conclusion: for $\beta < k$ the global energy minimum is $lw(\beta)$ and is attained at the configuration $u(x) = \beta x$, while for $\beta > k$ the infimum of the energy is given by the right-hand side of (1.15) and is not attained in the class of piecewise smooth displacement fields. The curve of the energy minima as functions of β is shown in Fig. 2a. Note that the branch $\beta > k$ is the half-line tangent to the curve $E = lw(\beta)$ at $\beta = k$.

The corresponding stress, given by (1.5),

$$\sigma(\beta) = \begin{cases} w'(\beta) & \text{for } \beta \leq k\,, \\ w'(k) & \text{for } \beta > k\,, \end{cases} \tag{1.16}$$

is represented in Fig. 2b.

Let us see whether, for $\beta > k$, it is possible to define equilibrium configurations in some generalized sense. In the minimizing sequence $n \mapsto u_n$ constructed above, nothing was said about the location of the jumps. Thus, we are free of locating them anywhere in the bar. In particular, as suggested by Fig. 3, given any smooth function $x \mapsto u_0(x)$ with

$$u_0(0) = 0\,, \quad u_0(l) = l\beta\,, \qquad u_0'(x) \geq k \quad \forall x \in (0,l)\,,$$

it is possible to locate the jumps in such a way that $n \mapsto u_n$ converges uniformly to u_0.

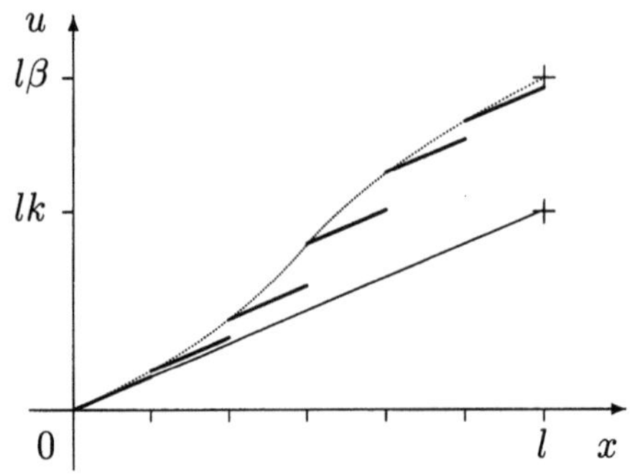

Figure 3.
Graphs of an element u_n of a minimizing sequence for $\beta > k$ (solid line), and of its limit u_0 (dotted line).

This does not imply that u_0 is an energy minimizer. Indeed, u_0 is smooth and satisfies (1.2) and (1.3). Therefore, by (1.15),

$$E(u_0) > \lim_{n\to\infty} E(u_n)\,. \tag{1.17}$$

This inequality tells us that, in spite of uniform convergence, the displacement field u_0 is inadequate to describe the limit of the sequence $n \mapsto u_n$ from the viewpoint of energy. The inadequacy is due to the presence of a very large number of very small jumps which, in the limit, affect the total energy of the bar, and yet cannot be detected from the observation of the macroscopic deformation.

Inequality (1.17) can be interpreted as a sign that a *structural change* occurs in the bar when subject to elongations $\beta > k$. Thus, for $\beta > k$ there are energy minimizers, but each of them is more complex than a piecewise smooth displacement field, since its description has to include some information about the process of diffusion of the jumps. In our example this can be done in purely kinematical terms, by specifying the displacement field u_0 and the strain $\epsilon = k$. In this way, at any point x we know how much of the macroscopic deformation $u_0'(x)$ is due to smooth deformation and how much is due to diffusion of the jumps: the two contributions are k and $u_0'(x) - k$, respectively.

The pair (u_0, k) is called a *structured deformation.* The purpose of the next lectures is to extend the ideas introduced above to more complex situations: first, to situations in which ε is not constant over the bar, second, to bodies with dimension greater than one.

2 Classical Deformations

2.1 Admissible Deformations

In continuum mechanics, a *deformation* is a mapping $f : \mathcal{E} \to \mathcal{E}$, where $\mathcal{E}$ is a Euclidean point space of dimension $N \leq 3$.[3] It is generally assumed that f is

(A1) injective,
(A2) continuous,
(A3) orientation-preserving.

These requirements are intended to rule out the possibility of interpenetrations, fractures, and reflections, respectively. A set of *admissible deformations* is a set $\mathcal{D}$ of deformations which, besides (A1)-(A3), satisfy the two additional requirements

(A4) if f belongs to $\mathcal{D}$, then its inverse f^{-1} belongs to $\mathcal{D}$,
(A5) if f and g belong to $\mathcal{D}$ and if $f(\mathcal{E}) = \mathcal{E}$, then the composition $f \circ g$ belongs to $\mathcal{D}$.

If we combine (A2) and (A4), we see that every admissible deformation is a continuous mapping with a continuous inverse, i.e., a *homeomorphism.* Supplementary regularity assumptions characterize more restricted classes of admissible deformations. For example, deformations are frequently required to be not only continuous, but also continuously differentiable. Assumption (A4) then implies that f^{-1} be continuously differentiable as well, and a function f with these properties is called a C^1*-diffeomorphism.*

Both homeomorphisms and C^1-diffeomorphisms satisfy the properties (A1), (A2), (A4), (A5). Then the set of all orientation-preserving homeomorphisms of $\mathcal{E}$ and the set of all orientation-preserving C^1-diffeomorphisms of $\mathcal{E}$ are examples of sets of admissible deformations.

Let us recall that if f is differentiable at x then the orientation-preserving assumption is expressed by the inequality

$$\det \nabla f(x) \ > \ 0\,. \tag{2.1}$$

where $\nabla f(x)$ is the derivative of f at x and $\det\nabla f(x)$ is its determinant.

[3] In the literature there is some confusion between the terms *Euclidean space* and *Euclidean point space.* The first should denote a vector space with inner product, see e.g. Halmos (1974) or Kolmogorov and Fomin (1975), and the second should denote a metric space with a Euclidean metric, see e.g. Noll (1973).

2.2 Fit Regions

Continuum mechanics deals with bodies that occupy regions of a Euclidean point space. Such regions "should meet certain requirements which ensure that what is usually done in continuum physics can be justified in precise mathematical terms".[4] According to Noll and Virga (1988), each collection of "fit regions"

- should satisfy the axioms of a *material universe*,[5]
- should be invariant under admissible deformations,[6]
- each region in the collection should have a *surface-like boundary*, for which a form of the Integral-Gradient (Gauss-Green) Theorem should be valid.

The same authors add:

> "It is also desirable that the class of fit regions include all that can possibly be imagined by an engineer but exclude those that can be dreamt up *only* by an ingenious mathematician".

The class of fit regions proposed by Noll and Virga is the set of all regions $\mathcal{A}$ with the following properties:

- $\mathcal{A}$ is bounded,
- $\mathcal{A}$ is regularly open,
- $\mathcal{A}$ has finite perimeter,
- $\mathcal{A}$ has negligible boundary.

A region $\mathcal{A}$ is *regularly open* if

$$\text{int clo}\mathcal{A} = \mathcal{A}\,,$$

where *int* and *clo* denote the topological interior and closure of $\mathcal{A}$, respectively. The boundary of $\mathcal{A}$ is *negligible* if its $N-$dimensional Lebesgue measure[7] is zero. The *perimeter* of a set can be identified with the $(N-1)$-dimensional Hausdorff measure of the essential boundary.[8]

Recently I proposed a simpler class of fit regions, whose definition does not involve any measure-theoretic concept other than the Hausdorff measure.[9]

[4] Noll and Virga (1988), Sect. 1.

[5] See e.g. Noll (1973) or Truesdell (1991).

[6] In other words, the image of every fit region under an admissible deformation should be a fit region.

[7] That is, volume if $N = 3$, area if $N = 2$, length if $N = 1$.

[8] That is, area if $N = 3$, length if $N = 2$, counting measure if $N = 1$. For a definition of Hausdorff's measure see e.g. Vol'pert and Hudjaev (1985), and for a definition of the essential boundary see Section 5.1 below. The perimeter of a set is a basic concept in Geometric Measure Theory. For its definition and for its identification with the $(N-1)$-dimensional Hausdorff measure of the essential boundary see Ambrosio et al. (2000).

[9] Del Piero (2003).

According to this definition, a region $\mathcal{A}$ of $\mathcal{E}$ is fit if

(B1) $\mathcal{A}$ is bounded,
(B2) $\mathcal{A}$ has regularly open interior,
(B3) $\mathcal{A}$ has regularly closed closure,
(B4) $\mathcal{A}$ has finite boundary.

Assumptions (B2), (B3) mean that

$$\text{int}\mathcal{A} = \text{int clo}\mathcal{A}, \qquad \text{clo}\mathcal{A} = \text{clo int}\mathcal{A}\,,$$

respectively, and *finite boundary* means that the $(N-1)$-dimensional Hausdorff measure of the boundary is finite. Assumptions (B2), (B3) are illustrated in Fig. 4, where a two-dimensional region $\mathcal{A}$ is shown together with the regions int$\mathcal{A}$, clo$\mathcal{A}$, int clo$\mathcal{A}$, clo int$\mathcal{A}$. From the picture it is clear that (B2) states that a fit region cannot have missing lines or points, while (B3) excludes the presence of isolated lines or points.

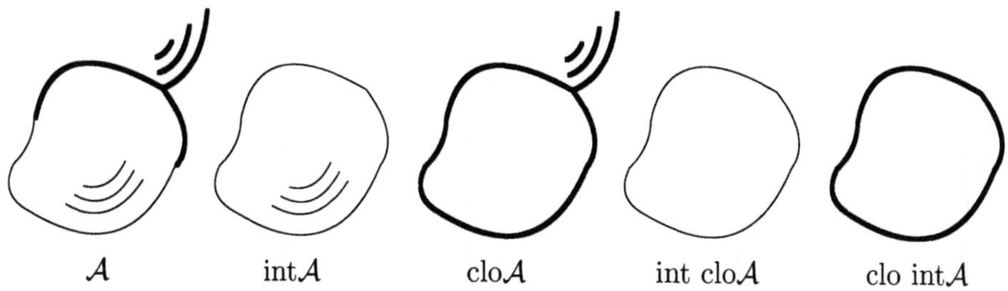

Figure 4.
A two-dimensional region $\mathcal{A}$ and the regions int$\mathcal{A}$, clo$\mathcal{A}$, int clo$\mathcal{A}$, clo int$\mathcal{A}$.
Thick lines are included in the region, thin lines are not included.

A fit region need not be open or closed. The set of all fit regions with the same interior forms an equivalence class. If we denote by $\tilde{\mathcal{A}}$ the equivalence class containing the fit region $\mathcal{A}$, we can prove that $\tilde{\mathcal{A}}$ has exactly one open representative, int$\mathcal{A}$, and one closed representative, clo$\mathcal{A}$. If we compare the set of all open representatives with the set of all fit regions in the sense of Noll and Virga, we see that the first set is strictly included in the second. In Del Piero (2003) I give the following example of a fit region in the sense of Noll and Virga which does not satisfy the assumption (B4). In a plane, choose a system of Cartesian coordinates (x, y). Take a positive number l and for each d in $(0, l)$ consider the region

$$\mathcal{S}(y,d) := \bigcup_{h=1}^{\infty} \bigcup_{k=1}^{2^{h-1}} \{\mathcal{B}(x_{h,p}, y+y_h; r_h)\,|\, p = 2k-1, x_{h,p} = \tfrac{lp}{2^h},\ y_h = \tfrac{d}{2^h}, r_h = \tfrac{d}{4^h}\}\,,$$

where $\mathcal{B}(x, y; r)$ denotes the open ball centered at (x, y) and with radius r. This region is fit both in the sense of Noll and Virga and in the sense of the definition (B1)-(B4). In particular, its perimeter is the sum of the lengths of the boundaries of the balls which form the region, and this sum is equal to πd. The boundary of $\mathcal{S}(y,d)$ consists of all boundaries of the balls plus the segment $((0,l)\times\{y\})$, and its length is $\pi d + l$.

Now consider the region

$$\mathcal{A}_l := \bigcup_{q=1}^{\infty} \{\, \mathcal{S}(y_q, d_q) \mid y_q = (1-\tfrac{1}{2^q})l,\ d_q = \tfrac{1}{2^q} l \,\}\,.$$

As shown in Fig. 5b, all regions $\mathcal{S}(y_q, d_q)$ are disjoint. Then the perimeter of $\mathcal{A}_l$ and the length of its boundary are the sums of the perimeters and of the lengths of the boundaries of the regions $\mathcal{S}(y_q, d_q)$, respectively. A quick computation shows that the perimeter of $\mathcal{A}_l$ is πl and that the length of the boundary is infinite.

Thus, $\mathcal{A}_l$ does not satisfy the requirement (B4) for fit regions. On the other hand, it can be easily checked that it satisfies all requirements in Noll and Virga's definition. In my opinion, this is a good example of a region which can be "dreamt up by an ingenious mathematician" but never "imagined by an engineer".

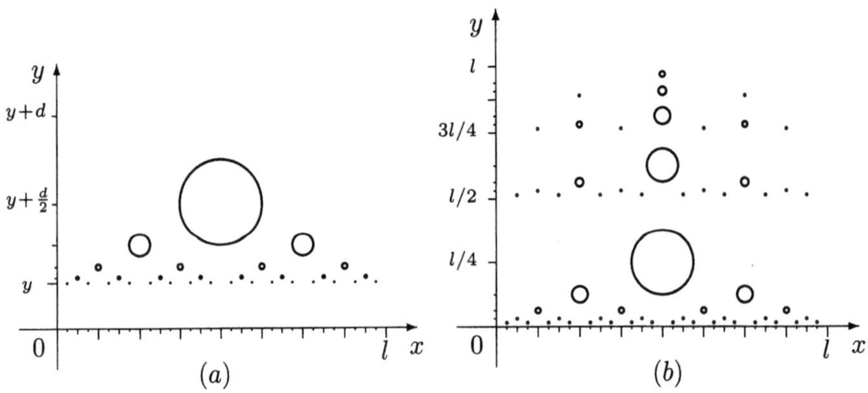

Figure 5.
The region $\mathcal{S}(y,d)$ (a) and the region $\mathcal{A}_l$ (b).

2.3 Classical Deformations

In Del Piero (2003) it is proved that the class of all regions which satisfy the conditions (B1)-(B4) satisfies the axioms of a material universe and is invariant under C^1-diffeomorphisms, and that each region in this class has a surface-like boundary. It can be easily verified that the same properties are enjoyed by the class of all *open* regions which satisfy the conditions (B1)-(B4). Thus, it makes sense to choose

- as *fit regions*, the *open* sets of $\mathcal{E}$ which satisfy the conditions (B1)-(B4),
- as *admissible deformations*, the orientation-preserving C^1-diffeomorphisms of $\mathcal{E}$.

It can also be verified that the class of fit regions chosen above is a subclass of the class of all fit regions in the sense of Noll and Virga (1988), and that in the one-dimensional case the two classes coincide. Indeed, according to both definitions a one-dimensional fit region is a finite union of bounded open intervals whose closures are pairwise disjoint.

With fit regions and admissible deformations selected as above, a *classical deformation* is defined as a mapping $f : \mathcal{A} \to \mathcal{E}$ such that[10]

(C1) $\mathcal{A}$ is a fit region,
(C2) f is the restriction to $\mathcal{A}$ of an admissible deformation over $\mathcal{E}$.

Every classical deformation is differentiable and orientation-preserving. Then the determinant of f is continuous and satisfies inequality (2.1) at every point of $\mathcal{A}$. By the boundedness of $\mathcal{A}$, it follows that there are positive constants m, M, depending on f and $\mathcal{A}$, such that

$$m \leq \det \nabla f(x) \leq M \qquad \forall x \in \mathcal{A} . \tag{2.2}$$

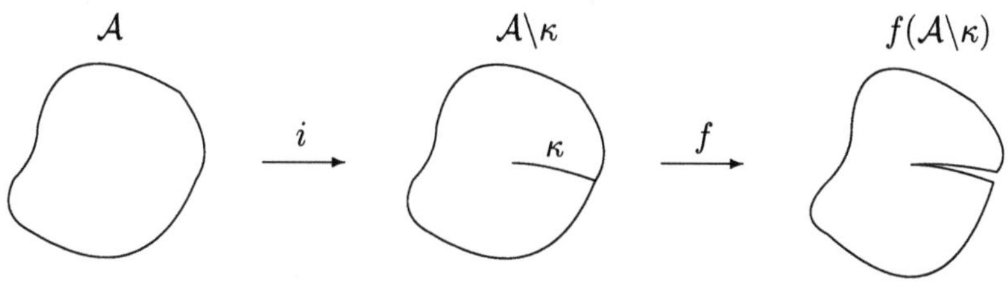

Figure 6.
Formation and opening of a crack.

3 Simple Deformations

3.1 Piecewise Fit Regions

Classical deformations are inadequate to describe the phenomenon of the formation and opening of a fracture. As shown in Fig. 6, this phenomenon can be modeled as the creation in $\mathcal{A}$ of a *crack site* κ without any change in the positions of the points of $\mathcal{A}$ (formation), followed by a deformation f of the cracked region $\mathcal{A}\backslash\kappa$ (opening). Crack sites will be assumed to be closed sets with zero Lebesgue measure. Consequently, the cracked region $\mathcal{A}\backslash\kappa$ has the same closure as the uncracked region $\mathcal{A}$, and if $\mathcal{A}$ is a fit region then from the property (B2) of fit regions we have

$$\mathcal{A} = \text{int}\mathcal{A} = \text{int clo}\mathcal{A} = \text{int clo}(\mathcal{A}\backslash\kappa) .$$

Then $\mathcal{A}\backslash\kappa$ cannot be a fit region. Indeed, κ is included in $\mathcal{A} = \text{int clo}(\mathcal{A}\backslash\kappa)$ but not in $\mathcal{A}\backslash\kappa = \text{int}(\mathcal{A}\backslash\kappa)$. Then $\text{int}(\mathcal{A}\backslash\kappa) \neq \text{int clo}(\mathcal{A}\backslash\kappa)$, in contradiction to (B2).

[10]Del Piero and Owen (1993), Section 2.

Thus, regions more general than fit regions are required to describe the formation and opening of a fracture. A convenient generalization is provided by piecewise fit regions. A *piecewise fit region* is a finite union of fit regions.[11] Relevant properties of piecewise fit regions are listed in Del Piero and Owen (1993). Here I establish a relationship between piecewise fit regions $\mathcal{A}$ and the corresponding regions int clo$\mathcal{A}$.

Let $\mathcal{A}$ be a piecewise fit region. Then,

(D1) $\mathcal{A} \subset \text{int clo}\mathcal{A}\,,$

(D2) $\text{clo}\mathcal{A} = \text{clo int clo}\mathcal{A}\,,$

(D3) $\mathcal{A}$ and int clo$\mathcal{A}$ have finite boundary,

(D4) int clo$\mathcal{A}$ is a fit region,

(D5) int clo$\mathcal{A}$ is the smallest fit region including $\mathcal{A}$.

Property (D1) holds because $\mathcal{A}$ open implies $\mathcal{A} = \text{int}\mathcal{A} \subset \text{int clo}\mathcal{A}$, and property (D2) is proved by observing that the inclusion clo int clo$\mathcal{A} \subset$ clo$\mathcal{A}$ holds for every region $\mathcal{A}$, and that the opposite inclusion holds for all open regions. To prove (D3) we observe that, by (D1) and (D2),

$$\text{bdy int clo}\mathcal{A} \;\subset\; \text{bdy}\mathcal{A}\,,$$

and that

$$\text{bdy}\mathcal{A} \;=\; \text{bdy}\,(\,\bigcup_i \mathcal{A}_i) \;\subset\; \bigcup_i\; (\text{bdy}\mathcal{A}_i)$$

because $\mathcal{A}$ is a finite union of fit regions $\mathcal{A}_i$. Then each $\mathcal{A}_i$ has finite boundary; therefore, both $\mathcal{A}$ and int clo$\mathcal{A}$ have finite boundary.

To prove (D4), it is sufficient to check the items (B1)–(B4) in the definition of a fit region. We see that $\mathcal{A}$ is bounded because it is a finite union of bounded regions. Then its closure is bounded, and so is int clo$\mathcal{A}$. Thus, int clo$\mathcal{A}$ satisfies (B1). It also satisfies (B2), which is a consequence of (D2), and (B3), which is true for all open regions. Finally, (B4) is implied by (D3).

Thus, int clo$\mathcal{A}$ is a fit region, and it includes $\mathcal{A}$ by (D1). Now let $\mathcal{B}$ be another fit region including $\mathcal{A}$. Then int clo$\mathcal{A} \subset$ int clo$\mathcal{B}$, and from the property (B2) of fit regions it follows that

$$\mathcal{A} \;\subset\; \text{int clo}\mathcal{A} \;\subset\; \text{int clo}\mathcal{B} \;=\; \mathcal{B}\,.$$

Then int clo$\mathcal{A}$ is the smallest fit region including $\mathcal{A}$. We may express this property by saying that int clo$\mathcal{A}$ is the *fit envelope* of the piecewise fit region $\mathcal{A}$.

[11] The union need not be disjoint. In fact, a finite union of pairwise disjoint fit regions is a fit region. An example of a two-dimensional piecewise fit region which is the union of non-disjoint fit regions is a disc without the radius $\theta = 0$. It is the union of the non-disjoint sectors $0 < \theta < \alpha$, $2\pi - \alpha < \theta < 2\pi$, with α any angle in $(\pi, 2\pi)$.

3.2 Piecewise Classical Deformations

In the example in Fig. 6, f maps points with arbitrarily small distance located on opposite sides of κ into points with a finite distance. Therefore, the deformation f cannot be extended, in general, to a C^1-diffeomorphism of $\mathcal{E}$.

Thus, just as done for fit regions, classical deformations need to be conveniently generalized. A *piecewise classical deformation* is a mapping f from a piecewise fit region $\mathcal{A}$, satisfying the following condition: there is a finite number of fit regions $\mathcal{A}_j$ such that: (i) their union is $\mathcal{A}$, (ii) the restriction of f to each $\mathcal{A}_j$ is a classical deformation.

To fix the ideas, consider the one-dimensional example represented in Fig. 7. Take a positive integer n, and set

$$\kappa_n := \bigcup_{h=1}^{n-1} \{\tfrac{h}{n}\}, \qquad \mathcal{A}_n := (0,1)\backslash\kappa_n . \tag{3.1}$$

The region $\mathcal{A}_n$ is not fit, because int clo$\mathcal{A}_n = (0,1)$ does not coincide with int$\mathcal{A}_n = (0,1)\backslash\kappa_n$. But it is piecewise fit, because it is the union of the fit regions

$$\mathcal{A}_{n,h} := (\tfrac{h-1}{n}, \tfrac{h}{n}), \qquad h \in \{1,..n\} . \tag{3.2}$$

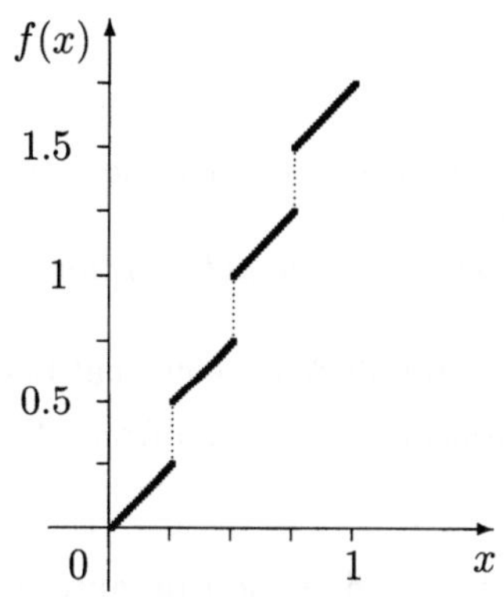

Figure 7.
The function f_n in (3.3), with $n = 4$.

The function f_n defined by

$$f_n(x) := f_{n,h}(x), \qquad x \in \mathcal{A}_{n,h} , \tag{3.3}$$

with

$$f_{n,h}(x) := x + \tfrac{h-1}{n} , \tag{3.4}$$

is a piecewise classical deformation, since each one of the mappings $f_{n,h}$ is a C^1-diffeomorphism of $\mathcal{E}$. However, f_n cannot be extended to a C^1-diffeomorphism of $\mathcal{E}$, because every extension fails to be continuous at κ_n.

In this example, the pair (κ_n, f_n) describes the formation of a crack site κ_n in the fit region $\mathcal{A} = (0,1)$ followed by the piecewise classical deformation f_n from the piecewise fit region $\mathcal{A}\backslash\kappa_n$. In fact, $\mathcal{A}$ need not be a fit region: the same example may describe the formation of a *new* crack site in any piecewise fit region $\mathcal{A} = (0,1)\backslash\gamma$, with γ a finite set disjoint from κ_n. In this case, the crack site for the piecewise classical deformation (κ_n, f_n) is the union of the pre-existing crack site γ and the new crack site κ_n.

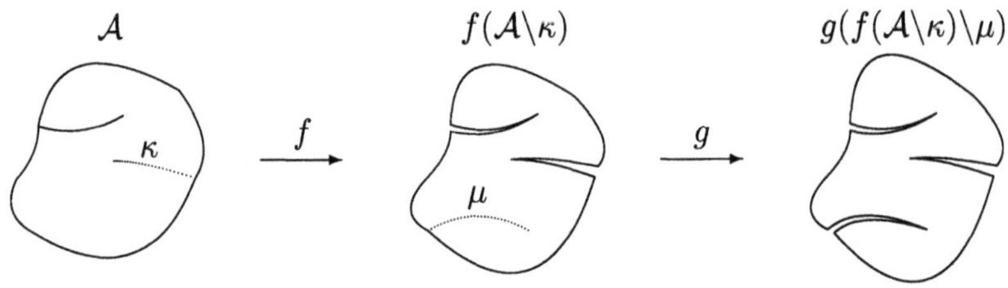

Figure 8.
Composition of two simple deformations (κ, f), (μ, g)
(dotted lines denote the new crack sites created in each simple deformation).

3.3 Simple Deformations

We are now ready to introduce a class of deformations appropriate to describe the formation and opening of fractures.

Let $\mathcal{A}$ be a piecewise fit region. A *simple deformation* from $\mathcal{A}$ is a pair (κ, f), where[12]

(E1) κ is a subset of $\mathcal{A}$ with null N-dimensional Lebesgue measure,

(E2) f is a piecewise classical deformation from $\mathcal{A}\backslash\kappa$,

(E3) f is injective.

In this definition, $\mathcal{A}\backslash\kappa$ is a piecewise fit region because it is the domain of a piecewise classical deformation. Moreover, $\mathcal{A}\backslash\kappa$ and $\mathcal{A}$ have the same fit envelope. Indeed, as already pointed out in Section 3.1, the property (E1) of κ implies that $\mathcal{A}\backslash\kappa$ and $\mathcal{A}$ have the same closure, and the equality int clo$(\mathcal{A}\backslash\kappa) =$ int clo$\mathcal{A}$ follows. Finally, the set κ has finite $(N-1)$-dimensional Hausdorff measure. Indeed, the relations

$$\kappa \subset \mathcal{A} \subset \text{clo}\mathcal{A} = \text{clo}(\mathcal{A}\backslash\kappa), \qquad \kappa \cap \text{int}(\mathcal{A}\backslash\kappa) \subset \kappa \cap (\mathcal{A}\backslash\kappa) = \emptyset$$

tell us that κ is included in the boundary of $\mathcal{A}\backslash\kappa$, which has finite $(N-1)$-dimensional Hausdorff measure by (D3). A one-dimensional example of a simple deformation is the pair (κ_n, f_n) defined by (3.1)–(3.4) and shown in Fig. 7.

[12]Del Piero and Owen (1993), Section 3.

Consider a piecewise fit region $\mathcal{A}$ and two simple deformations (κ, f) from $\mathcal{A}$ and (μ, g) from $f(\mathcal{A}\backslash\kappa)$. The *composition* of (μ, g) and (κ, f) is the pair

$$(\mu, g) \circ (\kappa, f) \;:=\; (\kappa \cup f^{-1}(\mu)\,,\, g \circ f|_{\mathcal{A}\backslash(\kappa \cup f^{-1}(\mu))})\,. \tag{3.5}$$

This pair is a simple deformation from $\mathcal{A}$.[13]

An inverse of a simple deformation cannot be defined in general. Indeed, an inverse of a simple deformation (κ, f) from $\mathcal{A}$ should be a simple deformation (μ, g) from $f(\mathcal{A}\backslash\kappa)$ such that

$$(\mu, g) \circ (\kappa, f) \;:=\; (\emptyset, i_{\mathcal{A}})\;, \tag{3.6}$$

where $\emptyset$ is the empty set and $i_{\mathcal{A}}$ is the restriction to $\mathcal{A}$ of the identity mapping $i(x) = x$. Then $\kappa \cup f^{-1}(\mu) = \emptyset$ by (3.5), and this is possible only if $\kappa = \mu = \emptyset$. The same rule (3.5) requires that $g \circ f = i_{\mathcal{A}}$, so that $g = f^{-1}$. In conclusion, a simple deformation (κ, f) is invertible if and only if $\kappa = \emptyset$, i.e., if and only if it does not involve the creation of a new crack site. For an invertible simple deformation $(\emptyset, f)$ from $\mathcal{A}$, the inverse $(\emptyset, f^{-1})$ is unique and is a simple deformation from $f(\mathcal{A})$.[14]

Every simple deformation (κ, f) from $\mathcal{A}$ admits the factorization

$$(\kappa, f) \;:=\; (\emptyset, f) \circ (\kappa, i_{\mathcal{A}\backslash\kappa}) \tag{3.7}$$

into a deformation describing the creation of a new crack site without changes in the positions of the points of $\mathcal{A}$, followed by a deformation not involving the creation of new cracks. This is precisely the factorization shown in Fig. 6. Both factors are simple deformations.[15] They can be called a *pure cracking* and a *simple deformation without cracking*, respectively. The fact that a pure cracking is not invertible reflects the irreversibility of the process of formation of a fracture, a basic assumption made in many theories of fracture.

3.4 Limits of Simple Deformations

We have seen that simple deformations provide a geometrical basis appropriate to the study of the formation and opening of fractures. Now we shall see that a surprisingly richer variety of phenomena can be described by considering sequences of simple deformations and by investigating their limit properties. In this section I consider limits in the sense of Del Piero and Owen (1993); a weaker notion of limit will be considered later, in Section 6.3.

Let $\mathcal{A}$ be a piecewise fit region, let κ be a subset of $\mathcal{A}$, let g be a map from $\mathcal{A}\backslash\kappa$ into $\mathcal{E}$, and let G be a tensor field on $\mathcal{A}\backslash\kappa$. A sequence $n \mapsto (\kappa_n, f_n)$ of simple deformations from $\mathcal{A}$ is said to converge to the triple (κ, g, G) if

[13] Del Piero and Owen (1993), Proposition 3.4.
[14] Del Piero and Owen (1993), Proposition 3.7.
[15] Del Piero and Owen (1993), Proposition 3.5.

$$\liminf_{n\to\infty} \kappa_n = \kappa\,, \tag{3.8}$$

$$\lim_{n\to\infty} \sup_{x\in\mathcal{A}\setminus(\kappa_n\cup\kappa)} |f_n(x) - g(x)| = 0\,, \tag{3.9}$$

$$\lim_{n\to\infty} \sup_{x\in\mathcal{A}\setminus(\kappa_n\cup\kappa)} |\nabla f_n(x) - G(x)| = 0\,, \tag{3.10}$$

where

$$\liminf_{n\to\infty} \kappa_n := \bigcup_{p=1}^{\infty} \bigcap_{n=p}^{\infty} \kappa_n\,. \tag{3.11}$$

According to this definition, the limit set κ is made of points which belong to all sets κ_n with n greater than some positive integer p. The notation

$$(\kappa_n, f_n) \to (\kappa, g, G) \tag{3.12}$$

denotes the type of convergence just defined. A triple (κ, g, G) which is a limit of a sequence of simple deformations from $\mathcal{A}$ is called a *limit of simple deformations* from $\mathcal{A}$. A basic example is provided by the sequence $n \mapsto (\kappa_n, f_n)$ defined by (3.1)–(3.4). It converges to the triple

$$\kappa = \emptyset\,, \quad g(x) = 2x\,, \quad G(x) = 1\,. \tag{3.13}$$

To get some insight into the nature of limits of simple deformations, let us discuss this example in detail. Assume that the simple deformations (κ_n, f_n) describe the axial extension of a bar of initial unit length. Each of them describes a deformation obtained by cutting the bar into n pieces of length $1/n$ and translating each piece rigidly a distance $1/n$ from the neighboring ones. We see that, with increasing n, the cracks *diffuse* throughout $\mathcal{A}$, that is, their number increases indefinitely with n and they spread over the whole region $\mathcal{A}$.

In the limit element (κ, g, G), the occurring of crack diffusion cannot be detected from the first two items in the triple. Indeed, the pair (κ, g) in (3.13) could well describe a homogeneous extension of a bar in which the initial length is doubled. Thus, some sign of the presence of crack diffusion is expected to come from the third item G. The fact that G is the limit of the deformation gradients ∇f_n and that the gradients are local descriptors of the deformation, suggests the interpretation of G as a local measure of deformation in the limit (κ, g, G). In our example, $G(x) = 1$ conveys the information that no part of the bar has actually been stretched, and this in spite of the fact that the total length has been doubled. The apparent contradiction is explained by observing that in each term of the approximating sequence the elongation is obtained by opening the cracks rather than by stretching the uncracked parts of the bar. Although there are no macroscopic cracks in the limit element, the presence of diffused cracks is revealed by the fact that the limit G of the deformation gradients is not equal to the gradient of the *macroscopic deformation* g. Thus, the inequality $\nabla g \neq G$ is a sign that some *structural change* has occurred in the deformation described by the triple (κ, g, G).

To make clear the correlation between the inequality $\nabla g \neq G$ and the occurrence of crack diffusion, let us first give a precise definition of the diffusion zone. For every sequence $n \mapsto (\kappa_n, f_n)$ of simple deformations from $\mathcal{A}$ and for every positive integer p, consider the set

$$\Lambda_p(n \mapsto \kappa_n) \; := \; \text{clo}_{\mathcal{A}}(\bigcup_{n=p}^{\infty} \kappa_n)$$

obtained by taking the closure in $\mathcal{A}$ of the set of all points which belong to at least one of the κ_n with $n \geq p$. Each Λ_p includes the limit set κ. Indeed, by (3.11), each point in κ belongs to all κ_n with n greater than some p. Moreover, it is clear from the preceding definition that Λ_{p+1} is included in Λ_p for all p. The set

$$\Lambda_\infty(n \mapsto \kappa_n) \; := \; \bigcap_{p=1}^{\infty} \Lambda_p(n \mapsto \kappa_n)$$

is the set of all points x in $\mathcal{A}$ for which there is a subsequence of $n \mapsto \kappa_n$, and for each κ_n in the subsequence there is a point x_n such that $x_n \to x$. Thus, $\Lambda_\infty(n \mapsto \kappa_n)$ is the *diffusion zone* for the sequence $n \mapsto (\kappa_n, f_n)$.

It may happen that different sequences of simple deformations with different diffusion zones converge to the same limit (κ, g, G). In this case, the structural changes described by the limit element are expected to occur in the *intersection* of the diffusion zones. The *minimum* diffusion zone

$$\Phi(\kappa, g, G) \; := \; \bigcap \{\Lambda_\infty(n \mapsto \kappa_n) \mid (\kappa_n, f_n) \to (\kappa, g, G)\}$$

was called the *fractured zone* in Del Piero and Owen (1993). It includes the set κ, because it is the intersection of regions Λ_p all including κ, and it is closed in $\mathcal{A}$, because all Λ_p are closed in $\mathcal{A}$. Then the set $\mathcal{A}\backslash\Phi(\kappa, g, G)$ of all points in $\mathcal{A}$ which are away from the diffusion zone is open. For each one of such points, there are an open ball $\mathcal{B}(x_0, \delta)$ in $\mathcal{A}$, centered at x_0 and with radius $\delta > 0$, and a sequence $n \mapsto (\kappa_n, f_n)$ converging to (κ, g, G) such that

$$\kappa \cap \mathcal{B}(x_0, \delta) \; = \; \emptyset \, , \qquad \kappa_n \cap \mathcal{B}(x_0, \delta) \; = \; \emptyset \quad \forall n \in \mathbb{N} \, .$$

By the fact that each (κ_n, f_n) is a simple deformation from $\mathcal{A}$, the restriction of each f_n to $\mathcal{B}(x_0, \delta)$ is continuously differentiable. Due to the uniform convergence of $n \mapsto f_n$ to g and of $n \mapsto \nabla f_n$ to G, standard theorems of differential calculus ensure that g is continuously differentiable and that $\nabla g = G$ in $\mathcal{B}(x_0, \delta)$.[16] Then we have $\nabla g = G$ at all points of $\mathcal{A}\backslash\Phi(\kappa, g, G)$, and because this set is open we conclude that

$$\mathcal{A} \setminus \Phi(\kappa, g, G) \; \subset \; \text{int} \, \{x \in \mathcal{A}\backslash\kappa \mid \nabla g(x) \; = \; G(x)\} \, . \tag{3.14}$$

[16]See e.g. Cartan (1967), Theorems 3.6.1 and 3.6.2.

Consequently, all points at which $\nabla g \neq G$ belong to the fractured zone. Thus, the inequality $\nabla g \neq G$ characterizes points at which crack diffusion occurs in all sequences $n \mapsto (\kappa_n, f_n)$ which converge to (κ, g, G), and therefore structural changes occur in the limit element.

Structural changes may be of different types, including dislocations and slips in crystalline solids, changes of orientation of the molecules in liquid crystals, incorporation of voids in soils, as well as some types of plastic deformation.[17] In Owen (1995), the term *disarrangements* was proposed as a collective name for structural changes due to both diffused (*microdisarrangements*) and concentrated (*macrodisarrangements*) discontinuities, and the fractured zone was called the *disarrangement zone*. This terminology has become standard in the subsequent literature. Here the three items in (κ, g, G) will be called the *new macrodisarrangement site*,[18] the *macroscopic deformation*, and the *deformation tensor without disarrangements*, respectively.

All limits of simple deformations have the following properties[19]

(F1) κ is a set with zero Lebesgue measure,
(F2) g and G are continuous on $\mathcal{A}\backslash\kappa$,
(F3) at every point x of $\mathcal{A}\backslash\kappa$ at which g is differentiable,

$$\det \nabla g(x) \;\geq\; \det G(x)\,. \tag{3.15}$$

By contrast, some nice properties of simple deformations are not preserved in the limit. For example,

- $\mathcal{A}\backslash\kappa$ need not be piecewise fit,
- g need not be differentiable,
- g need not be injective.

Thus, the first two items in a limit of simple deformations do not form, in general, a simple deformation. An example of a limit of simple deformations from $\mathcal{A}$ with $\mathcal{A}\backslash\kappa$ not piecewise fit is provided by the sequence $n \mapsto (\kappa_n, f_n)$ defined in (3.1)–(3.4). Indeed, if we take the subsequence $n \mapsto (\kappa_{2^n}, f_{2^n})$, we see that it converges to the limit (κ, g, G), where g and G are the same as in (3.13), and κ is the set of all dyadic rationals

$$\{\, \tfrac{p}{2^m},\ \ m \in \mathbb{N},\ \ p \in \{1,..\, 2^m - 1\}\}\ .$$

This κ is not closed; consequently, $\mathcal{A}\backslash\kappa$ is not open, and therefore it is not piecewise fit.

To construct an example of a limit of simple deformations which is neither differentiable nor injective, consider a function $f : [0,1] \to \mathbb{R}$ continuous, nondecreasing, and

[17]See Del Piero and Owen (1993), Sections 6 and 7. The sequence defined by (3.1)-(3.4) is a minimizing sequence for the problem considered in Lecture 1, with a uniform distribution of jumps across the bar. The structural changes occurring in that specific problem may be interpreted as plastic deformations.

[18]See point (vi) in Section 5.1 below.

[19]Del Piero and Owen (1993), Theorems 4.10 and 4.17.

with $f(0)=0,\ f(1)=1$. For each positive integer n, consider the points

$$x_{n,h} \;=\; \min\,\{x \in [0,1] \mid f(x) = \tfrac{h}{n}\,,\ \ h \in \{0,1,..\,n-1\}\}\,,$$

and the simple deformations (κ_n, f_n) from $(0,1)$ defined by

$$\begin{aligned} \kappa_n \;&=\; \{x_{n,h}\,,\ h \in \{1,..\,n-1\}\}\,, \\ f_n(x) \;&=\; \tfrac{h}{n} + \tfrac{x-x_{n,h}}{n} \qquad \text{for } \ x \in (x_{n,h},\, x_{n,h+1}),\ \ h \in \{0,1,..\,n-1\}\,. \end{aligned}$$

The sequence $n \mapsto (\kappa_n, f_n)$ converges to $(\emptyset, f|_{(0,1)}, 0)$. This is a limit of simple deformations, in which f can be any continuous nondecreasing function. For example, f can be the Cantor function, which is neither differentiable nor injective.[20]

4 Structured Deformations

4.1 Definition

As remarked in the preceding section, limits of simple deformations may lack some physically desirable properties. For example, they may involve deformations that are not injective. The following list of "physically reasonable" properties defines triples (κ, g, G) called *structured deformations* from $\mathcal{A}$.[21]

(G1) (κ, g) is a simple deformation from $\mathcal{A}$,

(G2) G is a continuous tensor field on $\mathcal{A}\backslash\kappa$, and there is a finite collection of fit regions $\mathcal{A}_j$ whose union is $\mathcal{A}\backslash\kappa$, such that G has a continuous extension to the closure of each $\mathcal{A}_j$,

(G3) There is a positive constant m such that

$$m \;<\; \det G(x) \;\leq\; \det \nabla g(x) \qquad \forall x \in \mathcal{A}\backslash\kappa\,. \tag{4.1}$$

Notice that (G1) implies that both $\mathcal{A}$ and $\mathcal{A}\backslash\kappa$ are piecewise fit regions, that κ is a subset of $\mathcal{A}$ with zero Lebesgue measure, and that g is piecewise classical and injective. Then from (G2) and (G3) it follows that structured deformations have the properties (F1)–(F3) enjoyed by limits of simple deformations. For the three items in (κ, g, G) we will use the same names adopted for limits of simple deformations.

While there is no satisfactory notion of composition for limits of simple deformations, the composition of a structured deformation (κ, g, G) from $\mathcal{A}$ with a structured deformation (μ, h, H) from $g(\mathcal{A}\backslash\kappa)$ may be defined as

$$(\mu, h, H) \circ (\kappa, g, G) \;:=\; (\kappa \cup g^{-1}(\mu),\ h \circ g|_{\mathcal{A}\backslash(\kappa\cup g^{-1}(\mu))},\ (H \circ g)G|_{\mathcal{A}\backslash(\kappa\cup g^{-1}(\mu))})\,, \tag{4.2}$$

[20] Del Piero and Owen (1993), Section 4.
[21] Del Piero and Owen (1993), Section 5.

where the first two items in the triple on the right-hand side correspond to the simple deformation $(\mu, h) \circ (\kappa, g)$ and the third item is the pointwise composition of the tensor fields $(H \circ g)$ and G. The composition of two structured deformations is a structured deformation.[22]

A structured deformation (κ, g, G) is invertible if and only if

$$\kappa = \emptyset, \qquad \det G = \det \nabla g\,. \tag{4.3}$$

If this is the case, the unique inverse is $(\emptyset, g^{-1}, G^{-1} \circ g^{-1})$; it is a structured deformation.[23]

Every structured deformation (κ, g, G) from $\mathcal{A}$ admits the factorization

$$(\kappa, g, G) = (\emptyset, i_{g(\mathcal{A}\setminus\kappa)}, (G \circ g^{-1})\nabla(g^{-1})) \circ (\kappa, g, \nabla g)\,, \tag{4.4}$$

in which both factors are structured deformations.[24] In the factor $(\kappa, g, \nabla g)$, the fact that the third item is the derivative of the second tells us that there are no microdisarrangements, while in the remaining factor the fact that the first two items are the empty set and the identical transformation tells us that there are no macrodisarrangements and no macroscopic deformation. The two factors $(\kappa, g, \nabla g)$, $(\emptyset, i_{g(\mathcal{A}\setminus\kappa)}, (G \circ g^{-1})\nabla(g^{-1}))$ can be called a *structured deformation without microdisarrangements* and a *pure microdisarrangement*, respectively. The latter will also be called a *purely microscopic deformation.*

The factor $(\kappa, g, \nabla g)$ can be regarded as the limit of the constant sequence of simple deformations $n \mapsto (\kappa, g)$; therefore, it can be identified with the simple deformation (κ, g). It can be further decomposed according to the law

$$(\kappa, g, \nabla g) = (\emptyset, g, \nabla g) \circ (\kappa, i_{\mathcal{A}\setminus\kappa}, I_{\mathcal{A}\setminus\kappa})\,, \tag{4.5}$$

which reflects the factorization (3.7) of a simple deformation into a pure cracking followed by an invertible simple deformation.

4.2 Approximation Theorem

A one-dimensional example of a structured deformation is the limit (3.13) of the sequence of simple deformations defined by (3.1)–(3.4). Now we address to the question of whether all structured deformations are limits of simple deformations. The answer is affirmative, and the proof of this statement, the Approximation Theorem, was given in Del Piero and Owen (1993).

The theorem is based on a lemma concerning purely microscopic deformations, whose proof is technically complicated. Here I sketch the proof of the much simpler one-dimensional case. I recall that, in a one-dimensional purely microscopic deformation $(\emptyset, i, H)$ from $\mathcal{A} = (0,1)$, H is a continuous scalar field with a continuous extension to $[0,1]$. Also, the condition (4.1) on the determinant reduces to

$$m < H(x) \leq 1 \qquad \forall x \in (0,1)\,. \tag{4.6}$$

[22] Del Piero and Owen (1993), Proposition 5.3.

[23] Del Piero and Owen (1993), Propositions 5.4 and 5.5.

[24] Del Piero and Owen (1993), Proposition 5.6.

Approximation Lemma. For every purely microscopic deformation $(\emptyset, i|_{(0,1)}, H)$ from $(0,1)$ there exists a sequence $n \mapsto (\kappa_n, f_n)$ of simple deformations from $(0,1)$ such that $(\kappa_n, f_n) \to (\emptyset, i_{(0,1)}, H)$.

Sketch of proof. For each positive integer n, subdivide $(0,1)$ into n intervals of equal length $1/n$, and take

$$\kappa_n = \bigcup_{h=1}^{n-1} \{\tfrac{h}{n}\},$$

$$f_n(x) = \tfrac{h}{n} + (x - \tfrac{h}{n})\, H(\tfrac{h}{n}) \qquad \text{for } \tfrac{h}{n} < x < \tfrac{h+1}{n}, \quad h \in \{0, 1, .. n-1\}.$$

The gradient of f_n is

$$\nabla f_n(x) = H(\tfrac{h}{n}) \qquad \text{for } \tfrac{h}{n} < x < \tfrac{h+1}{n},$$

and the jump of f_n at $x = h/n$ is

$$[f_n](\tfrac{h}{n}) = \tfrac{h}{n} - \tfrac{h-1}{n} - (\tfrac{h}{n} - \tfrac{h-1}{n})\, H(\tfrac{h-1}{n}) = \tfrac{1}{n}(1 - H(\tfrac{h-1}{n})).$$

Both are positive by (4.6). Then f_n is injective, and because the remaining conditions (E1), (E2) in the definition of a simple deformation are satisfied, each pair (κ_n, f_n) is a simple deformation from (0,1). The proof then reduces to checking that the sequence $n \mapsto (\kappa_n, f_n)$ converges to $(\emptyset, i_{(0,1)}, H)$.[25] □

Now consider a three-dimensional piecewise fit region $\mathcal{A}$ and a structured deformation (κ, g, G) from $\mathcal{A}$. Taking for granted that the Approximation Lemma holds for all three-dimensional purely microscopic deformations, we have the following

Approximation Theorem. For each structured deformation (κ, g, G) from a piecewise fit region $\mathcal{A}$ there is a sequence $n \mapsto (\kappa_n, f_n)$ of simple deformations from $\mathcal{A}$ such that $(\kappa_n, f_n) \to (\kappa, g, G)$.

Sketch of proof. By (4.4), (κ, g, G) can be decomposed into the simple deformation $(\kappa, g, \nabla g)$ from $\mathcal{A}$ followed by the purely microscopic deformation $(\emptyset, i_{g(\mathcal{A}\backslash\kappa)}, (G \circ g^{-1})\nabla(g^{-1}))$ from $g(\mathcal{A}\backslash\kappa)$. Moreover, the Approximation Lemma ensures that there is a sequence $n \mapsto (\kappa_n, f_n)$ of simple deformations from $g(\mathcal{A}\backslash\kappa)$ which converges to the purely microscopic deformation. Then the proof reduces to checking that the sequence

$$n \mapsto (\kappa_n, f_n) \circ (\kappa, g)$$

of simple deformations from $\mathcal{A}$ converges to (κ, g, G). □

---

[25]For details in the proof of this Lemma and of the following Approximation Theorem see Del Piero and Owen (2000), Part One, Section 4.

A consequence of the Approximation Theorem is that the definitions of diffusion zone and of disarrangement zone given in Section 3.4 for limits of simple deformations apply to structured deformations as well. Moreover, the additional structure enjoyed by structured deformations provides an improvement of the characterization (3.14) of the disarrangement zone of a limit of simple deformations.

Theorem.[26] For a structured deformation, the inclusion in (3.14) is replaced by equality. □

Thus, we have the following explicit expression for the disarrangement zone of a structured deformation (κ, g, G):

$$\Phi(\kappa, g, G) \;=\; \mathcal{A} \setminus \mathrm{int}\,\{\, x \in \mathcal{A} \setminus \kappa \mid \nabla g(x) = G(x)\,\}\,. \tag{4.7}$$

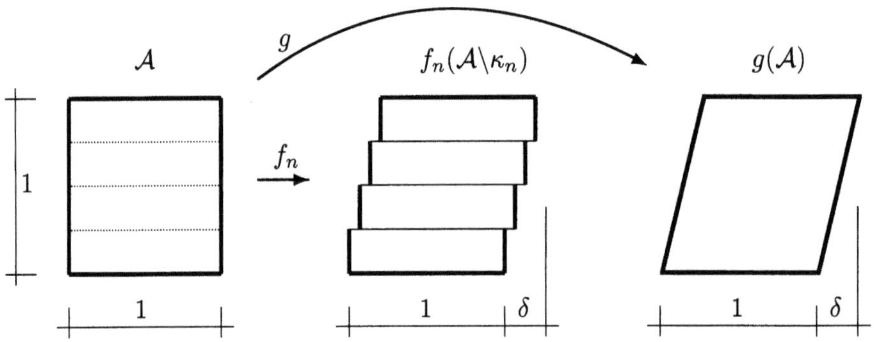

Figure 9. The deck of cards.

4.3 Examples

Basic two-dimensional examples of structured deformations are the following.

(i) The deck of cards

Let $\mathcal{A}$ be the square $(0,1) \times (0,1)$. For fixed $\delta > 0$ and $n \in \mathbb{N}$, consider the simple deformation (κ_n, f_n) with

$$\begin{aligned} \kappa_n &= (0,1) \times \{\, \tfrac{h}{n} \mid h \in \{1, 2, .. n-1\}\}\,, \\ f_n(x,y) &= (x + \tfrac{h\delta}{n}, y) \quad \text{for} \;\; y \in (\tfrac{h-1}{n}, \tfrac{h}{n})\,, \qquad h \in \{1, 2, .. n\}\,. \end{aligned}$$

Each f_n is a piecewise rigid deformation, with $\nabla f_n = I$ and with slips of intensity δ/n concentrated at the lines which form the crack site κ_n. The sequence $n \mapsto (\kappa_n, f_n)$ converges to the structured deformation $(\emptyset, g, I)$, with

$$g(x,y) \;=\; (x + y\delta,\, y)\,.$$

[26] Del Piero and Owen (1993), Theorem 5.10.

The expression of the gradient of g

$$(\nabla g(x,y)) = \begin{pmatrix} 1 & \delta \\ 0 & 1 \end{pmatrix}$$

tells us that the macroscopic deformation g is a homogeneous simple shear, and the fact that $\nabla g(x,y) \neq I$ at all points of $\mathcal{A}$ tells us that the disarrangement zone (4.7) equals $\mathcal{A}$, so that the disarrangements diffuse throughout the whole region.

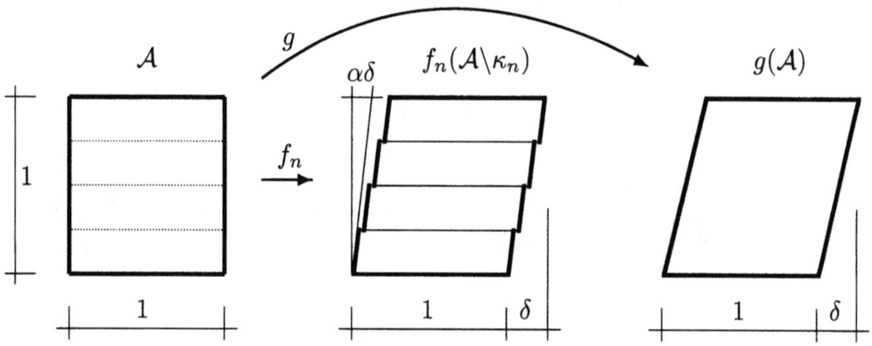

Figure 10. Two-level shear.

(ii) Two-level shear

Let $\mathcal{A}$, δ, and κ_n be as in the preceding example, and for a fixed $\alpha \in \mathbb{R}$ take

$$f_n(x,y) = (x + (1-\alpha)\tfrac{h\delta}{n} + \alpha\delta y,\ y) \quad \text{for } y \in (\tfrac{h-1}{n}, \tfrac{h}{n}),\ \ h \in \{1, 2, .. n\}.$$

Then

$$(\nabla f_n(x,y)) = \begin{pmatrix} 1 & \alpha\delta \\ 0 & 1 \end{pmatrix},$$

and because the right-hand side does not depend on n we have $G = \nabla f_n$ for all n. The approximating sequence $n \mapsto (\kappa_n, f_n)$ converges to the structured deformation $(\emptyset, g, G)$, with the same g as in the preceding example and with G as above. Thus, the macroscopic deformation g is the same as in the deck of cards, but, as shown in Fig. 10, each deformation f_n is a combination of simple shear and of slips of intensity $(1-\alpha)\delta/n$ along the lines forming the crack site κ_n. In particular, for $\alpha = 0$ we recover the deck of cards, while for $\alpha = 1$ we have the homogeneous simple shear $(\emptyset, g, \nabla g)$, that is, a deformation without any disarrangement.

(iii) Incorporation of voids

Let $\mathcal{A}$ be the square $(0,1)\times(0,1)$. For a fixed $\delta > 0$, for each $n \in \mathbb{N}$ consider the simple deformation (κ_n, f_n) from $\mathcal{A}$ given by

$$\kappa_n = \Big(\bigcup_{h=1}^{n-1} (\{\tfrac{h}{n}\} \times (0,1)) \Big) \cup \Big(\bigcup_{k=1}^{n-1} ((0,1) \times \{\tfrac{k}{n}\}) \Big),$$

$$f_n(x,y) = (x + \tfrac{h\delta}{n},\, y + \tfrac{k\delta}{n}) \quad \text{for } x \in (\tfrac{h-1}{n}, \tfrac{h}{n}),\ y \in (\tfrac{k-1}{n}, \tfrac{k}{n}),\ \ h,k \in \{1,2,..n\}.$$

The sequence $n \mapsto (\kappa_n, f_n)$ converges to the structured deformation $(\emptyset, g, I)$, with macroscopic deformation

$$g(x,y) = ((1+\delta)x,\, (1+\delta)y)\,.$$

This a homogeneous, uniform dilatation. The fact that $G = \nabla f_n = I$ tells us that the area increase associated with g is not due to the smooth deformation of the individual pieces of $\mathcal{A}\backslash\kappa_n$, but to the incorporation of the voids enclosed in the regions $f_n(\mathcal{A}\backslash\kappa_n)$.

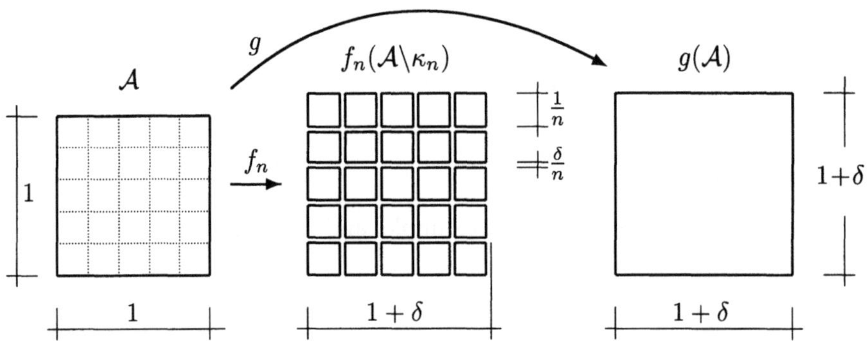

Figure 11. Incorporation of voids.

4.4 A Review of Classes of Deformations

At this point it is convenient to review the classes of deformations considered so far. Our starting point was the definition of *classical deformations* given in Section 2.3. *Piecewise classical deformations* were defined in Section 3.2. From the definitions it is clear that classical deformations are special injective piecewise classical deformations, and pairs (κ, f) consisting of a crack site κ and of an injective piecewise classical deformation were called *simple deformations* in Section 3.3.

In the same section we showed that a simple deformation (κ, f) is invertible if and only if κ is the empty set. Now we observe that, if f is an injective piecewise classical deformation, then the pair $(\emptyset, f)$ is an invertible simple deformation. Therefore, each invertible simple deformation can be identified with an injective piecewise classical deformation and *vice-versa.*

In Section 4.1, *structured deformations* from a piecewise fit region $\mathcal{A}$ were defined as triples (κ, g, G), with (κ, g) a simple deformation from $\mathcal{A}$, and the factorization (4.4) provided two basic subclasses of structured deformations:

$(\kappa, g, \nabla g)$	structured deformations without microdisarrangements,
$(\emptyset, i_{\mathcal{A}}, G)$	purely microscopic deformations.

The first subclass subclass was identified with the set of all simple deformations. In particular, since a simple deformation (κ, g) is invertible if and only if κ is the empty set, a structured deformation of the type $(\emptyset, g, \nabla g)$ can be identified with an invertible simple deformation. Finally, the Approximation Theorem in Section 4.2 established that all structured deformations are limits of simple deformations. Then the inclusion relations shown in Fig. 12 hold between the classes of deformations defined above. In the figure, Cld, Sid and Std denote the sets of all classical, simple, and structured deformations, respectively, Inj Pcd is the set of all injective piecewise classical deformations, Inv denotes sets of invertible deformations, and Lim Sid is the set of all limits of simple deformations.

$$\text{Cld} \subset \text{Inj Pcd} = \text{Inv Sid} \subset \left\{ \begin{array}{c} \text{Sid} \\ \text{Inv Std} \end{array} \right\} \subset \text{Std} \subset \text{Lim Sid}$$

Figure 12. Inclusions between classes of deformations.

5 Trace Theorem and Integral-Gradient Formula

5.1 Some Concepts from Geometric Measure Theory

The Trace Theorem and the Integral-Gradient Formula for structured deformations have been proved in Del Piero and Owen (1995). They involve some concepts from geometric measure theory, which I recall briefly. For a more detailed introduction to the subject I refer to Vol'pert and Hudjaev (1985).

(i) Essential Boundary

Let x be a point of $\mathcal{E}$, let $\mathcal{A}$ be a subset of $\mathcal{E}$, and let $\mathcal{B}(x, \delta)$ be the ball centered at x and with radius δ. We say that a point x of $\mathcal{E}$ is a *point of density* for $\mathcal{A}$ if the limit[27]

$$\lim_{\delta \to 0+} \frac{|\mathcal{A} \cap \mathcal{B}(x,\delta)|}{|\mathcal{B}(x,\delta)|} \tag{5.1}$$

[27]Here and in the following, $|\cdot|$ denotes the N-dimensional Lebesgue measure and $\mathcal{H}^{N-1}(\cdot)$ denotes the $(N-1)$-dimensional Hausdorff measure.

is equal to one, and that it is a *point of rarefaction* if the same limit is equal to zero. The *essential boundary* $\text{eby}\mathcal{A}$ of $\mathcal{A}$ is the set of all points of $\mathcal{E}$ which are neither density points nor points of rarefaction for $\mathcal{A}$.

All points in the interior of $\mathcal{A}$ are points of density and all points in the exterior of $\mathcal{A}$ are points of rarefaction. Consequently, all points in the essential boundary belong to the boundary of $\mathcal{A}$:

$$\text{eby}\mathcal{A} \subset \text{bdy}\mathcal{A} .$$

(ii) Reduced Boundary

Let ν be a unit vector, and consider the hemiball

$$\mathcal{K}(x,\delta;\nu) := \{ y \in \mathcal{B}(x,\delta) \mid (y-x)\cdot\nu > 0 \} .$$

If

$$\lim_{\delta\to 0+} \frac{|\mathcal{A}\cap\mathcal{K}(x,\delta;\nu)|}{|\mathcal{K}(x,\delta;\nu)|} = 1 , \qquad \lim_{\delta\to 0+} \frac{|\mathcal{A}\cap\mathcal{K}(x,\delta;-\nu)|}{|\mathcal{K}(x,\delta;-\nu)|} = 0 , \tag{5.2}$$

we say that ν is the (measure theoretic) *inward normal*, and that $-\nu$ is the *outward normal* to $\mathcal{A}$ at x. The set of all points of $\mathcal{E}$ for which the inward normal to $\mathcal{A}$ exists is called the *reduced boundary* of $\mathcal{A}$. For points in the reduced boundary, the limit (5.1) is equal to 1/2. Thus, all points in the reduced boundary belong to the essential boundary:

$$\text{rby}\mathcal{A} \subset \text{eby}\mathcal{A} .$$

If $\mathcal{H}^{N-1}(\text{eby}\mathcal{A})$ is finite, then reduced boundary and essential boundary differ by a set of $(N-1)$-dimensional measure zero:[28]

$$\mathcal{H}^{N-1}(\text{rby}\mathcal{A}) = \mathcal{H}^{N-1}(\text{eby}\mathcal{A}) .$$

(iii) Approximate Limit

Consider a mapping $f:\mathcal{E}\to\mathcal{E}$. We say that a point z in $\mathcal{E}$ is the *approximate limit* of f at x, and we write

$$z = \lim_{y\to x} f(y)$$

if, for every $\epsilon > 0$,

$$\lim_{\delta\to 0+} \frac{|\{ y\in\mathcal{E} \mid |f(y)-z| < \epsilon \} \cap \mathcal{B}(x,\delta)|}{|\mathcal{B}(x,\delta)|} = 1 .$$

For any subset $\mathcal{A}$ of $\mathcal{E}$, we say that z is the approximate limit of f at x *relative to* $\mathcal{A}$, and we write

[28] Vol'pert and Hudjaev (1985), p.174.

$$z = \lim_{\substack{y \to x \\ y \in \mathcal{A}}} f(y)$$

if

$$\lim_{\delta \to 0+} \frac{|\{y \in \mathcal{A} \mid |f(y) - z| < \epsilon\} \cap \mathcal{B}(x,\delta)|}{|\mathcal{A} \cap \mathcal{B}(x,\delta)|} = 1\,.$$

For example, let $\mathcal{E} = \mathbb{R}$ and let f be the map

$$f(x) = \begin{cases} 1 & \text{if } x \text{ is irrational and belongs to } (0,1)\,, \\ 0 & \text{otherwise}\,. \end{cases}$$

Then the approximate limit of f at x is equal to one for all x in $(0,1)$, to zero for all x outside $[0,1]$, and does not exist at $x = 0$ and at $x = 1$. At $x = 1$, the approximate limit of f relative to $(0,1)$ is equal to one, and the approximate limit of f relative to $(1,2)$ is equal to zero.

(iv) Jumps

Let x be a point of $\mathcal{E}$ and let ν be a unit vector. Consider the half-space $\Pi(x,\nu)$ through x with inward normal ν

$$\Pi(x,\nu) := \{\, y \in \mathcal{E} \mid (y - x) \cdot \nu > 0 \,\}\,,$$

and for any $f : \mathcal{E} \to \mathcal{E}$ set

$$f_\nu(x) := \lim_{\substack{y \to x \\ y \in \Pi(x,\nu)}} f(y)\,.$$

We say that x is a *regular point* for f if there is a pair $(\nu, -\nu)$ of unit vectors for which the limits $f_\nu(x)$, $f_{-\nu}(x)$ exist. If the two limits are equal, their common value is the approximate limit at x, and x is said to be a point of *approximate continuity* for f. If f is summable, then the set of all points which are not points of approximate continuity for f has N-dimensional Lebesgue measure zero,[29] and if f is a function of bounded variation then the set of all points which are not regular points for f has $(N-1)$-dimensional Hausdorff measure zero.[30]

A regular point x at which the two limits $f_\nu(x)$, $f_{-\nu}(x)$ are not equal is a *jump point* for f. At a jump point, the pair $(\nu, -\nu)$ is unique to within interchange of entries; it is called the pair of *defining vectors.* The vector

$$[f](x) := f_\nu(x) - f_{-\nu}(x) \tag{5.3}$$

is the *jump* of f at x. The set of all jump points for f is the *jump set* for f; it will denoted by $S(f)$.

[29]Vol'pert and Hudjaev (1985), p. 169.
[30]Vol'pert and Hudjaev (1985), p. 178.

Notice that the definition (5.3) depends on the orientation of ν: if we interchange ν and $-\nu$, the jump changes sign. On the contrary, the *jump tensor*

$$Jf(x) \;:=\; [f](x) \otimes \nu \tag{5.4}$$

is independent on the orientation of ν.

For example, let $\mathcal{E} = \mathbb{R}^2$ and let f be the map

$$f(x_1, x_2) \;=\; \begin{cases} (x_1+1, x_2) & \text{if } x_1 > 0,\ |x_2| < x_1, \\ (x_1, x_2) & \text{otherwise}. \end{cases}$$

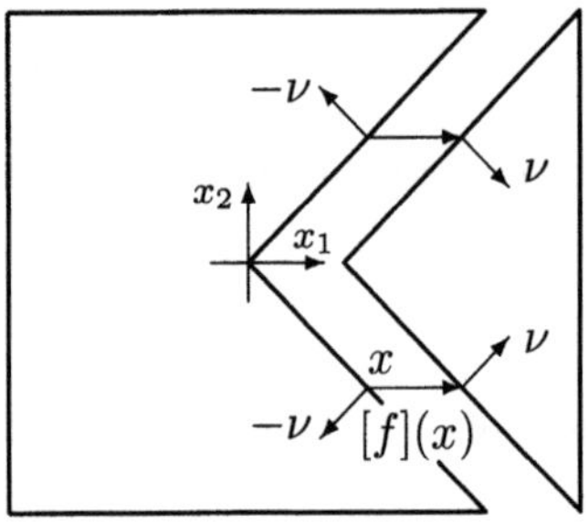

Figure 13.
Defining vectors $(\nu, -\nu)$ and jump $[f](x)$ at x.

Then all points in $\mathcal{E}$ are regular points, except the origin. The jump set consists of the two half-lines

$$x_2 \;=\; x_1 \;>\; 0\,, \qquad -x_2 \;=\; x_1 \;>\; 0\,,$$

and for them we take $\nu = (1/\sqrt{2}, -1/\sqrt{2})$ and $\nu = (1/\sqrt{2}, 1/\sqrt{2})$, respectively, as shown in Fig. 13. At points $x = (x_1, x_2)$ in both half-lines, the approximate limits $f_\nu(x)$, $f_{-\nu}(x)$ are

$$f_\nu(x) \;=\; (x_1+1,\, x_2)\,, \qquad f_{-\nu}(x) \;=\; (x_1, x_2)\,,$$

and the jump is

$$[f](x) \;=\; (1,\, 0)\,.$$

The jump tensors for the two half-lines are

$$\begin{pmatrix} 1/\sqrt{2} & -1/\sqrt{2} \\ 0 & 0 \end{pmatrix}, \qquad \begin{pmatrix} 1/\sqrt{2} & 1/\sqrt{2} \\ 0 & 0 \end{pmatrix},$$

respectively.

In the one-dimensional case, the tensor $Jf(x)$ degenerates into the scalar $(f_\nu(x) - f_{-\nu}(x))\,\nu$, and ν is a scalar which can only take the values 1 and -1. In both cases, we have

$$Jf(x) \;=\; f(x+) - f(x-)\,, \tag{5.5}$$

with $f(x+)$ and $f(x-)$ the approximate limits of f at x from the right and from the left, respectively. It is usual to choose $\nu = 1$; with this choice, we have $f_\nu(x) = f(x+)$, and therefore $Jf = [f]$.

(v) Inward Trace

Let $\mathcal{A}$ be a subset of $\mathcal{E}$, let x be a point in the essential boundary of $\mathcal{A}$, and let f be a mapping from $\mathcal{E}$ into $\mathcal{E}$. If the approximate limit

$$f^+(x) \;:=\; \lim_{\substack{y\to x\\ y\in\mathcal{A}}} f(y)$$

exists, it is called the *inward trace* of f at x. If x is a point in the reduced boundary of $\mathcal{A}$, i.e., if the inward normal ν exists at x, then the inward trace exists if and only if the approximate limit f_ν exists, and in this case the two limits coincide:

$$f^+(x) \;=\; f_\nu(x)\,.$$

(vi) The Jump Set of a Simple Deformation

Let $\mathcal{A}$ be a subset of $\mathcal{E}$, and let f be a mapping of $\mathcal{A}$ into $\mathcal{E}$. At every x in $\mathcal{A}$, the definition of approximate limit of f relative to $\mathcal{A}$, given above for functions defined over the whole space $\mathcal{E}$, is meaningful for f as well. Moreover, if x is a density point for $\mathcal{A}$, the approximate limit for f at x relative to $\mathcal{A}$, if it exists, coincides with the approximate limit at x of *any* extension of f to $\mathcal{E}$. This motivates the definition

$$\lim_{y\to x} f(y) \;:=\; \lim_{\substack{y\to x\\ y\in\mathcal{A}}} f(y) \qquad \forall x \in \mathrm{dns}\mathcal{A}\,,$$

where dns$\mathcal{A}$ denotes the set of all density points of $\mathcal{A}$. Similarly, one may define the jump vector as in (5.3), after setting

$$f_\nu(x) \;:=\; \lim_{\substack{y\to x\\ y\in\Pi(x,\nu)\cap\mathcal{A}}} f(y) \qquad \forall x \in \mathrm{dns}\mathcal{A}\,.$$

Accordingly, the *jump set* of $f : \mathcal{A} \to \mathcal{E}$ can be defined as the set of all density points of $\mathcal{A}$ at which the jump vector exists and is different from zero.

Let now $\mathcal{A}$ be a piecewise fit region, and let f be a piecewise classical deformation from $\mathcal{A}$. Because f is continuous on $\mathcal{A}$ by definition, $S(f)$ is now a subset of dns$\mathcal{A}\backslash\mathcal{A}$. Then $S(f)$ is a set with finite $(N-1)$-dimensional Hausdorff measure. Indeed, dns$\mathcal{A}$ is included in $\mathrm{clo}\mathcal{A}$, and therefore $\mathrm{dns}\mathcal{A}\backslash\mathcal{A}$ is included in $\mathrm{clo}\mathcal{A}\backslash\mathcal{A} = \mathrm{bdy}\mathcal{A}$, and the boundary of a piecewise fit region has finite $(N-1)$-dimensional Hausdorff measure by (D3) in Section 3.1. Note that, because the set of density points and the essential boundary are disjoint, we have

$$S(f) \;\subset\; \mathrm{dns}\mathcal{A}\backslash\mathcal{A} \;\subset\; \mathrm{bdy}\mathcal{A}\backslash\mathrm{eby}\mathcal{A}\,,$$

that is, $S(f)$ is included in that part of the boundary which does not belong to the essential boundary.

A useful property of piecewise fit regions, to be added to those in (D1)–(D5) in Section 3.1, is that the density set coincides with the density set of the fit envelope:

$$\mathrm{dns}\mathcal{A} \;=\; \mathrm{dns}\,(\mathrm{int}\;\mathrm{clo}\mathcal{A})\,.$$

Indeed, since the density set is not changed by adding or subtracting sets of zero N-dimensional Lebesgue measure, the inclusion $\mathcal{A} \subset \mathrm{int}\;\mathrm{clo}\mathcal{A} \subset \mathrm{clo}\mathcal{A}$ and the fact that $\mathcal{A}$ and $\mathrm{clo}\mathcal{A}$ differ by a set with zero N-dimensional Lebesgue measure imply that $\mathcal{A}$, $\mathrm{int}\;\mathrm{clo}\mathcal{A}$ and $\mathrm{clo}\mathcal{A}$ all have the same density set.

If (κ, f) is a simple deformation from $\mathcal{A}$, then f is a piecewise classical deformation from the piecewise fit region $\mathcal{A}\backslash\kappa$, and therefore $S(f)$ is included in $\mathrm{dns}(\mathcal{A}\backslash\kappa)\setminus(\mathcal{A}\backslash\kappa)$. But the density set of $\mathcal{A}\backslash\kappa$ is the same as the density set of its fit envelope, and the fit envelope of $\mathcal{A}\backslash\kappa$ coincides with the fit envelope of $\mathcal{A}$ as shown in Section 3.3. Then $S(f)$ belongs to $\mathrm{dns}\mathcal{A}\setminus(\mathcal{A}\backslash\kappa)$, and the equality

$$\mathrm{dns}\mathcal{A}\setminus(\mathcal{A}\backslash\kappa) \;=\; (\mathrm{dns}\mathcal{A}\setminus\mathcal{A})\cup\kappa\,,$$

which follows from the inclusion $\kappa \subset \mathcal{A} \subset \mathrm{dns}\mathcal{A}$, tells us that

$$S(f) \;\subset\; (\mathrm{dns}\mathcal{A}\setminus\mathcal{A})\cup\kappa\,. \tag{5.6}$$

One may interpret $\mathrm{dns}\mathcal{A}\backslash\mathcal{A}$ as the site of all pre-existing cracks, and κ as the site of the new cracks created in the simple deformation (κ, f). Their union is the crack site after the deformation, and the difference $((\mathrm{dns}\mathcal{A}\backslash\mathcal{A})\cup\kappa)\setminus S(f)$ is the site of all unopened cracks.

5.2 Trace Theorem

In a simple deformation (κ, f) from $\mathcal{A}$, the inward trace of f exists at all points in the essential boundary of $\mathcal{A}$, except at most at a set with $(N-1)$-dimensional Hausdorff measure zero.[31] The Trace Theorem[32] states that, if $n \mapsto (\kappa_n, f_n)$ is a sequence of simple deformations from $\mathcal{A}$ converging to a limit (κ, g, G), then the sequence $n \mapsto f_n^+(x)$ of the inward traces converges uniformly to the inward trace $g^+(x)$ almost everywhere, in the sense of the $(N-1)$-dimensional Hausdorff measure, in the essential boundary of $\mathcal{A}$.

[31] Del Piero and Owen (1995), Lemma 3.1.

[32] Del Piero and Owen (1995), Theorem 4.1.

The following proof, taken from Del Piero and Owen (2000), holds whenever $\kappa_n \to \kappa$ and $f_n \to g$ in the sense of definitions (3.8), (3.9). Therefore, it applies to objects more general than limits of simple deformations.

Trace Theorem. Let $\mathcal{A}$ be a piecewise fit region in $\mathcal{E}$, and let $n \mapsto (\kappa_n, f_n)$ be a sequence of simple deformations from $\mathcal{A}$ such that $\kappa_n \to \kappa$ and $f_n \to g$ in the sense defined in (3.8), (3.9). Then there is a subset $\mathcal{C}$ of eby$\mathcal{A}$ with $\mathcal{H}^{N-1}(\mathcal{C}) = \mathcal{H}^{N-1}(\text{eby}\mathcal{A})$ such that: (i) the inward trace $g^+(x)$ of g exists at all points x of $\mathcal{C}$, (ii) the sequence $n \mapsto f_n^+$ of the inward traces of the functions f_n converges uniformly to g^+ in $\mathcal{C}$.

Proof. The convergence $f_n \to g$ implies that for every $\epsilon > 0$ there is a positive integer n_ϵ such that

$$|f_n(y) - g(y)| < \epsilon \qquad \forall n > n_\epsilon\,,\ \forall y \in \mathcal{A}\backslash(\kappa_n \cup \kappa)\,. \tag{5.7}$$

Moreover, if we call $\mathcal{C}_n$ the set of all points of eby$\mathcal{A}$ at which the inward trace f_n^+ exists, then, by the definition of the inward trace, for all x in $\mathcal{C}_n$ we have

$$\lim_{\delta \to 0+} \frac{|\{\, y \in \mathcal{A} \mid |f_n(y) - f_n^+(x)| < \epsilon \,\} \cap \mathcal{B}(x,\delta)\,|}{|\mathcal{A} \cap \mathcal{B}(x,\delta)|} = 1\,. \tag{5.8}$$

Thus, for sufficiently small δ, for all $m, n > n_\epsilon$ and for all $x \in \mathcal{C}_m \cap \mathcal{C}_n$ there is a subset $\mathcal{S}$ of $\mathcal{B}(x,\delta)$ with N-dimensional Lebesgue measure $|\mathcal{S}| > 0$, such that $|f_n(y) - f_n^+(x)| < \epsilon$ and $|f_m(y) - f_m^+(x)| < \epsilon$ for every y in $\mathcal{S}$. Because $|\kappa_n \cup \kappa_m \cup \kappa| = 0$, the difference $\mathcal{S}\backslash(\kappa_n \cup \kappa_m \cup \kappa)$ is non-empty, and by (5.7), (5.8) for every y in the difference we have

$$\begin{aligned} &|f_m^+(x) - f_n^+(x)| \\ \le\ & |f_m^+(x) - f_m(y)| + |f_m(y) - g(y)| + |g(y) - f_n(y)| + |f_n(y) - f_n^+(x)| \\ <\ & 4\epsilon\,. \end{aligned} \tag{5.9}$$

Denote by $\mathcal{C}$ the intersection of all $\mathcal{C}_n$. By Lemma 3.1 in Del Piero and Owen (1995), each $\mathcal{C}_n$ differs from eby$\mathcal{A}$ by a set with $(N-1)$-dimensional Hausdorff measure zero. Then $\mathcal{C}$ differs from eby$\mathcal{A}$ at most by a countable union of sets with $(N-1)$-dimensional Hausdorff easure zero. Therefore, $\mathcal{H}^{N-1}(\mathcal{C}) = \mathcal{H}^{N-1}(\text{eby}\mathcal{A})$.

Moreover, inequality (5.9) tells us that at all $x \in \mathcal{C}$ the sequence $n \mapsto f_n^+(x)$ is a Cauchy sequence, and therefore it converges to a limit $\varphi(x)$. Then for $m \to \infty$ we have

$$|f_n^+(x) - \varphi(x)| \le 4\epsilon \qquad \forall n > n_\epsilon,\ \forall x \in \mathcal{C}\,.$$

This proves that the convergence $f_n^+(x) \to \varphi$ is uniform in $\mathcal{C}$; it remains to prove that φ is the inward trace of g. For this purpose we consider the inequality

$$|g(y) - \varphi(x)| \le |g(y) - f_n(y)| + |f_n(y) - f_n^+(x)| + |f_n^+(x) - \varphi(x)|\,,$$

and from (5.7) we get

$$|g(y) - \varphi(x)| \;<\; 5\epsilon + |f_n(y) - f_n^+(x)| \qquad \forall x \in \mathcal{C},\ \forall y \in \mathcal{A}\backslash(\kappa_n \cup \kappa)\,.$$

Then

$$\{\, y \in \mathcal{A}\backslash(\kappa_n \cup \kappa) \mid\ |f_n(y) - f_n^+(x)| < \epsilon \,\} \;\subset\; \{\, y \in \mathcal{A}\backslash(\kappa_n \cup \kappa) \mid\ |g(y) - \varphi(x)| < 6\epsilon \,\},$$

and

$$\begin{aligned} &|\,\{\, y \in \mathcal{A}\backslash(\kappa_n \cup \kappa) \mid\ |f_n(y) - f_n^+(x)| < \epsilon \,\} \cap \mathcal{B}(x,\delta)\,| \\ &\qquad \leq\ |\,\{\, y \in \mathcal{A}\backslash(\kappa_n \cup \kappa) \mid\ |g(y) - \varphi(x)| < 6\epsilon \,\} \cap \mathcal{B}(x,\delta)\,|\,. \end{aligned}$$

By (5.8) and because $|\kappa_n \cup \kappa| = 0$, this implies

$$\lim_{\delta \to 0+} \frac{|\,\{\, y \in \mathcal{A} \mid |g(y) - \varphi(x)| < 6\epsilon \,\} \cap \mathcal{B}(x,\delta)\,|}{|\,\mathcal{A} \cap \mathcal{B}(x,\delta)|} \;=\; 1$$

for all x in $\mathcal{C}$, and from the definition of the inward trace we conclude that $\varphi(x) = g^+(x)$ at all x in $\mathcal{C}$. □

5.3 Integral-Gradient Formula

Let $\mathcal{A}$ be a piecewise fit region of $\mathcal{E}$, and let (κ, f) be a simple deformation from $\mathcal{A}$. The Integral-Gradient Formula

$$\int_{\mathcal{A}} \nabla f(x)\, dx \;+\; \int_{S(f)} Jf(x)\, d\mathcal{H}^{N-1} \;=\; \int_{\mathrm{eby}\mathcal{A}} (f^+(x) - o) \otimes \nu(x)\, d\mathcal{H}^{N-1} \tag{5.10}$$

was proved in Del Piero and Owen (1995), equation (4.3). Here o is an arbitrary point of $\mathcal{E}$, $f^+(x)$ and $\nu(x)$ are the inward trace and the *outward* normal at a point x in the essential boundary of $\mathcal{A}$, respectively, and the jump tensor $Jf(x)$ and the jump set $S(f)$ are defined as in Section 5.1, point (vi).

Here I prove a one-dimensional version of the formula.

Integral-Gradient Formula for One-Dimensional Simple Deformations. For every one-dimensional piecewise fit region $\mathcal{A}$ and for every simple deformation (κ, f) from $\mathcal{A}$,

$$\int_{\mathcal{A}} \nabla f(x)\, dx \;+\; \sum_{x \in S(f)} (f(x+) - f(x-)) \;=\; \sum_{x \in \mathrm{eby}\mathcal{A}} f^+(x) \nu(x)\,. \tag{5.11}$$

Proof. Consider first the case in which $\mathcal{A}$ is the interval (a, b). Then f is piecewise C^1 in (a, b), and the equality

$$\int_a^b \nabla f(x)\, dx \;+\; \sum_{x \in S(f)} (f(x+) - f(x-)) \;=\; f(b-) - f(a+) \tag{5.12}$$

follows directly from the Fundamental Theorem of Calculus. Moreover, the essential boundary consists of the two points $x = a$, $x = b$, with outward normals $\nu(a) = -1$ and $\nu(b) = 1$. Therefore,

$$\sum_{x \in \text{eby}\mathcal{A}} f^+(x)\nu(x) = f^+(b)\,\nu(b) + f^+(a)\,\nu(a) = f(b-) - f(a+)\,,$$

and equation (5.11) follows. Let now $\mathcal{A}$ be any one-dimensional piecewise fit region. Then $\mathcal{A}$ is a finite union of one-dimensional fit regions and, because a one-dimensional fit region is a finite union of bounded open intervals whose closures are pairwise disjoint, $\mathcal{A}$ is a finite union of bounded open intervals (a_i, b_i) whose closures need not be disjoint. Thus, the set $\text{dns}\mathcal{A} \setminus \mathcal{A}$ consists of those of the endpoints a_i, b_i which are common to the closures of two intervals and, by (5.6), $S(f)$ is included in the union of $\text{dns}\mathcal{A} \setminus \mathcal{A}$ and κ. Then writing equation (5.12) for each interval and summing over all intervals we get (5.11). □

Suppose that (κ, f) describes the axial deformation of a bar. Then the right-hand side of equation (5.12) expresses the total length of the deformed bar, and the same equation tells us that this length is the sum of the total length of the unfractured pieces of the bar, plus the total opening of the cracks. By extrapolating to higher dimension, one may interpret the right-hand side of (5.10) as the *total deformation* of the body, and the two integrals on the left as the *total deformation without macrodisarrangements* and the *total deformation due to macrodisarrangements*, respectively. In the next section this decomposition is used to determine a local measure of deformation due to microdisarrangements.

5.4 A Volume Measure for the Deformation Due to Microdisarrangements

Consider a piecewise fit region $\mathcal{A}$ of $\mathcal{E}$, a structured deformation (κ, g, G) from $\mathcal{A}$, and a sequence $n \mapsto (\kappa_n, f_n)$ of simple deformations from $\mathcal{A}$ converging to (κ, g, G). For each (κ_n, f_n), we may write the integral-gradient formula (5.10):

$$\int_{\mathcal{A}} \nabla f_n(x)\,dx + \int_{S(f_n)} Jf_n(x)\,d\mathcal{H}^{N-1} = \int_{\text{eby}\mathcal{A}} f_n^+(x) \otimes \nu(x)\,d\mathcal{H}^{N-1}.$$

By (3.10), when $n \to \infty$ the sequence $n \mapsto \nabla f_n$ converges uniformly to G almost everywhere in the sense of the N-dimensional Lebesgue measure. Moreover, by the trace theorem, $n \mapsto f_n^+$ converges uniformly to g^+, $\mathcal{H}^{N-1}$-almost eveywhere on the essential boundary of $\mathcal{A}$, and $n \mapsto Jf_n$ converges uniformly to Jg, $\mathcal{H}^{N-1}$-almost eveywhere on $S(g)$. In the limit, we get

$$\int_{\mathcal{A}} G(x)\,dx + \int_{S(g)} Jg(x)\,d\mathcal{H}^{N-1} + \lim_{n\to\infty} \int_{S(f_n)\setminus S(g)} Jf_n(x)\,d\mathcal{H}^{N-1}$$
$$= \int_{\text{eby}\mathcal{A}} g^+(x) \otimes \nu(x)\,d\mathcal{H}^{N-1}. \qquad (5.13)$$

On the other hand, we may write the formula (5.10) for the simple deformation (κ, g):

$$\int_{\mathcal{A}} \nabla g(x)\, dx + \int_{S(g)} Jg(x)\, d\mathcal{H}^{N-1} = \int_{\text{eby}\mathcal{A}} g^{+}(x) \otimes \nu(x)\, d\mathcal{H}^{N-1}\,, \tag{5.14}$$

and subtracting from the preceding equation we get

$$\int_{\mathcal{A}} (\nabla g(x) - G(x))\, dx = \lim_{n\to\infty} \int_{S(f_n)\setminus S(g)} Jf_n(x)\, d\mathcal{H}^{N-1}\,. \tag{5.15}$$

In the preceding section, the three terms in equation (5.14) have been interpreted as the total deformation without macrodisarrangements, the total deformation due to macrodisarrangements, and the total deformation of the body, respectively. Now, the limit in (5.15) may be interpreted as the *total deformation due to microdisarrangements*, because it keeps track of all disarrangements in the approximating sequence which diffuse away from the macrodisarrangement site κ of the limit element.

Equation (5.15) tells us that this deformation can be expressed as a volume integral, so that the integrand function $\nabla g - G$ is a *volume density of microdisarrangements.*[33] In particular, at any point x in $\mathcal{A}$ away from $S(g)$ we have

$$\nabla g(x) - G(x) = \lim_{\delta\to 0+} \lim_{n\to\infty} \frac{\int_{S(f_n)\,\cap\,\mathcal{B}(x,\delta)} Jf_n(x)\, d\mathcal{H}^{N-1}}{|\mathcal{B}(x,\delta)|}\,. \tag{5.16}$$

This result completes the discussion in Section 3.4 about the correlation between crack diffusion and the inequality $\nabla g \neq G$. Now we know that the difference $\nabla g - G$ is a density of deformation due to microdisarrangements. Because this difference is a tensor, not only does it measure the intensity of the deformation, but it also provides information on the nature of the microdisarrangements. For instance, in the examples discussed in Section 4.3, a tensor of pure shear corresponds to microslips, and a spherical tensor corresponds to incorporation of voids.

6 Energy

6.1 The Energy of a Structured Deformation

For a simple deformation (κ, f) from a piecewise fit region $\mathcal{A}$, the total energy is the sum of two terms: a bulk term which measures the energy away from the crack site $(\text{dns}\mathcal{A}\setminus\mathcal{A})\cup\kappa$, and an interfacial term which specifies the energy required for the formation and opening of the cracks:

$$E_{Sid}(\kappa, f) = \int_{\mathcal{A}} w(\nabla f(x))\, dx + \int_{(\text{dns}\mathcal{A}\setminus\mathcal{A})\cup\kappa} \theta(Jf(x))\, d\mathcal{H}^{N-1}\,. \tag{6.1}$$

The *bulk energy density* w and the *interfacial energy density* θ are assumed to be continuous, non-negative functions satifying

$$w(I) = 0\,, \qquad \theta(O) = 0\,, \tag{6.2}$$

[33]Del Piero and Owen (1995), Section 5.

with I the identity tensor and O the null tensor. While the first assumption is standard (it corresponds to assigning a null energy to the undeformed configuration), the second introduces a substantial restriction. Indeed, $\theta(O) = 0$ means that no energy is required for crack formation and that, consequently, the whole interfacial energy is spent in the process of crack opening.

A consequence of this assumption may be deduced from the diagram in Fig. 14, where the one-dimensional interfacial energies assumed in the fracture models of Griffith and of Barenblatt are compared. In the first model, the whole energy is required for crack formation, while in the second there no energy cost to be paid for crack formation and the whole energy is needed for crack opening. Thus, the assumption $\theta(O) = 0$ rules out Griffith's model from the present analysis.[34]

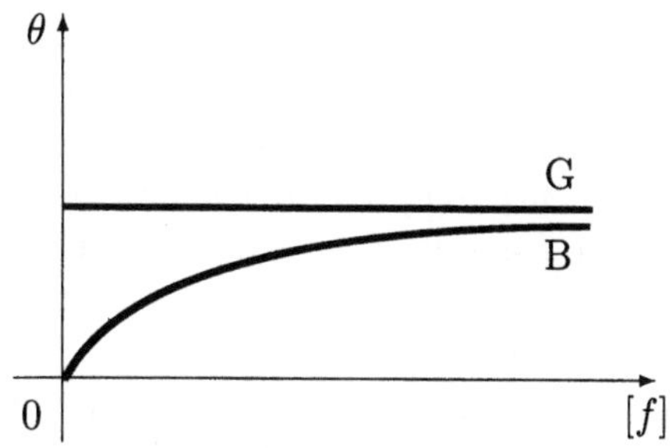

Figure 14. Forms of the interfacial energy density θ in Griffith's theory (G) and in Barenblatt's theory (B).

Another consequence of the same assumption is that points in the site of all unopened cracks $((\text{dns}\mathcal{A}\backslash\mathcal{A}) \cup \kappa)\backslash S(f)$ do not contribute to the energy. Accordingly, $(\text{dns}\mathcal{A}\backslash\mathcal{A}) \cup \kappa$ can be replaced by $S(f)$ in the expression (6.1). Then the energy is determined by the second item in the pair (κ, f), and it can be rewritten in the form

$$E_{Sid}(f) \;=\; \int_{\mathcal{A}} w(\nabla f(x))\, dx \;+\; \int_{S(f)} \theta(Jf(x))\, d\mathcal{H}^{N-1}\,. \tag{6.3}$$

For a structured deformation (κ, g, G), an obvious candidate energy is

$$E_{Std}(g, G) \;=\; \lim_{n\to\infty} \; E_{Sid}(f_n)\,, \tag{6.4}$$

with $n \mapsto (\kappa_n, f_n)$ any sequence of simple deformations converging to (κ, g, G). This definition makes sense only if the limit on the right-hand side exists and is independent of the approximating sequence. In what follows I consider the one-dimensional case, and I show that the two above conditions may or may not be true, depending on the properties of θ.

[34] A more general theory including Griffith's model is developed in Del Piero and Truskinovsky (in preparation).

More precisely, I consider the two following cases:

(H1) θ is defined on $\mathbb{R}$, has right and left derivatives $\theta'(0+), \theta'(0-)$ at the origin, and $\theta'(0+) > 0 > \theta'(0-)$,

(H2) θ is defined on $[0, +\infty)$, has a right derivative $\theta'(0+)$ at the origin, and $\theta'(0+) > 0$.

Recalling that the derivative of θ represents the stress transmitted across the crack, we see that both assumptions require a non-zero activation stress for crack opening.[35] As to assumption (H2), the restriction of the domain of θ to the positive jumps is justified by the observation that this is a necessary and sufficient condition for preserving the orientation of the vector joining any two points of $\mathcal{A}$. Indeed, for a simple deformation the derivative $\nabla f(x)$ is positive by the one-dimensional version of inequality (2.1), and, by (5.5), a positive jump at x means that $f(x+) > f(x-)$; then a simple deformation with positive jumps is increasing, and an increasing function is orientation-preserving. Conversely, the jumps of an orientation-preserving simple deformation must be positive. Indeed, if x is a jump point for f and if $f(x+) - f(x-) < 0$, then by the continuity of f there is an $\epsilon > 0$ such that $f(x+\epsilon) < f(x-\epsilon)$, and therefore the orientation of the pair $(x-\epsilon, x+\epsilon)$ is not preserved in the deformation.

Now consider a structured deformation (κ, g, G) from a one-dimensional piecewise fit region $\mathcal{A}$, and a sequence $n \mapsto (\kappa_n, f_n)$ of simple deformations from $\mathcal{A}$ converging to (κ, g, G). For each f_n consider the energy $E_{Sid}(f_n)$ given by (6.3), and examine the limit behavior of the sequence $n \mapsto E_{Sid}(f_n)$ when $n \to \infty$. The equalities[36]

$$\lim_{n\to\infty} \int_{\mathcal{A}} w(\nabla f_n(x))\, dx = \int_{\mathcal{A}} w(G(x))\, dx\,, \tag{6.5}$$

$$\lim_{n\to\infty} \sum_{x\in S(g)} \theta([f_n](x)) = \sum_{x\in S(g)} \theta([g](x))\,, \tag{6.6}$$

follow from the continuity assumed for w and θ, from the uniform convergence of $n \mapsto \nabla f_n$ to G, and from the uniform convergence of the sequences $n \mapsto [f_n](x)$ to $[g](x)$ for all $x \in S(g)$.[37]

To evaluate the remaining limit

$$\lim_{n\to\infty} \sum_{x\in S(f_n)\setminus S(g)} \theta([f_n](x))\,, \tag{6.7}$$

let us prove a preliminary result.

---

[35] See e.g. Del Piero (1998), Section 4.

[36] In the one-dimensional case I identify Jf with $[f]$, see the comment following equation (5.5).

[37] Proofs of (6.5) and (6.6) may be found in Del Piero and Owen (2000), Part Two, Sections 2.2 and 2.3, respectively.

Lemma. Let $\mathcal{A}$ be a one-dimensional piecewise fit region, let (κ, g, G) be a structured deformation from $\mathcal{A}$, and let $n \mapsto (\kappa_n, f_n)$ be a sequence of simple deformations from $\mathcal{A}$ converging to (κ, g, G). Then

$$\lim_{n\to\infty} \sup_{x\in\mathcal{S}(f_n)\backslash\mathcal{S}(g)} |[f_n](x)| = 0 . \tag{6.8}$$

Proof. By the uniform convergence of $n \mapsto f_n$ to g we have

$$\lim_{n\to\infty} \sup_{x\in\mathcal{S}(f_n)\backslash\mathcal{S}(g)} |f_n(x+) - g(x+)| = 0 ,$$

$$\lim_{n\to\infty} \sup_{x\in\mathcal{S}(f_n)\backslash\mathcal{S}(g)} |f_n(x-) - g(x-)| = 0 .$$

By (5.5), $f_n(x+) - f_n(x-) = [f_n(x)]$ and $g(x+) - g(x-) = 0$ on $\mathcal{S}(f_n)\backslash\mathcal{S}(g)$. Thus,

$$\sup_{x\in\mathcal{S}(f_n)\backslash\mathcal{S}(g)} |[f_n](x)| \leq \sup_{x\in\mathcal{S}(f_n)\backslash\mathcal{S}(g)} |f_n(x+) - g(x+)| + \sup_{x\in\mathcal{S}(f_n)\backslash\mathcal{S}(g)} |g(x-) - f_n(x-)| ,$$

and (6.8) follows when $n \to \infty$. □

Then we establish the following lower bound for the limit (6.7).

Proposition 6.1. Let $\mathcal{A}$, (κ, g, G), and $n \mapsto (\kappa_n, f_n)$ be as in the preceding Lemma. If θ satisfies the assumption (H1), then

$$\lim_{n\to\infty} \sum_{x\in\mathcal{S}(f_n)\backslash\mathcal{S}(g)} \theta([f_n](x)) \geq \theta'(0+) \int_{\mathcal{A}} (\nabla g(x) - G(x))\, dx , \tag{6.9}$$

and if θ satisfies (H2) the above inequality holds as an equality.

Proof. By assumption (H1) on the existence of the derivatives $\theta'(0+), \theta'(0-)$, for sufficiently small $\epsilon > 0$ there is a $\delta > 0$ such that

$$|\theta(a) - a\,\theta'(0+)| < \epsilon\, a \qquad \text{if } 0 < a < \delta , \tag{6.10}$$

$$|\theta(a) - a\,\theta'(0-)| < -\epsilon\, a \qquad \text{if } -\delta < a < 0 . \tag{6.11}$$

Denote by a^+, a^- the positive part and the negative part of a:

$$a^+ := \tfrac{1}{2}(|a| + a) , \qquad a^- := \tfrac{1}{2}(|a| - a) .$$

Then (6.10) and (6.11) can be collected into the single inequality

$$|\,\theta(a) - \theta'(0+)\, a^+ + \theta'(0-)\, a^-| < \epsilon\, |a| ,$$

which can be further transformed into

$$|\,\theta(a) - \theta'(0+)\, a - (\theta'(0+) - \theta'(0-))\, a^-| < \epsilon\, (a + 2a^-) . \tag{6.12}$$

By the preceding lemma, with δ we may associate a positive integer n_δ such that $|[f_n](x)| < \delta$ for all $n > n_\delta$ and for all x in $S(f_n)\backslash S(g)$. Then for a fixed $n > n_\delta$, in (6.12) we may replace a by any one of the jumps $[f_n](x)$, and summing with respect to x we get

$$\sum_{x\in S(f_n)\backslash S(g)} \theta([f_n](x)) > (\theta'(0+) - \epsilon) \sum_{x\in S(f_n)\backslash S(g)} [f_n](x)$$
$$+ (\theta'(0+) - \theta'(0-) - 2\epsilon) \sum_{x\in S(f_n)\backslash S(g)} [f_n](x)^- . \tag{6.13}$$

For sufficiently small ϵ the last term is non-negative. Thus, for $n \to \infty$ from the relation (5.15) we get

$$\lim_{n\to\infty} \sum_{x\in S(f_n)\backslash S(g)} \theta([f_n](x)) \geq (\theta'(0+) - \epsilon) \int_{\mathcal{A}}(\nabla g(x) - G(x))\, dx \,,$$

and (6.9) follows from the arbitrariness of ϵ.

If assumption (H2) holds, then we may set $a^- = 0$ and $a^+ = [f_n](x)$ in inequality (6.12), and summing with respect to x we obtain

$$\sum_{x\in S(f_n)\backslash S(g)} |\, \theta([f_n](x)) - \theta'(0+)[f_n](x)\,| < \epsilon \sum_{x\in S(f_n)\backslash S(g)} [f_n](x) \,.$$

Using again the limit relation (5.15) and the arbitrariness of ϵ, for $n \to \infty$ we get (6.9) with the equality sign. □

By summing inequality (6.9) and equations (6.5), (6.6) we get

$$\lim_{n\to\infty} E_{Sid}(f_n) \geq \int_A \left(w(G(x)) + \theta'(0+)(\nabla g(x) - G(x)) \right) dx + \sum_{x\in S(g)} \theta([g](x)) \tag{6.14}$$

for all sequences $n \mapsto (\kappa_n, f_n)$ converging to the structured deformation (κ, g, G). This inequality is satisfied as an equality if assumption (H2) holds. Thus, in the case of θ defined on $[0, +\infty)$ the right-hand side of (6.14) can be taken as the energy of the structured deformation (κ, g, G).

If θ is defined over the whole real line, the limit of $n \mapsto E_{Sid}(f_n)$ may not exist, or may depend on the approximating sequence. For example, the sequence $n \mapsto (\kappa_n, f_n)$ of simple deformations from (0,1) defined by

$$\kappa_n = \{\tfrac{h}{n},\quad h = 1, 2, .. n-1\}\,,$$
$$f_n(x) = \begin{cases} x + 1/n & \text{for } x \in [\,\frac{h}{n}, \frac{h+1}{n})\,, \ h \text{ even}\,, \\ x - 1/n & \text{for } x \in [\,\frac{h}{n}, \frac{h+1}{n})\,, \ h \text{ odd}\,, \end{cases}$$

converges to the structured deformation (κ, g, G) with

$$\kappa = \emptyset, \quad g(x) = x, \quad G(x) = 1.$$

Each f_n has $(n-1)$ jumps of alternating sign, all of intensity $2/n$. The case $n = 4$ is shown in Fig. 15a. Recalling that $w(1) = 0$ by assumption $(6.2)_1$ and assuming the interface energy

$$\theta(a) = |a|,$$

the energy of f_n is

$$E_{Sid}(f_n) = (n-1)\theta(\tfrac{2}{n}) = 2\,\tfrac{n-1}{n},$$

so that

$$\lim_{n\to\infty} E_{Sid}(f_n) = 2.$$

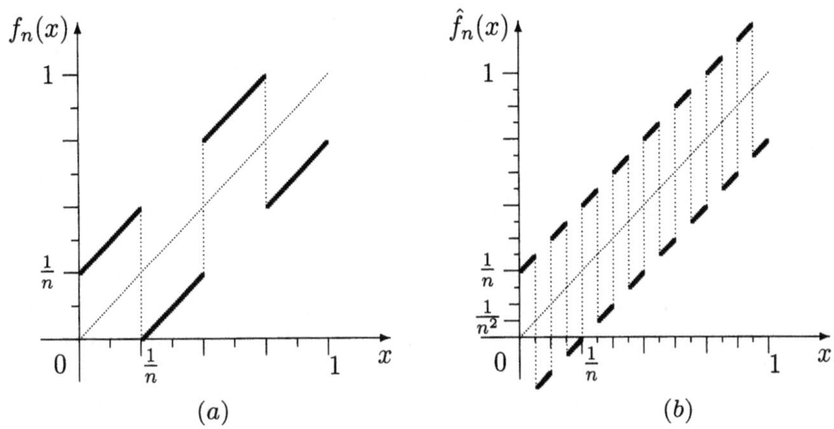

Figure 15.
Two approximating sequences for the same structured deformation, one with converging energies (a), and one with diverging energies (b).

Now consider the constant sequence $(\tilde{\kappa}_n, \tilde{f}_n) = (\emptyset, i)$. It converges to the same structured deformation, but $E_{Sid}(\tilde{f}_n) = 0$ for all n. Thus, we have two different limits for two sequences converging to the same structured deformation.

A sequence $(\hat{\kappa}_n, \hat{f}_n)$ converging to the same structured deformation and with diverging energies is obtained by taking

$$\hat{\kappa}_n = \{\tfrac{h}{n^2}, \quad h = 1, 2, .. \, n^2 - 1\},$$

$$\hat{f}_n(x) = \begin{cases} x + 1/n & \text{for } x \in [\frac{h}{n^2}, \frac{h+1}{n^2}), \quad h \text{ even}, \\ x - 1/n & \text{for } x \in [\frac{h}{n^2}, \frac{h+1}{n^2}), \quad h \text{ odd}. \end{cases}$$

In each $\hat{f}_n$ there are (n^2-1) jumps of alternating sign, all of intensity $2/n$. The case $n=4$ is shown in Fig. 15b. Taking $\theta(a)=|a|$ as above, the energy of $\hat{f}_n$ is

$$E_{Sid}(\hat{f}_n) \;=\; (n^2-1)\,\theta(\tfrac{2}{n}) \;=\; 2\,\tfrac{n^2-1}{n}\,,$$

and diverges for $n\to\infty$.

In situations in which the limit energy depends on the approximating sequence, it is standard to take as the energy of the structured deformation the *relaxed energy*, defined as the lowest energy which can be attained in the limit by an approximating sequence:

$$E_{Std}(g,G) \;:=\; \inf_{(\kappa_n,f_n)\to(\kappa,g,G)}\ \liminf_{n\to\infty}\ E_{Sid}(f_n)\,, \tag{6.15}$$

A lower bound for the relaxed energy is provided by the right-hand side of (6.14); this bound becomes the exact expression of the relaxed energy if for every structured deformation there is an approximating sequence for which (6.14) is satisfied as an equality. By Proposition 6.1, this is done by all sequences which have positive jumps away from $S(g)$. Thus the question comes, whether every structured deformation has such an approximating sequence. The answer is affirmative.

Proposition 6.2. Let (κ,g,G) be a structured deformation from a one-dimensional piecewise fit region $\mathcal{A}$. Then there is a sequence $n\mapsto(\kappa_n,f_n)$ of simple deformations from $\mathcal{A}$ with positive jumps outside $S(g)$ and converging to (κ,g,G).

Proof. It is sufficient to prove the statement in the case in which $\mathcal{A}$ is an interval (a,b) and κ is the empty set. If $(\emptyset,g,G)$ is a structured deformation, then $\nabla g(x)\geq G(x)>0$ at all x in (a,b) by the one-dimensional version of the determinant inequality (4.1). Let h be the primitive of $\nabla g-G$ with $h(a)=0$. Then h is continuous and non-decreasing, and $h(b)>0$ except in the trivial case $\nabla g=G$. For each n in $\mathbb{N}$ there are points $x_{r,n},\ r\in\{0,1,..\,n\}$, at which

$$h(x_{r,n}) \;=\; \tfrac{r}{n}\,h(b)\,.$$

In particular, $x_{0,n}=a$ and $x_{n,n}=b$ for all n. Consider the piecewise constant function

$$\varphi_n(x) \;:=\; \tfrac{r}{n}\,h(b) \qquad \text{for } x\in(x_{r,n},x_{r+1,n})\,,\ r\in\{0,1,..\,n-1\}\,.$$

It has positive jumps, and the sequence $n\mapsto\varphi_n$ converges uniformly to h.

Now consider the simple deformations (κ_n,f_n), with

$$\kappa_n \;=\; S(\varphi_n(x))\,, \qquad f_n \;=\; \varphi_n+g-h\,.$$

Each f_n has positive jumps because, g and h being continuous, f_n has the same jumps as φ_n. Moreover, $\nabla\varphi_n=0$ implies $\nabla f_n=\nabla g-\nabla h=G$.

The limit of $n\mapsto\kappa_n$ is the empty set, because no point in $S(\varphi_n)$ belongs to $S(\varphi_m)$ for all $m>n$. Moreover, the constant sequence $n\mapsto\nabla f_n$ converges to G, and $n\mapsto f_n$ converges uniformly to g, due to the uniform convergence of $n\mapsto\varphi_n$ to h. Therefore, $n\mapsto(\kappa_n,f_n)$ has all the desired properties. □

This result motivates the choice of the right-hand side of (6.14) as the expression of the energy of a structured deformation[38]

$$E_{Std}(g,G) := \int_{\mathcal{A}}\big(w(G(x)) + \theta'(0+)(\nabla g(x) - G(x))\big)\,dx + \sum_{x\in S(g)} \theta([g](x)) \qquad (6.16)$$

when either of assumptions (H1), (H2) is satisfied.

By comparing the energy of (κ, g, G) with the energy of the simple deformation (κ, g) given by (6.3), we have

$$E_{Std}(g,G) \; - \; E_{Sid}(g) \; = \; \int_{\mathcal{A}}\big(w(G(x)) - w(\nabla g(x)) + \theta'(0+)(\nabla g(x) - G(x))\big)\,dx\,.$$

In particular, when the structured deformation is a simple deformation $(\kappa, g, \nabla g)$ we get

$$E_{Std}(g, \nabla g) \; = \; E_{Sid}(g)\,. \qquad (6.17)$$

Let me close with a comment on the energy minimization performed in Lecture 1. There, the energy (1.1) is the energy of the simple deformation (κ, f) from $(0, l)$, with $f(x) = u(x) + x$ and $\kappa = S(f) = S(u)$. The right-hand side of (1.15) is the energy of the structured deformation (κ, g, G) with $\kappa = \emptyset$, $g(x) = x + \beta x$, and $G(x) = 1 + k$. The sequence $n \mapsto (\kappa_n, f_n)$, with $f_n(x) = x + u_n(x)$, $\kappa_n = S(u_n)$, and with the u_n shown in Fig. 3, is an approximating sequence for (κ, g, G). The computation made in Section 1 shows that, for $\beta > k$, the global energy minimum is attained at the structured deformation (κ, g, G).

6.2 The Non-Homogeneous Case

In the preceding section, the independence of θ on the point was crucial in proving the formula (6.16) for the energy of a structured deformation. Here I consider the case of a smooth dependence of w and θ on x, and for this *non-homogeneous case* I prove a generalized version of (6.16). First of all, I write without proof the counterparts of the limit relations (6.5), (6.6):[39]

$$\lim_{n\to\infty} \int_{\mathcal{A}} w(\nabla f_n(x), x)\,dx \; = \; \int_{\mathcal{A}} w(G(x), x)\,dx\,, \qquad (6.18)$$

$$\lim_{n\to\infty} \sum_{x\in S(g)} \theta([f_n](x), x) \; = \; \sum_{x\in S(g)} \theta([g](x), x)\,. \qquad (6.19)$$

[38] Del Piero and Owen (2000), Part 2, Section 2.4. In (*ibidem*, Sect. 2.6) it is shown that assuming a more general form for E_{Sid} leads to an explicit expression of E_{Std} in which the dependence upon $\nabla g - G$ is non-linear.

[39] The proofs are obvious extensions of those given in Del Piero and Owen (2000), Part 2, for the homogeneous case.

To evaluate the limit of the interfacial energy over $S(f_n)\backslash S(g)$, I begin with a generalization of the limit formula (5.15). Let $\mathcal{A}$ be a one-dimensional piecewise fit region, let (κ, f) be a simple deformation from $\mathcal{A}$, and let h be a smooth, positive scalar field over $\mathcal{A}$. Then the Integral-Gradient Formula (5.10) can be applied to the product (hf):

$$\int_{\mathcal{A}} \nabla(hf)(x)\, dx + \sum_{x\in S(hf)} [hf](x) = \sum_{x\in \mathrm{eby}\mathcal{A}} (hf)^+(x)\nu(x)\,,$$

and in view of the smoothness of h we have

$$\int_{\mathcal{A}} (h(x)\nabla f(x) + \nabla h(x) f(x))\, dx + \sum_{x\in S(f)} h(x)\,[f](x) = \sum_{x\in \mathrm{eby}\mathcal{A}} h(x) f^+(x)\nu(x)\,. \tag{6.20}$$

Consider a sequence $n \mapsto (\kappa_n, f_n)$ of simple deformations from $\mathcal{A}$ converging to the structured deformation (κ, g, G), and write (6.20) for each f_n. Then in the limit as $n \to \infty$,

$$\int_{\mathcal{A}} (h(x)G(x) + \nabla h(x) g(x))\, dx + \sum_{x\in S(g)} h(x)\,[g](x)$$

$$+ \lim_{n\to\infty} \sum_{x\in S(f_n)\backslash S(g)} h(x)\,[f_n](x) = \sum_{x\in \mathrm{eby}\mathcal{A}} h(x)\, g^+(x)\,\nu(x)\,,$$

and comparing with equation (6.20) for g we have

$$\lim_{n\to\infty} \sum_{x\in S(f_n)\backslash S(g)} h(x)\,[f_n](x) = \int_{\mathcal{A}} h(x)(\nabla g(x) - G(x))\, dx\,. \tag{6.21}$$

This is the desired generalization of (5.15) in the presence of a smooth field h.

Now I prove a generalized version of inequality (6.9) in Proposition 6.1. Assume that θ varies smoothly with x, and that either (H1) or (H2) is satisfied at all x in $[0, l]$, with θ replaced by $\theta(\cdot, x)$. Set

$$\mu(a, x) := \left| \frac{\theta(a,x)}{a} - \theta'(0+, x) \right|.$$

This function is continuous in $(0, +\infty) \times [0, l]$, and if either (H1) or (H2) holds it has a continuous extension to $\{0\} \times [0, l]$ as

$$\mu(0, x) = 0\,. \tag{6.22}$$

Then μ is uniformly continuous in every compact subset of $[0, +\infty) \times [0, l]$.

For a fixed $\varepsilon > 0$ and for each x in $[0, l]$ consider the number

$$a_\varepsilon(x) := \sup\{\, a > 0 \mid \mu(z, x) < \varepsilon \quad \forall z \in (0, a)\,\}\,, \tag{6.23}$$

and let me prove that

$$\mu(a_\varepsilon(x), x) = \varepsilon\,. \tag{6.24}$$

Indeed, by the uniform continuity of μ, for every $\eta > 0$ there is a $\zeta > 0$ such that $|a_2 - a_1| < \zeta$ implies $|\mu(a_2, x) - \mu(a_1, x)| < \eta$. Then fix $\eta > 0$ and take a_1, a_2 such that

$$a_1 \ < \ a_\varepsilon(x) \ < \ a_2 \ < a_1 + \zeta \, .$$

Then,

$$\mu(a_2, x) - \eta \ < \ \mu(a_\varepsilon(x), x) \ < \ \mu(a_1, x) + \eta \, .$$

But $\mu(a_1, x) < \varepsilon$ because $a_1 < a_\varepsilon(x)$. Moreover, by (6.23), a_2 can be chosen so that $\mu(a_2, x) \geq \varepsilon$, no matter how small ζ is. Therefore,

$$\varepsilon - \eta \ < \ \mu(a_\varepsilon(x), x) \ < \ \varepsilon + \eta \, ,$$

and (6.24) follows from the arbitrariness of η. Now define

$$\bar{a}_\varepsilon \ := \ \inf_{x \in [0,l]} \ a_\varepsilon(x) \, ,$$

and let me show that $\bar{a}_\varepsilon$ is strictly positive. Indeed, assume that $\bar{a}_\varepsilon = 0$. Then for every $\delta > 0$ there is an x in $[0, l]$ such that $a_\varepsilon(x) < \delta$. Take $\delta_n = 1/n$, and consider the sequence of the x_n such that $a_\varepsilon(x_n) < 1/n$ for each n. Because each x_n belongs to the closed interval $[0, l]$, the sequence $n \mapsto x_n$ has a converging subsequence. For each x_n in the subsequence, we have $\mu(a_\varepsilon(x_n), x_n) = \varepsilon$ by (6.24). Then,

$$\lim_{n \to \infty} \ \mu(a_\varepsilon(x_n), x_n) \ = \ \varepsilon \, .$$

Denote by $\bar{x}$ the limit element of the subsequence. By the continuity of μ, $a_\varepsilon(x_n) \to 0$ and $x_n \to \bar{x}$ implies $\mu(0, \bar{x}) = \varepsilon$, in contradiction with (6.22). Then the premise $\bar{a}_\varepsilon = 0$ is false and we have proved that, for every $\varepsilon > 0$,

$$\bar{a}_\varepsilon \ := \ \inf_{x \in [0,l]} \ \sup \left\{ a > 0 \ \middle| \ \left| \tfrac{\theta(a,x)}{a} - \theta'(0+, x) \right| < \varepsilon \right\} \ > \ 0 \, .$$

In quite a similar way it can be proved that

$$\tilde{a}_\varepsilon \ := \ \inf_{x \in [0,l]} \ \inf \left\{ a < 0 \ \middle| \ \left| \tfrac{\theta(a,x)}{a} - \theta'(0-, x) \right| < \varepsilon \right\} \ > \ 0 \, ,$$

and after taking $\delta = \min \{\bar{a}_\varepsilon, \tilde{a}_\varepsilon\}$ we have that for all a with $|a| < \delta$ inequality (6.12) holds, with $\theta(a), \theta'(0+), \theta'(0-)$ replaced by $\theta(a, x), \theta'(0+, x), \theta'(0-, x)$, respectively.

At this point we may proceed as in the proof of Proposition 6.1, with this choice of δ and with θ replaced by $\theta(\cdot, x)$. In place of (6.13), we get the inequality

$$\sum_{x \in S(f_n) \setminus S(g)} \theta([f_n](x), x) \ > \ \sum_{x \in S(f_n) \setminus S(g)} (\theta'(0+, x) - \epsilon) \, [f_n](x) \, ,$$

and for $n \to \infty$ from equation (6.21) with $h(x) = \theta'(0+, x)$ and from the arbitrariness of ϵ we obtain

$$\lim_{n\to\infty} \sum_{x\in S(f_n)\setminus S(g)} \theta([f_n](x), x) \;\geq\; \int_{\mathcal{A}} \theta'(0+, x)(\nabla g(x) - G(x))\, dx\,, \tag{6.25}$$

with the equality sign for all $n \mapsto f_n$ with positive jumps. Recalling the limit relations (6.18), (6.19) and Proposition 6.2, we conclude that in the non-homogeneous case the expression of the energy of a structured deformation is

$$E_{Std}(g, G) \;=\; \int_{\mathcal{A}} (w(G(x), x) + \theta'(0+, x)(\nabla g(x) - G(x)))\, dx \;+\; \sum_{x\in S(g)} \theta([g](x), x)\,, \tag{6.26}$$

whenever either assumption (H1) or (H2) is satisfied at all x in $[0, l]$.

6.3 The Theory of Choksi and Fonseca

In Choksi and Fonseca (1997), a broader class of deformations and a weaker type of convergence are taken into consideration. In their theory, $\mathcal{A}$ is an open bounded region of $\mathcal{E}$, and simple deformations from $\mathcal{A}$ are elements of the space $SBV(\mathcal{A})$ of all *special functions of bounded variation.* A function $f : \mathcal{A} \to \mathcal{E}$ belongs to $SBV(\mathcal{A})$ if it is integrable over $\mathcal{A}$, if its derivative ∇f is integrable away from the jump set $S(f)$, and if the jump tensor Jf has bounded total variation:[40]

$$\int_{S(f)} |Jf(x)|\, d\mathcal{H}^{N-1} \;<\; +\infty\,.$$

Simple deformations in Choksi and Fonseca's theory are less regular than those defined here in Section 3.3, which are C^1 away from $S(f)$. Moreover, no injectivity is required, so that interpenetration is allowed, and no difference is made between deformations without cracks and deformations with unopened cracks, so that Griffith's model is excluded *a priori.*

In Choksi and Fonseca's theory, a structured deformation from $\mathcal{A}$ is a pair (g, G) with g in $SBV(\mathcal{A})$ and G in $L^1(\mathcal{A})$. No determinant inequality such as (4.1) is assumed. A sequence $n \mapsto f_n$ of simple deformations from $\mathcal{A}$ is said to converge to the structured deformation (g, G) if:

- f_n converges to g strongly in $L^1(\mathcal{A})$,
- ∇f_n converges to G weak* in the sense of measures,

that is, if

$$\lim_{n\to\infty} \int_{\mathcal{A}} |f_n(x) - g(x)|\, dx \;=\; 0\,,$$

$$\lim_{n\to\infty} \int_{\mathcal{A}} (\nabla f_n(x) - G(x))\, \varphi(x)\, dx \;=\; 0 \qquad \forall \varphi \in C_c(\mathcal{A})\,,$$

[40]See e.g. Ambrosio (1998) or Ambrosio et al. (2000).

where $C_c(\mathcal{A})$ is the set of all continuous functions over $\mathcal{A}$ with compact support. This type of convergence will be denoted by

$$f_n \rightharpoonup (g, G)\,. \tag{6.27}$$

Choksi and Fonseca show that a celebrated theorem of Alberti (1991) provides the Approximation Theorem for structured deformations in the new context: for every pair (g, G) in $SBV(\mathcal{A}) \times L^1(\mathcal{A})$ there is a sequence $n \mapsto f_n$ in $SBV(\mathcal{A})$ such that $f_n \rightharpoonup (g, G)$.

The energy of a simple deformation is still defined by (6.3), and the energy of a structured deformation is still the relaxed energy (6.15), with the convergence (3.12) replaced by the weaker convergence (6.27):

$$E^*_{Std}(g, G) \;=\; \inf_{f_n \rightharpoonup (g,G)} \;\liminf_{n\to\infty}\; E_{Sid}(f_n)\,. \tag{6.28}$$

In the one-dimensional case there is an explicit expression for this energy:[41]

$$\begin{aligned} E^*_{Std}(g, G) \;=\; & \textstyle\int_{\mathcal{A}} (w_c(G(x)) + \theta_s'(0+)(\nabla g(x) - G(x))^+ - \theta_s'(0-)(\nabla g(x) - G(x))^-)dx \\ & + \sum_{x \in S(g)} \theta_s([g](x))\,. \end{aligned} \tag{6.29}$$

Here w_c is the convex envelope of w, θ_s is the subadditive envelope of θ, and a^+, a^- are the positive and the negative part of a. Recall that a real-valued function φ is *subadditive* if

$$\varphi(a+b) \;\leq\; \varphi(a) + \varphi(b) \qquad \forall a, b \in \mathbb{R}\,, \tag{6.30}$$

and that the *subadditive envelope* θ_s of θ is the greatest subadditive function φ such that $\varphi \leq \theta$, just as the *convex envelope* w_c of w is the greatest convex function ϖ such that $\varpi \leq w$. In general, E^*_{Std} is lower than the energy E_{Std} in (6.16), because in Choksi and Fonseca's theory the infimum (6.28) is taken over a larger number of approximating sequences.

Let us show by examples that the greater freedom allowed in the SBV-theory of Choksi and Fonseca makes it possible to describe some physically relevant phenomena, not captured by the L^∞-theory illustrated in the preceding sections.

Example 1: w non-convex. Let w be concave on an interval (β, γ) of the real line and convex in $(-\infty, \beta)$ and in $(\gamma, +\infty)$. Then the convex envelope of w is given by

$$w_c(G) \;=\; \begin{cases} w(G) & \text{for all } G \text{ outside } (\alpha, \delta)\,, \\ \lambda w(\alpha) + (1-\lambda) w(\delta) & \text{for } G = \lambda\alpha + (1-\lambda)\delta\,,\;\; \lambda \in (0,1)\,, \end{cases} \tag{6.31}$$

[41] Del Piero (2001).

with α and δ as shown in Fig. 16a. Let $\mathcal{A}$ be the interval $(0,l)$, and let (g,G) be the structured deformation

$$g(x) \;=\; \gamma\, x\,, \qquad G(x) \;=\; \lambda\alpha + (1-\lambda)\delta\,, \tag{6.32}$$

with λ fixed in $(0,1)$ and $\gamma > \delta$, so that

$$\nabla g - G \;=\; \gamma - \lambda\alpha - (1-\lambda)\delta \;>\; 0\,.$$

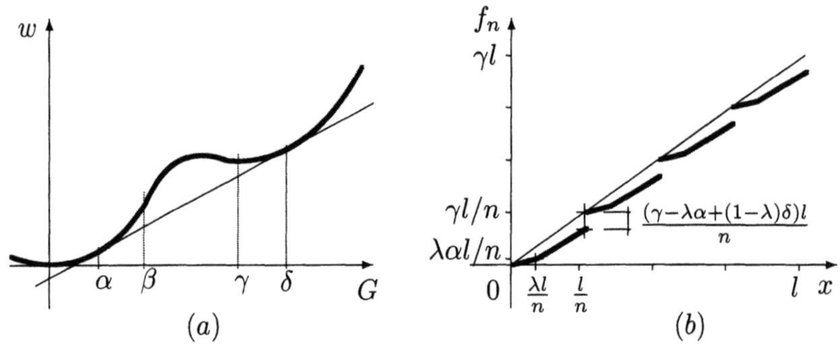

Figure 16.
A non-convex bulk energy density (a),
and an approximating sequence for the structured deformation (6.32) (b).

Assume that θ is subadditive, so that $\theta_s = \theta$ and, in particular, $\theta_s'(0+) = \theta'(0+)$. Then, by (6.16) and (6.29),

$$E_{Std}(g,G) \;=\; l\,(w(G) + \theta'(0+)(\gamma - G))\,, \qquad E^*_{Std}(g,G) \;=\; l\,(w_c(G) + \theta'(0+)(\gamma - G))\,.$$

Therefore $E_{Std}(g,G) > E^*_{Std}(g,G)$, because G is included in the interval (α,δ) in which $w > w_c$.

Some information on the nature of the structural changes involved in the structured deformation (g,G) is provided by the approximating sequences which achieve the infimum in (6.28). For example, consider the sequence $n \mapsto f_n$ with

$$f_n(x) = \begin{cases} \frac{\gamma h l}{n} + \alpha(x - \frac{hl}{n}) & \text{for } x \in [\frac{hl}{n}, \frac{(h+\lambda)l}{n})\,, \quad h \in \{0,1,..\,n-1\}, \\ (\frac{\gamma h + \alpha\lambda)l}{n} + \delta(x - \frac{(h+\lambda)l}{n}) & \text{for } x \in [\frac{(h+\lambda)l}{n}, \frac{(h+1)l}{n}),\ h \in \{0,1,..\,n-1\}. \end{cases}$$

As shown in Fig. 16b, each f_n has $(n-1)$ jumps of intensity

$$[f_n] \;=\; (\gamma - \lambda\alpha - (1-\lambda)\delta)\,\tfrac{l}{n} \;=\; (\nabla g - G)\,\tfrac{l}{n}\,, \tag{6.33}$$

and $|g(x) - f_n(x)| < [f_n]$ for all x. Then f_n converges to g uniformly, and therefore in $L^1(0,l)$. Moreover, ∇f_n is piecewise constant and takes the values α and δ in such a proportion that the average over any interval of length l/n is exactly G:

$$\int_y^{y+l/n} (\nabla f_n(x) - G(x))\, dx \;=\; 0 \qquad \text{for all } \; y \in (0, l - l/n)\,.$$

Because the sequence $n \mapsto \nabla f_n$ is uniformly bounded in L^∞, the above equation implies the weak convergence to G in L^1,[42] and therefore the weak* convergence to G in the sense of measures. On the other hand, the convergence is not uniform, because

$$\sup_{x\in(0,l)} \; |\nabla f_n(x) - G(x)| \;=\; \max\{\delta - G,\, G-\alpha\} \;=\; \max\{1-\lambda, \lambda\}\,(\delta-\alpha)$$

for all n. Thus, $n \mapsto f_n$ converges to (g, G) in the sense of (6.27) but not in the sense of (3.12). Moreover, by (6.3),(6.31) and (6.33),

$$\begin{aligned} E_{Sid}(f_n) \;&=\; \lambda l\, w(\alpha) + (1-\lambda) l\, w(\delta) + (n-1)\,\theta((\nabla g - G)\tfrac{l}{n}) \\ &=\; l w_c(G) + (n-1)(\theta'(0+)(\nabla g - G)\tfrac{l}{n} + o(\tfrac{1}{n}))\,, \end{aligned}$$

so that

$$\lim_{n\to\infty} \; E_{Sid}(f_n) \;=\; l w_c(G) + l\,\theta'(0+)(\nabla g - G) \;=\; E^*_{Std}(g, G)\,,$$

that is, $n \mapsto f_n$ achieves the infimum in (6.28). When $n \to \infty$, all jumps $[f_n]$ tend to disappear, and the derivatives ∇f_n oscillate between the values α and δ with higher and higher frequency. The limit element (g, G) can be interpreted as a *fine mixture*, in which the *constituent phases* α and δ enter in the proportion λ and $(1-\lambda)$, respectively. Thus, the SBV-theory is able to describe structured deformations which are fine mixtures of two constituent phases, a type of microstructure which cannot be described within the L^∞-theory.[43]

Example 2: θ non-subadditive. Let θ be a convex-concave energy satisfying the assumption (H2) and the inequality

$$\theta([f]) \;>\; \theta'(0+)\,[f] \qquad \text{for all}\,[f] > 0\,. \tag{6.34}$$

An energy of this type is shown in Fig. 1. The function θ is not subadditive; indeed, by repeated use of inequality (6.30) with $a = b = [f]/n$, for any subadditive function φ we have

$$\varphi([f]) \;\leq\; n\,\varphi(\tfrac{[f]}{n})$$

[42] Dacorogna (1989), Lemma 1.4.

[43] Fine mixtures of deformations instead of deformation gradients can be described by limits of simple deformations called *mixing deformations*, see Del Piero and Owen (1993), Section 7d.

for all $[f] > 0$ and for all $n \in \mathbb{N}$, and if $\varphi(0) = 0$ we have

$$\varphi(\tfrac{[f]}{n}) \;=\; \varphi'(0+)\,\tfrac{[f]}{n} + o(\tfrac{[f]}{n})\,.$$

For $n \to \infty$, from the preceding inequality we get

$$\varphi([f]) \;\leq\; \varphi'(0+)\,[f] \qquad \text{for all } [f] > 0\,. \tag{6.35}$$

Then comparison with (6.34) shows that θ cannot be subadditive.

Now consider the function

$$\psi([f]) \;:=\; \theta'(0+)\,[f]\,, \qquad [f] > 0\,. \tag{6.36}$$

This function is subadditive and satisfies $\psi \leq \theta$ by (6.34). In fact, ψ is the greatest function with these two properties, that is, ψ is the subadditive envelope of θ. Assume, indeed, that there exists a subadditive function φ such that $\varphi([f]) > \psi([f]) = \theta'(0+)[f]$ at some $[f]$. Then from (6.35) it follows that $\varphi'(0+) > \theta'(0+)$, and therefore the condition $\varphi \leq \theta$ is violated in a right neighborhood of the origin.

For $\mathcal{A} = (0, l)$ consider the structured deformation

$$g(x) \;=\; \begin{cases} x & \text{for } x \in (0, l/2) \\ x + a & \text{for } x \in [\,l/2, l) \end{cases}\,, \qquad G(x) \;=\; 1\,, \tag{6.37}$$

with a a positive constant. Assume that w is convex and that $w(1) = 0$ as in $(6.2)_1$. Then $w_c(G(x)) = w(G(x)) = 0$, and from (6.16), (6.29) we have

$$E_{Std}(g, G) \;=\; \theta(a)\,, \qquad E^*_{Std}(g, G) \;=\theta_s(a)\,, \tag{6.38}$$

so that $E_{Std}(g, G) > E^*_{Std}(g, G)$.

Take the sequence of simple deformations

$$f_n(x) \;=\; \begin{cases} x & \text{for } x \in (0, l/2)\,, \\ x + \frac{ha}{n} & \text{for } x \in [\frac{l}{2} + \frac{hl}{n^2}, \frac{l}{2} + \frac{(h+1)l}{n^2})\,, \quad h \in \{0, 1, .. \, n-1\}\,, \\ x + a & \text{for } x \in [\frac{l}{2} + \frac{l}{n}, l)\,. \end{cases} \tag{6.39}$$

The function f_n with $n = 4$ is shown in Fig. 17.

The sequence $n \to f_n$ converges to g strongly in L^1 but not in the sense of (3.9).

Indeed,

$$\sup_{x\in(0,l)\setminus(S(f_n)\cup S(g))} |f_n(x) - g(x)| \;=\; |f_n(l/2) - g(l/2)| \;=\; a(1-\tfrac{1}{n})\,.$$

Moreover, $n \mapsto \nabla f_n$ is the constant sequence $n \mapsto G$, because $\nabla f_n(x) = G(x)$ for all x and for all n. Thus, we have the convergence $f_n \rightharpoonup (g, G)$ but *not* the convergence $f_n \to (g, G)$.

Each f_n has n jumps of intensity a/n, and $w(\nabla f_n) = w(1) = 0$. Then,

$$E_{Sid}(f_n) \;=\; n\theta(\tfrac{a}{n}) \;=\; \theta'(0+)\,a + n\,o(\tfrac{1}{n})\,.$$

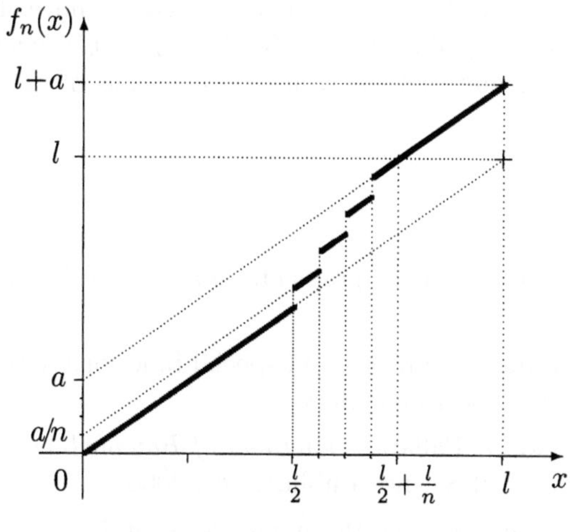

Figure 17.
The approximating sequence (6.39) for the structured deformation (6.37).

Recalling that $\theta'(0+)a = \theta_s(a)$, in the limit we have

$$\lim_{n\to\infty} E_{Sid}(f_n) \;=\; \theta'(0+)\,a \;=\; \theta_s(a) \;=\; E^*_{Std}(g, G)\,.$$

With growing n, the functions f_n exhibit jumps of smaller and smaller intensity, concentrated on a smaller and smaller interval. In the limiting structured deformation, there is only one jump at $x = l/2$. The approximating sequence (6.39) tells us that, among all sequences which converge to (g, G) in the sense of (6.27), one which achieves the infimum in (6.28) involves the coalescence at a single point of an infinite number of jumps of infinitesimal intensity. The jump in the limit element (g, G) was called a *complex fracture* in Del Piero (1989) and a *concentrated microfracture* in Del Piero and Owen (1990). Again, these are structured deformations which cannot be described within the L^∞-theory.

In conclusion, the theory of Choksi and Fonseca broadens the range of applicability of structured deformations, by including such objects as fine mixtures of solid phases and concentrated microfractures. In Del Piero (2001) it was pointed out that there may well be classes of continua undergoing structured deformations of different regularity, and that the regularity assumed for simple deformations and the definition of convergence for sequences of simple deformations are in some sense *constitutive assumptions* on the nature of the continuum. Here this observation is made more precise, by saying that regularity and type of convergence determine classes of continua undergoing specific microstructures.

For example, according to the convergence (3.12) two configurations with jumps of finite amplitude are close to each other only if their jump sets coincide, while this is not required in the convergence (6.24). Consequently, the second type of convergence seems to be more appropriate to describe materials in which defects represented by discontinuous displacements, such as dislocations, can easily move. It is then reasonable that, as suggested by the inequality $E_{Std}(g,G) \geq E^*_{Std}(g,G)$, such materials be able of storing less energy than materials in which propagation of defects is restrained or forbidden.

References

G. Alberti. A Lusin type theorem for gradients. *Journal of Functional Analysis.* 100: 110-118, 1991.

L. Ambrosio. Free discontinuity problems and special functions with bounded variation. *Progress in Mathematics*, 168:15-35, 1998.

L. Ambrosio, N. Fusco and D. Pallara. *Functions of Bounded Variation and Free Discontinuity Problems.* Oxford Science Publications, 2000.

H. Cartan. *Cours de calcul différentiel.* Hermann, Paris 1967.

R. Choksi and I. Fonseca. Bulk and interfacial energy densities for structured deformations of continua. *Archive for Rational Mechanics and Analysis*, 138:37-103, 1997.

R. Choksi, G. Del Piero, I. Fonseca and D.R. Owen. Structured deformations as energy minimizers in models of fracture and hysteresis. *Mathematics and Mechanics of Solids*, 4:321-356, 1999.

B. Dacorogna. *Direct methods in the Calculus of Variations.* Springer-Verlag, 1989.

G. Dal Maso and R. Toader. A model for the quasi-static growth of a brittle fracture: existence and approximation results. *Archive for Rational Mechanics and Analysis*, 162:101-135, 2002.

G. Del Piero. A new function space for the mathematical theory of plasticity. In A.S. Khan and M. Tokuda, editors, *Proceedings of the Second International Symposium on Plasticity and its Current Applications.* Pergamon Press, 1989.

G. Del Piero. Towards a unified approach to fracture, yielding, and damage. In E. Inan and K.Z. Markov, editors, *Proceedings of the 9th International Symposium "Continuum Models and Discrete Systems".* World Scientific, Singapore, 1998.

G. Del Piero. The energy of a one-dimensional structured deformation. *Mathematics and Mechanics of Solids*, 6:387-408, 2001.

G. Del Piero. A class of fit regions and a universe of shapes for continuum mechanics. *Journal of Elasticity*, 70:175-195, 2003.

G. Del Piero and D.R. Owen. New concepts in the mechanics of fractured continua. *Proceedings of the 10th National Congress AIMETA*, Pisa 1990.

G. Del Piero and D.R. Owen. Structured deformations of continua. *Archive for Rational Mechanics and Analysis*, 124:99-155, 1993.

G. Del Piero and D.R. Owen. Integral-gradient formulae for structured deformations. *Archive for Rational Mechanics and Analysis*, 131:121-138, 1995.

G. Del Piero and D.R. Owen. *Structured Deformations.* Quaderni I.N.d.A.M., Gruppo Nazionale Fisica Matematica no. 58, Centro Stampa 2p, Firenze 2000.

G. Del Piero and L. Truskinovsky. Elastic bars with cohesive energy. In preparation.

G. Francfort and J.J. Marigo. Revisiting brittle fracture as an energy minimization problem. *Journal of the Mechanics and Physics of Solids*, 46:1319-1342, 1998.

P.R. Halmos. *Finite-Dimensional Vector Spaces.* Springer-Verlag 1974.

A.N. Kolmogorov and S.V. Fomin. *Introductory Real Analysis.* Dover Publications, 1975.

W. Noll. Lectures on the foundations of Continuum Mechanics and Thermodynamics. *Archive for Rational Mechanics and Analysis*, 52:62-92, 1973.

W. Noll, E.G. Virga. Fit regions and functions of bounded variation. *Archive for Rational Mechanics and Analysis*, 102:1-21, 1988.

D.R. Owen. Disarrangements in continua and the geometry of microstructure. In K. Rajagopal, editor, *Recent Advances in Elasticity, Viscoelasticity, and Inelasticity*, pp. 67–81. World Scientific, 1995.

C. Truesdell. *A First Course in Rational Continuum Mechanics*, Vol.1. Academic Press, 1991.

A.I. Vol'pert and S.I. Hudjaev. *Analysis of Classes of Discontinuous Functions and Equations of Mathematical Physics.* Nijhoff, 1985.

Second-Order Structured Deformations: Approximation Theorems and Energetics

Roberto Paroni

Dipartimento di Architettura e Pianificazione, Università degli Studi di Sassari, Alghero, Italy

1 Introduction

Recent research on geometrical changes that can occur at different length scales has led to the concept of a structured deformation (κ, g, G), see [Del Piero and Owen (1993)]. Structured deformations have been applied to describe geometrical changes associated with slips and the presence of defects in single crystals, with deformations of liquid crystals, with fracture, and with the mixing of different substances [Del Piero and Owen (1993)]. One limitation of structured deformations is that the effects on macroscopic deformation of jumps in the gradients ∇f_m of approximating simple deformations are not captured. Another limitation is the following: Structured deformations lead to the additive decomposition $\nabla g(x) = G(x) + M(x)$, where M is a deformation due to disarrangements associated with non-smooth changes at a smaller length scale, but do not reveal any decomposition for $\nabla^2 g(x)$ which puts into light the effects of deformations at smaller length scales. Finally, definitive refinements of kinematical quantities such as acceleration and stretching are not available through such structured deformations.

Our goal in these notes is to review the notion of "second-order structured deformation" (κ, g, G, Σ) that extends the scope of the multiscale geometry afforded by structured deformations of the original form (κ, g, G). In Section 2 we review the theory put forth by Owen and Paroni (2000), here called "L^∞ setting". In particular in Section 2.1 we recall the notions of classical, simple, and of second-order structured deformations. In this setting the additional entry Σ that appears in the symbol (κ, g, G, Σ) for a second-order structured deformation is a piecewise continuous, third-order tensor field possessing the same symmetries as a second gradient, but not being necessarily a second gradient itself. The remaining entries form a triple (κ, g, G) that turns out to be a first-order structured deformation in which g is piecewise twice continuously differentiable and G is piecewise continuously differentiable. Section 2.1 also contains a brief discussion of compositions of second-order structured deformations. The Approximation Theorem for second-order structured deformations is presented, and partly proved, in Section 2.2. The main difficulties in the analysis of second-order structured deformations are the necessity of constructing extensions of certain quadratic maps to the entire space as C^2-diffeomorphisms (this is done in Section 2.2.1) and the necessity of constructing injective mappings starting from a family of restrictions of the constructed C^2-diffeomorphisms. In Section 2.3 we establish decompositions and identification relations for $\nabla^2 g$. Section 2.4 is devoted to two examples, including the bending of a rectangular block through distortions at a smaller length scale without slips ("bending through simple shears"), and the bending of a block through distortions *and* slips at smaller length scales ("bending of a deck of cards").

A study of the energetics of first-order structured deformations has led to a more general setting for defining and analyzing structured deformations [Choksi and Fonseca (1997), Del Piero, Lecture 6 in this volume] and a corresponding energetic analysis and extension for second-order structured deformations is in progress [Fonseca et al. (2002), Fonseca et al. (2003), Fonseca et al. (to appear)]. This analysis has lead to generalizations of the notion of second-order structured deformation which is the principal topic of Section 3. After recalling the main properties of the spaces SBH and SBV^2 in Section 3.1, we introduce in Section 3.2 the concept of second-order structured deformation in the SBV^2 setting. These deformations are the natural generalization of the "L^∞" second-order structured deformations, but contrary to what happens in the L^∞ setting, in the SBV^2 setting the field Σ has no symmetries. The Approximation Theorem is easily proved by means of a second order version of "Alberti's Theorem" [Alberti (1991)], the proof of which follows immediately by applying "Alberti's Theorem" twice.

In Section 3.3 we introduce the concept of second-order structured deformations in the SBH setting. These deformations are different from the ones discussed above since slipping and fracture is not allowed: deformations generally are smoother in the SBH setting than in the SBV^2 setting. Again the Approximation Theorem is obtained by applying an appropriate second order version of "Alberti's Theorem". The proof of this version turns out to be quite challenging and is reported, in dimension $N = 2$ where the main ideas can be explained also by means of some figures, in Section 3.3.

The notes end with a short section in which the energetics of second-order structured deformations is discussed.

2 The L^∞ setting

In this section we recall the concept of *second-order structured deformations*, introduced by Owen and Paroni (2000). They are a generalization to the realm of second-order derivatives of the previously developed class of (first-order) structured deformations due to Del Piero and Owen (1993), and the presentation of this section follows closely the one in [Owen and Paroni (2000)].

We now introduce some useful notation. For each $p = 1, \cdots, N$ and $N \times N$ matrix A, we denote by $I_p(A)$ the $p-$th invariant of A, i.e., the sum of the determinants of all of the principal $p \times p$ submatrices of A. The symbol I alone denotes the $N \times N$ identity matrix $I = \mathrm{diag}(1, 1, \cdots, 1)$, and $\mathcal{B}(x, r)$ denotes the open ball centered at $x \in \mathbb{R}^N$ of radius $r > 0$. Also, $\mathbb{R}^{N\times N}$ denotes the set of real $N \times N$ matrices.

For each subset $\mathcal{A}$ of $\mathbb{R}^N$, we denote by $\overline{\mathcal{A}}$ the closure of $\mathcal{A}$, by $\partial\mathcal{A}$ the boundary of $\mathcal{A}$, and by $\mathcal{A}^\circ$ the interior of $\mathcal{A}$. A *C^2-diffeomorphism* of $\mathbb{R}^N$ is a bijective, C^2-mapping $f : \mathbb{R}^N \to \mathbb{R}^N$ whose inverse also is of class C^2.

2.1 Definitions and preliminaries

The regions that a deforming body can occupy are described here in terms of "fit regions" [Noll and Virga (1998), Del Piero Lecture 2 in this volume]: a fit region of $\mathbb{R}^N$ is a bounded, regularly open, set with finite perimeter whose boundary has measure zero. As for first-order structured deformations, the notion of a second-order structured deformation is based on the notions of "classical deformations" and "simple deformations", here defined simply by strengthening the smoothness requirements on mappings from "C^1", in the first-order case [Del Piero and Owen (1993)], to "C^2" in the present case.

Definition 2.1. Let $\mathcal{A}$ be a fit region in $\mathbb{R}^N$. A *classical deformation from* $\mathcal{A}$ is a mapping f from $\mathcal{A}$ into $\mathbb{R}^N$ satisfying:
(Cld 1) f can be extended to a C^2-diffeomorphism of $\mathbb{R}^N$;
(Cld 2) f is orientation preserving, i.e.,

$$\det\nabla f(x) > 0 \text{ for all } x \in \mathcal{A}.$$

We recall that a *piecewise fit region* [Del Piero and Owen (1993)] is a finite union of (possibly overlapping) fit regions.

Definition 2.2. Let $\mathcal{A}$ be a piecewise fit region in $\mathbb{R}^N$. A *simple deformation from* $\mathcal{A}$ is a pair (κ, f), where κ is a subset of $\mathcal{A}$ and f is mapping from $\mathcal{A} \setminus \kappa$ into $\mathbb{R}^N$ with the following properties:

(Sid 1) vol $\kappa = 0$;
(Sid 2) f is injective;
(Sid 3) $\mathcal{A} \setminus \kappa$ is the union of finitely many fit regions such that the restriction of f to each of the fit regions is a classical deformation.

Definition 2.3. Let $\mathcal{A}$ be a piecewise fit region in $\mathbb{R}^N$. A *second-order structured deformation from* $\mathcal{A}$ is a quadruple (κ, g, G, Σ) in which

(Std 1) (κ, g) is a simple deformation from $\mathcal{A}$;
(Std 2) $G : \mathcal{A} \setminus \kappa \to \text{Lin}\,(\mathbb{R}^N)$ is of class C^1, $\Sigma : \mathcal{A} \setminus \kappa \to \text{Lin}_2(\mathbb{R}^N, \mathbb{R}^N)$ is symmetric (i.e., for every $x \in \mathcal{A} \setminus \kappa$, $\Sigma(x)$ is symmetric) and continuous, G has a piecewise C^1-extension to $\overline{\mathcal{A}}$, and Σ has a piecewise continuous extension to $\overline{\mathcal{A}}$, in the sense that there exists a finite collection of fit regions $\{\mathcal{A}_j : j = 1, \ldots, J\}$ whose union is $\mathcal{A} \setminus \kappa$ such that, for each $j = 1, \ldots, J$, $G|_{\mathcal{A}_j}$ has a C^1-extension to $\overline{\mathcal{A}_j}$ and $\Sigma|_{\mathcal{A}_j}$ has a continuous extension to $\overline{\mathcal{A}_j}$.
(Std 3) There exists $m > 0$ such that $m < \det G(x) \leq \det \nabla g(x)$ for all $x \in \mathcal{A} \setminus \kappa$.

We note that the fields g, G, and Σ associated with a second-order structured deformation, as well as the fields f, ∇f, and $\nabla^2 f$ associated with a simple deformation, are bounded.

The next definition provides a precise sense in which one can assert that a sequence of simple deformations approaches a second-order structured deformation.

Definition 2.4. Let $\mathcal{A}$ be a piecewise fit region in $\mathbb{R}^N$. We say that the sequence $m \mapsto (\kappa_m, f_m)$ of simple deformations from $\mathcal{A}$ *determines* the second-order structured deformation (κ, g, G, Σ) from $\mathcal{A}$, if the following conditions are satisfied:

$$\kappa = \liminf_{m\to\infty} \kappa_m := \bigcup_{m=1}^{\infty} \bigcap_{j=m}^{\infty} \kappa_j, \tag{2.1}$$

$$\lim_{m\to\infty} \|g - f_m\|_{L^\infty(\mathcal{A},\mathbb{R}^N)} = 0, \tag{2.2}$$

$$\lim_{m\to\infty} \|G - \nabla f_m\|_{L^\infty(\mathcal{A},\mathrm{Lin}(\mathbb{R}^N,\mathbb{R}^N))} = 0, \tag{2.3}$$

$$\lim_{m\to\infty} \|\Sigma - \nabla^2 f_m\|_{L^\infty(\mathcal{A},\mathrm{Lin}_2(\mathbb{R}^N,\mathbb{R}^N))} = 0. \tag{2.4}$$

Our final definition embodies the idea that a body may undergo a simple deformation followed by a second-order structured deformation.

Definition 2.5. Let (κ, f) be a simple deformation from $\mathcal{A}$, and let (μ, h, H, Θ) be a second-order structured deformation from $f(\mathcal{A}\backslash\kappa)$. Then the *composition* of (μ, h, H, Θ) with (κ, f) is the quadruple defined by

$$\begin{aligned}(\mu, h, H, \Theta) \quad \circ \quad & (\kappa, f) = (\kappa \cup f^{-1}(\mu),\ h \circ f\,|_{\mathcal{A}\backslash(\kappa\cup f^{-1}(\mu))},\\ & (H \circ f\,|_{\mathcal{A}\backslash(\kappa\cup f^{-1}(\mu))})\nabla f\,|_{\mathcal{A}\backslash(\kappa\cup f^{-1}(\mu))},\\ & (\Theta \circ f\,|_{\mathcal{A}\backslash(\kappa\cup f^{-1}(\mu))})[\nabla f\,|_{\mathcal{A}\backslash(\kappa\cup f^{-1}(\mu))}, \nabla f\,|_{\mathcal{A}\backslash(\kappa\cup f^{-1}(\mu))}]\\ & +(H \circ f\,|_{\mathcal{A}\backslash(\kappa\cup f^{-1}(\mu))})\nabla^2 f\,|_{\mathcal{A}\backslash(\kappa\cup f^{-1}(\mu))}).\end{aligned} \tag{2.5}$$

It is a routine matter to verify that the composition $(\mu, h, H, \Theta) \circ (\kappa, f)$ is a second-order structured deformation from the piecewise fit region $\mathcal{A}$. Moreover, the next result shows that a determining sequence for (μ, h, H, Θ) yields immediately a determining sequence for $(\mu, h, H, \Theta) \circ (\kappa, f)$.

Proposition 2.6. *Let (κ, f) be a simple deformation from $\mathcal{A}$, and let (μ, h, H, Θ) be a second-order structured deformation from $f(\mathcal{A}\backslash\kappa)$. Let $m \mapsto (\mu_m, h_m)$ be a sequence of simple deformations from $f(\mathcal{A}\backslash\kappa)$ that determines (μ, h, H, Θ). Then the sequence of simple deformations*

$$m \mapsto (\mu_m, h_m) \circ (\kappa, f) := (\kappa \cup f^{-1}(\mu_m), h_m \circ f\,|_{\mathcal{A}\backslash(\kappa\cup f^{-1}(\mu_m))})$$

determines the second-order structured deformation $(\mu, h, H, \Theta) \circ (\kappa, f)$.

PROOF. We notice that

$$\begin{aligned}\|(\nabla^2 h_m \circ f)[\nabla f, \nabla f] + (\nabla h_m \circ f)\nabla^2 f \\ -(\Theta \circ f)[\nabla f, \nabla f] - (H \circ f)\nabla^2 f\|_{L^\infty(\mathcal{A})} \\ \leq \|\nabla^2 h_m - \Theta\|_{L^\infty(\mathcal{A}\backslash\kappa)} \|\nabla f\|^2_{L^\infty(\mathcal{A})} \\ + \|\nabla h_m - H\|_{L^\infty(\mathcal{A}\backslash\kappa)} \|\nabla^2 f\|_{L^\infty(\mathcal{A})} \to 0\end{aligned}$$

as $m \to \infty$. The rest of the proof is identical to the one given in [Del Piero and Owen (1993), Proposition 4.9]. □

2.2 The L^∞ Approximation Theorem

In this section we shall state and sketch the proof of the Approximation Theorem for second-order structured deformations (Theorem 2, below): every second-order structured deformation has a determining sequence. The existence of a determining sequence reveals by means of relations (2.1)-(2.4) geometrical information about each entry in a second-order structured deformation (κ, g, G, Σ) that is not apparent from Definition 3 and that is useful in a variety of situations.

In order to understand how the proof goes, let us restrict our attention to the one dimensional case, $N := 1$, and to $\mathcal{A} := (0, 1)$. Let us consider a purely submacroscopic second-order structured deformation $(\emptyset, id, G, \Sigma)$. Here G and Σ are real-valued continuosly differentiable and continuous functions on $\mathcal{A}$, respectively, with $0 < m \leq G \leq id' = 1$ and where id denotes the identity function on $(0, 1)$. We then want to construct a sequence of simple deformations (κ_n, f_n) from $\mathcal{A}$ that determines $(\emptyset, id, G, \Sigma)$. A natural first guess for the determining sequence would be to set $\kappa_n := \{x_i : i = 1, 2, \ldots, n-1\}$, for each positive integer n and where for $i = 0, 1, \ldots, n$ we put $x_i := i/n$, and

$$f_n(x) := \sum_{i=0}^{n-1} \left(x_i + G(x_i)(x - x_i) + \frac{1}{2}\Sigma(x_i)(x - x_i)^2 \right) \chi_{(x_i, x_{i+1})}(x).$$

Here $\chi_{(x_i, x_{i+1})}$ denotes the characteristic function of the interval (x_i, x_{i+1}). In general, as one can easily check, the pair (κ_n, f_n) is not a simple deformation because the functions f_n fails to be one-to-one. The injectivity problem is easily resolved in one dimension, while in higher dimensions it turns out to be quite a challenge, cf. [Owen and Paroni (2000)]. Here, for simplicity, we continue only to consider the one dimensional case. In order to get a one-to-one function, we define

$$f_n(x) := \sum_{i=0}^{n-1} \left(x_i + (G(x_i) - \|\Sigma\|_{L^\infty}/n)(x - x_i) + \frac{1}{2}\Sigma(x_i)(x - x_i)^2 \right) \chi_{(x_i, x_{i+1})}(x). \tag{2.6}$$

To show that (κ_n, f_n) is a simple deformation, it suffices to show that f_n is injective and, for every $i = 0, 1, \ldots, n-1$, that f_n restricted to (x_i, x_{i+1}) is a classical deformation . Note that for $n > 2\|\Sigma\|_{L^\infty}/m$ and $x \in (0, 1)\backslash\kappa_n$,

$$f_n'(x) \geq \sum_{i=0}^{n-1} (m - \|\Sigma\|_{L^\infty}/n - \|\Sigma\|_{L^\infty}/n)\, \chi_{(x_i, x_{i+1})}(x) \geq m - 2\frac{\|\Sigma\|_{L^\infty}}{n} > 0.$$

Thus, for n large enough, $f_n' > 0$, and, as will be shown in the Extension Theorem subsection, f_n restricted to (x_i, x_{i+1}) can be extended to a C^2-diffeomorphism of $\mathbb{R}$. We now show that f_n is injective. Since f_n is increasing on each interval (x_i, x_{i+1}) for n large enough, it suffices to check that

$$f_n(x_i^-) := \lim_{x \to x_i^-} f_n(x) \leq \lim_{x \to x_i+} f_n(x) = x_i, \tag{2.7}$$

for every $i = 1, \ldots, n-1$. Note that $f_n'(x) \leq 1 - \|\Sigma\|_{L^\infty}/n + \|\Sigma\|_{L^\infty}/n = 1$, so that for $x \in (x_{i-1}, x_i)$ we have

$$f_n(x) - x_{i-1} = \int_{x_{i-1}}^{x} f_n'(t)\, dt \leq x - x_{i-1},$$

and, therefore, $f_n(x) \leq x$, which implies (2.7) and the injectivity of f_n. Finally, the sequence of simple deformations $n \mapsto (\kappa_n, f_n)$ satisfies

$$\liminf_{n \to +\infty} \kappa_n = \emptyset,$$

$$\lim_{n \to \infty} \|f_n - id\|_{L^\infty} \leq \lim_{n \to \infty} \left(\frac{\|G\|_{L^\infty}}{n} + \frac{\|\Sigma\|_{L^\infty}}{n}\frac{1}{n} + \frac{1}{2}\|\Sigma\|_{L^\infty}\frac{1}{n^2} \right) = 0,$$

$$\lim_{n \to \infty} \|f_n' - G\|_{L^\infty} \leq \lim_{n \to \infty} \frac{2\|\Sigma\|_{L^\infty} + \|G'\|_{L^\infty}}{n} = 0,$$

and

$$\lim_{n \to \infty} \|f_n'' - \Sigma\|_{L^\infty} = 0.$$

Thus, assuming that f_n restricted to (x_i, x_{i+1}) can be extended to a C^2-diffeomorphism of $\mathbb{R}$, we have proved that $n \mapsto (\kappa_n, f_n)$ determines the second-order stuctured deformation $(\emptyset, id, G, \Sigma)$. In [Owen and Paroni (2000)] the corresponding result was called the Approximation Lemma for second-order structured deformations.

The factorization of the type (2.5)

$$(\kappa, g, G, \Sigma) = (\emptyset, id, H, \Theta) \circ (\kappa . g), \tag{2.8}$$

where

$$H(y) = (G(\nabla g)^{-1})(g^{-1}(y)) \tag{2.9}$$

and

$$\begin{aligned} \Theta(y)[a, b] &= \Sigma(g^{-1}(y))[(\nabla g)^{-1}(g^{-1}(y))a, (\nabla g)^{-1}(g^{-1}(y))b] \\ &\quad -(G(\nabla g)^{-1})(g^{-1}(y))\nabla^2 g(g^{-1}(y))[a, b] \end{aligned} \tag{2.10}$$

for every $y \in g(\mathcal{A} \backslash \kappa)$ and $a, b \in \mathbb{R}^N$, together with Proposition 1, permits us to deduce from the Approximation Lemma the existence of a determining sequence for each second-order structured deformation. For brevity the details of the argument are left out. We now state the Approximation Theorem in its most general form.

Theorem 2.7. ***Approximation Theorem for Second-Order Structured Deformations*** *Let $\mathcal{A}$ be a piecewise fit region. For each second-order structured deformation (κ, g, G, Σ) from $\mathcal{A}$, there exists a sequence $m \mapsto (\kappa_m, f_m)$ of simple deformations from $\mathcal{A}$ that determines (κ, g, G, Σ).*

It is convenient to use the term "disarrangement" to denote non-smooth geometrical changes due to slip and separation on internal surfaces and occurring possibly at more than one length scale. Because ∇f_m and $\nabla^2 f_m$ do not measure such geometrical changes, it is natural in view of the Approximation Theorem to call $G = \lim_{m\to\infty} \nabla f_m$ the *(first-order) deformation without disarrangements* and $\Sigma = \lim_{m\to\infty} \nabla^2 f_m$ the *(second-order) deformation without disarrangements* for the second-order structured deformation (κ, g, G, Σ). Because κ_m is a collection of points where f_m can fail to be smooth, we call κ_m the *disarrangement site* for the simple deformation (κ_m, f_m) and, in view of the Approximation Theorem, we call $\kappa = \liminf_{m\to\infty} \kappa_m$ the *(permanent) disarrangement site* for the second-order structured deformation (κ, g, G, Σ). Measures of deformation due to disarrangements for second-order structured deformations will be identified in Section 2.3.

Extension Theorem In this section we prove a theorem, used in the previous section, which provides smooth diffeomorphic extensions to the entire space for members of a particular class of quadratic functions. The proof will be carried out in the general case.

The following proposition, whose proof is given in [Owen and Paroni (2000)], provides elementary estimates needed in the Extension Theorem.

Proposition 2.8. *For each $A \in \mathbb{R}^{N\times N}$ there holds*

$$1 - \sum_{j=1}^{N} j! \binom{N}{j} \|A\|^j \leq \det(I + A) \leq 1 + \sum_{j=1}^{N} j! \binom{N}{j} \|A\|^j. \tag{2.11}$$

If $B \in \mathbb{M}^{N\times N}$ is a symmetric matrix whose eigenvalues are less than or equal to 1 in absolute value, then we also have

$$|I_j(AB)| \leq j! \binom{N}{j} \|A\|^j \qquad \text{for every } j = 1, \ldots, N. \tag{2.12}$$

The main result of this section is the following Extension Theorem.

Theorem 2.9. *Let $a \in \mathbb{R}^N, A \in \mathrm{Lin}(\mathbb{R}^N, \mathbb{R}^N)$, $\Xi \in \mathrm{Lin}_2(\mathbb{R}^N, \mathbb{R}^N)$, and positive real numbers m, M be given such that $m \leq \det A < M$ and Ξ is symmetric. There exists $r = r(M, \det A, \|A^{-1}\Xi\|) > 0$ depending continuously on $(M, \det A, \|A^{-1}\Xi\|)$ such that the function $f : \mathcal{B}(0, r) \to \mathbb{R}^N$ defined by*

$$f(x) = \frac{1}{2}\Xi[x, x] + Ax + a \tag{2.13}$$

can be extended to a C^2-diffeomorphism f^e of $\mathbb{R}^N$ that satisfies

$$\frac{m}{2} \leq \det \nabla f^e(x) < M \tag{2.14}$$

for every $x \in \mathbb{R}^N$.

PROOF.We put $\Lambda := A^{-1}\Xi$, i.e., $\Lambda_{ijk} = (A^{-1})_{ip}\Xi_{pjk}$ and choose $r_1 > 0$ such that

$$\begin{aligned} r_1 \quad < \quad \sup \Bigg\{ q > 0 : \frac{M}{\det A} > 1 + \sum_{j=1}^{N} j! \binom{N}{j} \|\Lambda\|^j s^j \\ \text{and } 1 - \sum_{j=1}^{N} j! \binom{N}{j} \|\Lambda\|^j s^j > \frac{1}{2} \text{ for all } s \in [0, q] \Bigg\}. \end{aligned}$$

Further, we choose a number r_2 satisfying $0 < r_2 \leq \min\{r_1, (7\|\Lambda\|)^{-1}\}$; for example, elementary estimates show that we may put

$$r_2 := \min\left\{ (7\|\Lambda\|)^{-1}, (N!(1 + \|\Lambda\|)^N)^{-1}\frac{1}{2}, (N!(1 + \|\Lambda\|)^N)^{-1}\frac{M - \det A}{\det A} \right\}. \quad (2.15)$$

We put $r := r_2/4$, and we choose a C^2-function $\psi : (0, \infty) \to \mathbb{R}$ satisfying $\psi = 1$ on $(0, r)$, $\psi = 0$ on (r_2, ∞) and $-2/r_2 \leq \psi'(x) \leq 0$ for every $x \in (0, \infty)$. We define $f^e : \mathbb{R}^N \to \mathbb{R}^N$ by

$$f^e(x) = \psi(\|x\|)\frac{1}{2}\Xi[x, x] + Ax + a,$$

and we notice that f^e is a C^2-extension of the function f defined in (2.13) from $\mathcal{B}(0, r)$ to $\mathbb{R}^N$. The gradient of f^e is

$$\begin{aligned} \nabla f^e(x) &= \psi'(\|x\|)\frac{1}{2}\Xi[x, x] \otimes \frac{x}{\|x\|} + \psi(\|x\|)\Xi[x, \cdot] + A \\ &= A\left(\frac{1}{2}\psi'(\|x\|)\Lambda[x, x] \otimes \frac{x}{\|x\|} + \psi(\|x\|)\Lambda[x, \cdot] + I\right) \\ &= A\left(\Lambda[x, \cdot]\left(\frac{1}{2}\psi'(\|x\|)x \otimes \frac{x}{\|x\|} + \psi(\|x\|)I\right) + I\right) \\ &= A(\Lambda[x, \cdot]B(x) + I), \end{aligned}$$

where the symmetric matrix $B(x)$ is defined by

$$B(x) = \frac{1}{2}\psi'(\|x\|)x \otimes \frac{x}{\|x\|} + \psi(\|x\|)I$$

and where $\otimes$ denotes the tensor product: $(c \otimes d)e = (d \cdot e)c$ for all $c, d, e \in \mathbb{R}^N$. We now show that the function f^e satisfies (2.14). For $\|x\| \geq r_2$ we have $f^e(x) = Ax + a$ and hence (2.14) is clearly satisfied. For $\|x\| < r$ we have that $\psi(\|x\|) = 1$ and hence

$B(x) = I$. By using (2.11) we obtain the relations

$$\begin{aligned}
\det \nabla f^e(x) &= \det A \det(I + \Lambda[x,\cdot]) \\
&\in \det A \left(1 - \sum_{j=1}^{N} j! \binom{N}{j} \|\Lambda[x,\cdot]\|^j, 1 + \sum_{j=1}^{N} j! \binom{N}{j} \|\Lambda[x,\cdot]\|^j\right) \\
&\subset \det A \left(1 - \sum_{j=1}^{N} j! \binom{N}{j} \|\Lambda\|^j \frac{r_2^j}{4^j}, 1 + \sum_{j=1}^{N} j! \binom{N}{j} \|\Lambda\|^j \frac{r_2^j}{4^j}\right),
\end{aligned}$$

and, in view of (2.15), (2.14) is satisfied since $r = r_2/4 \leq r_1$. Finally, for $r \leq \|x\| \leq r_2$ we notice that the absolute value of each of the eigenvalues of $B(x)$ is less than 1. In fact, for each $\xi \in \mathbb{R}^N$ we have

$$B(x)\xi \cdot \xi = -\frac{1}{2}|\psi'(\|x\|)|\|x\| \left(\frac{x \cdot \xi}{\|x\|}\right)^2 + \psi(\|x\|)\|\xi\|^2;$$

hence

$$B(x)\xi \cdot \xi \leq \psi(\|x\|)\|\xi\|^2 \leq \|\xi\|^2,$$

and

$$B(x)\xi \cdot \xi \geq -\frac{1}{2}|\psi'(\|x\|)|\|x\| \left(\frac{x \cdot \xi}{\|x\|}\right)^2 \geq -\frac{1}{2}\frac{2}{r_2}\|x\|\|\xi\|^2 \geq -\|\xi\|^2.$$

From this bound on the eigenvalues of $B(x)$, the last equation in the proof of Proposition 2 along with (2.12) and (2.11) yield

$$\begin{aligned}
\det \nabla f^e(x) &= \det A \det(I + \Lambda[x,\cdot]B(x)) \\
&\in \det A \left(1 - \sum_{j=1}^{N} |I_j(\Lambda[x,\cdot]B(x))|, 1 + \sum_{j=1}^{N} |I_j(\Lambda[x,\cdot]B(x))|\right) \\
&\subset \det A \left(1 - \sum_{j=1}^{N} j! \binom{N}{j} \|\Lambda[x,\cdot]\|^j, 1 + \sum_{j=1}^{N} j! \binom{N}{j} \|\Lambda[x,\cdot]\|^j\right) \\
&\subset \det A \left(1 - \sum_{j=1}^{N} j! \binom{N}{j} \|\Lambda\|^j r_2^j, 1 + \sum_{j=1}^{N} j! \binom{N}{j} \|\Lambda\|^j r_2^j\right),
\end{aligned}$$

and, keeping in mind that $r_2 \leq r_1$, we conclude that (2.14) is satisfied.

We now show that f^e is injective. Since for $\|x\| > r_2$ the function $x \mapsto f^e(x)$ is injective, it suffices to consider the relation $f^e(x) = f^e(\tilde{x})$ when $\|x\| \leq r_2$ and $\|\tilde{x}\|$ is unrestricted. If $\|x\| \leq r_2$ and $\|\tilde{x}\| \geq r_2$ then $f^e(x) = f^e(\tilde{x})$ implies $\tilde{x} = x + \frac{1}{2}\psi(\|x\|)\Lambda[x,x]$, so that

$$\|x - \tilde{x}\| \leq \frac{1}{2}\|\Lambda\| \|x\|^2 \leq \frac{1}{2}\|\Lambda\| r_2^2 \leq r_2.$$

Therefore, when $\|x\| \leq r_2$ and $\|\tilde{x}\| \geq r_2$, $f^e(x) = f^e(\tilde{x})$ implies that $\|\tilde{x}\| \leq 2r_2$. Thus, it suffices to establish the injectivity of $x \mapsto f^e(x)$ on $\overline{B(0, 2r_2)}$ the closed ball of radius

$2r_2$ centered at 0. Let $y \in \mathbb{R}^N$ be given and define $\xi(x) = x + A^{-1}(y - f^e(x))$ for each $x \in \mathbb{R}^N$. We note that $\xi(x) = x$ if and only if $f^e(x) = y$, so to prove the injectivity of f^e it suffices to show that ξ has at most one fixed point in $\overline{\mathcal{B}(0, 2r_2)}$. The Mean Value Theorem, the inequality $r_2 \leq 1/(7\|\Lambda\|)$ and the estimate

$$\begin{aligned}\|\nabla\xi(x)\| &= \|I - A^{-1}\nabla f^e(x)\| = \left\|\Lambda[x,\cdot]\left(\frac{1}{2}\psi'(\|x\|)x \otimes \frac{x}{\|x\|} + \psi(\|x\|)I\right)\right\| \\ &\leq \|\Lambda\|\|x\|\left(\frac{1}{2}\frac{2}{r_2}\|x\| + 1\right) \leq \|\Lambda\| 2r_2(2+1) = 6r_2\|\Lambda\|\end{aligned}$$

show that $\|\xi(x) - \xi(\tilde{x})\| < \|x - \tilde{x}\|$ for all $x, \tilde{x} \in \overline{\mathcal{B}(0, 2r_2)}$ with $x \neq \tilde{x}$. Hence ξ has at most one fixed point on that ball. Thus, f^e is an injective, C^2-function with $\det \nabla f^e > 0$, and the Inverse Function Theorem tell us that f^e is locally invertible with a local inverse that is of class C^2.

In order to show that $f^e : \mathbb{R}^N \to \mathbb{R}^N$ is a C^2-diffeomorphism, it remains to show that the range of f^e is equal to $\mathbb{R}^N$. Because $f^e(x) = Ax + a$ for all $x \in \mathbb{R}^N \backslash \mathcal{B}(0, r_2)$, the injectivity of f^e and the invertibility of $(x \mapsto Ax + a) : \mathbb{R}^N \to \mathbb{R}^N$ yield

$$f^e(\mathcal{B}(0, r_2)) \subset \mathcal{A}(r_2) := \{Ax + a : |x| < r_2\}$$

and

$$f^e(\mathbb{R}^N \backslash \mathcal{B}(0, r_2)) = \mathbb{R}^N \backslash \mathcal{A}(r_2).$$

However, the inclusion in the first relation above must be equality. In fact, suppose, to the contrary, that $f^e(\mathcal{B}(0, r_2)) \neq \mathcal{A}(r_2)$. Then we may choose $y \in \mathcal{A}(r_2) \backslash f^e(\mathcal{B}(0, r_2))$. If $y \in \partial f^e(\mathcal{B}(0, r_2))$ the continuity and injectivity of f^e imply that $y \in \partial f^e(\mathcal{B}(0, r_2)) = f^e(\partial\mathcal{B}(0, r_2))$, and therefore there exists an $\tilde{x} \in \partial\mathcal{B}(0, r_2)$ such that $f^e(\tilde{x}) = A\tilde{x} + a = y$. But this contradicts the fact that $y \in \mathcal{A}(r_2)$ and that $x \mapsto Ax + a$ is an injective mapping. Thus $y \notin \partial f^e(\mathcal{B}(0, r_2))$ and hence $f^e(\mathcal{B}(0, r_2))$ is closed in $\mathcal{A}(r_2)$, since we can find a small ball centered in y, open in $\mathcal{A}(r_2)$ and which does not intersect $\partial f^e(\mathcal{B}(0, r_2))$. On the other hand, the local invertibility of f^e implies that $f^e(\mathcal{B}(0, r_2))$ is open in $\mathcal{A}(r_2)$. However, $\mathcal{A}(r_2)$ is connected and $f^e(\mathcal{B}(0, r_2))$ is a non-empty open and closed subset of $\mathcal{A}(r_2)$, so $f^e(\mathcal{B}(0, r_2)) = \mathcal{A}(r_2)$, which contradicts our assumption. Thus $f^e(\mathcal{B}(0, r_2)) = \mathcal{A}(r_2)$,

$$f^e(\mathbb{R}^N) = f^e(\mathcal{B}(0, r_2)) \cup f^e(\mathbb{R}^N \backslash \mathcal{B}(0, r_2)) = \mathcal{A}(r_2) \cup \left(\mathbb{R}^N \backslash \mathcal{A}(r_2)\right) = \mathbb{R}^N,$$

and f^e is surjective. □

2.3 Decompositions and identification relations

We recall that the Approximation Theorem for first-order structured deformations and a Gauss-Green formula imply the "identification relation" [Del Piero and Owen (1995), Del Piero, Lecture 5 in this volume]:

$$\begin{aligned}M(x) &:= \nabla g(x) - G(x) \qquad (2.16) \\ &= \lim_{r \to 0} \lim_{m \to \infty} \frac{1}{vol(B(x,r))} \int_{B(x,r) \cap \Gamma(f_m)} [f_m](y) \otimes \nu_{\Gamma(f_m)}(y) \, dA_y\end{aligned}$$

where $\Gamma(f_m)$ denotes the jump-set for f_m, $[f_m](y)$ denotes the jump in f_m at $y \in \Gamma(f_m)$, $\nu_{\Gamma(f_m)}(y)$ denotes the normal at the point $y \in \Gamma(f_m)$, and $\mathcal{A}_y$ denotes N-dimensional area measure. The term "identification relation" is appropriate for the formula (2.16), because it identifies the difference $M(x) = \nabla g(x) - G(x) = \left(\nabla \lim_{m\to\infty} f_m - \lim_{m\to\infty} \nabla f_m\right)|_x$ as a measure of deformation due to a large number of small jumps in f_m. This identification relation justifies calling $\nabla g(x) - G(x)$ the *(first-order) deformation due to (micro)disarrangements* . For a second-order structured deformation (κ, g, G, Σ) we have, by the Approximation Theorem, $\nabla G = \nabla \lim_{m\to\infty} \nabla f_m$ and $\Sigma = \lim_{m\to\infty} \nabla^2 f_m = \lim_{m\to\infty} \nabla(\nabla f_m)$; the results in [Del Piero and Owen (1995)] when applied to ∇f_m, instead of to f_m, then yield a new identification relation:

$$\nabla G(x) - \Sigma(x) = \lim_{r\to 0}\lim_{m\to\infty} \frac{1}{vol(B(x,r))} \int_{B(x,r)\cap\Gamma(\nabla f_m)} [\nabla f_m](y) \otimes \nu_{\Gamma(\nabla f_m)}(y)\, d\mathcal{A}_y. \tag{2.17}$$

Here, for each $A \in \mathrm{Lin}(\mathbb{R}^N)$ and $v \in \mathbb{R}^N$, we define [1] the tensor product $A \otimes v \in \mathrm{Lin}(\mathbb{R}^N, \mathrm{Lin}(\mathbb{R}^N))$ by $(A\otimes v)w = (v\cdot w)A$ for all $w \in \mathbb{R}^N$. The identification relation (2.17) permits us to call $\nabla G(x) - \Sigma(x)$ a *(second-order) deformation due to disarrangements.* If we differentiate both sides of the identification relation (2.16) and add the resulting equation to the identification relation (2.17), we obtain the identification relation

$$\begin{aligned}\nabla^2 g(x) - \Sigma(x) &= \nabla G(x) - \Sigma(x) + \nabla M(x)\\ &= \lim_{r\to 0}\lim_{m\to\infty} \frac{1}{vol(B(x,r))} \int_{B(x,r)\cap\Gamma(\nabla f_m)} [\nabla f_m](y) \otimes \nu_{\Gamma(\nabla f_m)}(y) d\mathcal{A}_y\\ &+\nabla_x \lim_{r\to 0}\lim_{m\to\infty} \frac{1}{vol(B(x,r))} \int_{B(x,r)\cap\Gamma(f_m)} [f_m](y) \otimes \nu_{\Gamma(f_m)}(y) d\mathcal{A}_y.\end{aligned} \tag{2.18}$$

In this relation, we may call $\nabla^2 g(x) - \Sigma(x)$ a (second-order) deformation due to disarrangments, and this difference is revealed as the term $\nabla G(x) - \Sigma(x)$ arising only from jumps in ∇f_m plus the term $\nabla M(x)$ arising only from jumps in f_m. We may now write

$$\nabla g(x) = G(x) + M(x),$$
$$\nabla^2 g(x) = \nabla G(x) - \Sigma(x) + \nabla M(x) + \Sigma(x)$$

and use relation Taylor's formula to obtain the *refined quadratic approximation:*

$$\begin{aligned} g(x) - g(y) &= G(x)(y-x) + M(x)(y-x)\\ &+\frac{1}{2}(\nabla G(x) - \Sigma(x))[y-x, y-x] + \frac{1}{2}\nabla M(x)[y-x, y-x]\\ &+\frac{1}{2}\Sigma(x)[y-x, y-x] + o(|y-x|^2). \end{aligned} \tag{2.19}$$

[1]This definition of tensor product corrects a misprint appearing on p. 232 of Paroni and Owen, 2000, just below equation (46).

In this approximation, the terms $G(x)(y-x)$ and $\frac{1}{2}\Sigma(x)[y-x,y-x]$ represent *translations without disarrangements*, from limits of first gradients ∇f_m and second gradients $\nabla^2 f_m$, respectively. The terms $M(x)(y-x)$, $\frac{1}{2}(\nabla G(x)-\Sigma(x))[y-x,y-x]$, and $\frac{1}{2}\nabla M(x)[y-x,y-x]$ represent *translations due to disarrangements*: $M(x)(y-x)$ and $\frac{1}{2}\nabla M(x)[y-x,y-x]$ arise due to jumps in f_m, whereas $\frac{1}{2}(\nabla G(x)-\Sigma(x))[y-x,y-x]$ arises due to jumps in ∇f_m.

2.4 Two examples

Example 1. ("Bending through simple shears") Let $N=2$ and let $\mathcal{A}=(0,1)\times(0,1)$. For each positive integer n, put $y_i := i/n$ for $i=0,1,\ldots,n$, and $\kappa_n := \{(0,1)\times\{y_i\} : i=1,2,\ldots,n-1\}$. Let $h\in C^2([0,1);\mathbb{R})$, and for $i=0,1,\ldots,n-1$ define

$$m_i := \frac{h(y_{i+1})-h(y_i)}{y_{i+1}-y_i},$$

as well as

$$h_n(y) := \sum_{i=0}^{n-1}\left(m_i(y-y_i)+h(y_i)\right)\chi_{(y_i,y_{i+1})}(y).$$

Finally, we put $f_n(x,y) := (x+h_n(y),y)$ and $g(x,y) := (x+h(y),y)$. It is easily seen that the simple deformation (κ_n, f_n) determines the second-order structured deformation $(\emptyset, g, \nabla g, 0)$. Moreover, M vanishes, and the translations due to disarrangements in (2.19) arise solely from the term $\nabla G-\Sigma=\nabla^2 g$, whose only non-zero component is $\nabla^2 g(x,y)_{122}=h''(y)$. In view of (2.17), the translations due to disarrangments associated with $(\emptyset, g, \nabla g, 0)$ arise through the jumps in ∇f_n, specifically, the jumps in h_n'.

Example 2. ("Bending of a deck of cards")

Let $N=2$ and let $\mathcal{A}=(-1,1)\times(2,3)$. For each positive integer n, put $y_i := 2+\frac{i}{n}$ for $i=0,1,\ldots,n$, and $\kappa_n := \{(-1,1)\times\{y_i\} : i=1,2,\ldots,n-1\}$. We define

$$\rho(y) := \sqrt{1+y^2} \tag{2.20}$$

and

$$f_n(x,y) := \frac{\rho(y)}{\rho([ny]/n)}(x,\sqrt{\rho([ny]/n)^2-x^2}) \tag{2.21}$$

for every $(x,y)\in(-1,1)\times(2,3)$. Here, $[\cdot]$ denotes the greatest integer function, and relation (2.21) tells us that lines $y=y_0$ are mapped by f_n into circular arcs with center at the origin and radius $\rho(y_0)$, while lines $x=x_0$ are mapped into a collection of line segments, each of which, when extended, passes through the origin. The figure shows the effect of the mapping f_n on rectangles of the form $(-1,1)\times(y_i,y_{i+1})$: each is mapped into an annular region with inner radius $\rho(y_i)$ and outer radius $\rho(y_{i+1})$. (This deformation is reminiscent of the geometrical changes in a deck of cards when it is bent prior to shuffling.) Because $0\le y-[ny]/n\le 1/n$, it follows that $\rho([ny]/n)$ converges to $\rho(y)$ as n tends to infinity, uniformly in y. Consequently, the sequence $n\mapsto f_n$ converges uniformly on $(-1,1)\times(2,3)$ to the function g defined by

$$g(x,y) := (x,\sqrt{1+y^2-x^2}), \tag{2.22}$$

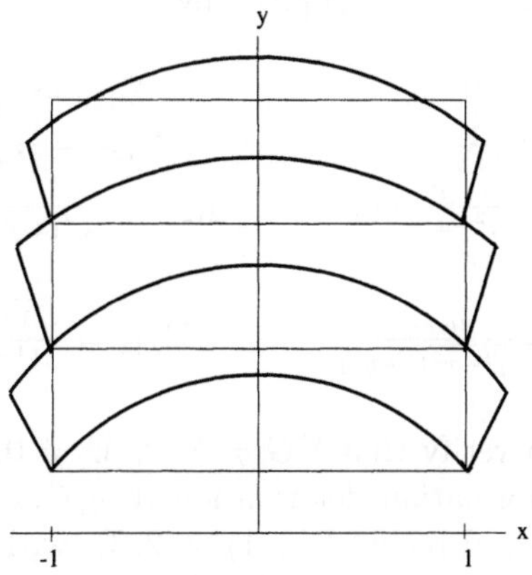

Figure 1. Microview of the deformation f_n of Example 2

and the gradients

$$\nabla f_n(x,y) = \begin{bmatrix} \frac{\rho(y)}{\rho([ny]/n)} & \frac{xy}{\rho([ny]/n)\rho(y)} \\ -\frac{\rho(y)x}{\rho([ny]/n)\sqrt{\rho([ny]/n)^2-x^2}} & \frac{y}{\rho([ny]/n)\rho(y)}\sqrt{\rho([ny]/n)^2-x^2} \end{bmatrix}$$

converge uniformly to

$$G(x,y) := \begin{bmatrix} 1 & \frac{xy}{1+y^2} \\ -\frac{x}{\sqrt{1+y^2-x^2}} & \frac{y\sqrt{1+y^2-x^2}}{1+y^2} \end{bmatrix}. \tag{2.23}$$

We easily find that

$$\nabla g(x,y) = \begin{bmatrix} 1 & 0 \\ -\frac{x}{\sqrt{1+y^2-x^2}} & \frac{y}{\sqrt{1+y^2-x^2}} \end{bmatrix} \tag{2.24}$$

and, therefore, on $(-1,1)\times(2,3)$, we have

$$\frac{2}{3} \leq \frac{y}{\sqrt{1+y^2-x^2}} = \det G(x,y) = \det \nabla g(x,y).$$

Consequently, the inequalities in (Std 3) of Definition 3 are satisfied by G and ∇g. Finally, the second gradients $\nabla^2 f_n$ are easily shown to converge uniformly to the tensor field Σ having components Σ_{ijk} at (x, y) given by

$$\Sigma_{111} = 0, \qquad \Sigma_{112} = \frac{y}{1+y^2},$$
$$\Sigma_{121} = \frac{y}{1+y^2}, \qquad \Sigma_{122} = \frac{x}{(1+y^2)^2},$$
$$\Sigma_{211} = -\frac{1+y^2}{(1+y^2-x^2)^{3/2}}, \qquad \Sigma_{212} = -\frac{xy}{\sqrt{1+y^2-x^2}(1+y^2)},$$

$$\Sigma_{221} = -\frac{xy}{\sqrt{1+y^2-x^2}(1+y^2)}, \qquad \Sigma_{222} = \frac{\sqrt{1+y^2-x^2}}{(1+y^2)^2},$$

and it is straightforward to verify that $\nabla G \neq \Sigma$, $\nabla M \neq 0$, and $\nabla^2 g \neq \Sigma$. Thus, the second-order structured deformation deformation $(\emptyset, g, G, \Sigma)$ represents a macroscopic bending of the two-dimensional region $(-1, 1) \times (2, 3)$, with accompanying disarrangements occuring due both to jumps in f_n and in ∇f_n.

3 The SBV^2 and SBH setting

In this Section we consider two frameworks for second-order structured deformations that encompass modern techniques in analysis. The material presented in this Section is mainly taken from [Fonseca et al. (to appear)].

3.1 Preliminaries

In what follows $\mathcal{L}^N$ and $\mathcal{H}^{N-1}$ are, respectively, the N dimensional Lebesgue measure and the $N-1$ dimensional Hausdorff measure in $\mathbb{R}^N$, and Ω is a bounded open subset of $\mathbb{R}^N$. When there is no possibility of confusion, if $E \subset \mathbb{R}^N$ is a Lebesgue measurable set then we abbreviate $|E| := \mathcal{L}^N(E)$. We denote by $\mathbb{R}^{N\times N}_{\text{sym}}$ the space of all symmetric $N \times N$ matrices endowed with the norm

$$|A| := \sum_{i,i=1}^{N} |A_{ij}| \quad \text{where } A = (A_{ij}) \quad i, j = 1, \cdots, N,$$

i.e., $A_{ij} = A\mathbf{e}_i \cdot \mathbf{e}_j$ where $\{\mathbf{e}_1, \ldots, \mathbf{e}_N\}$ is the standard orthonormal basis of $\mathbb{R}^N$.

We review briefly some facts about functions of bounded variation which will be useful in the sequel. A function $u \in L^1(\Omega; \mathbb{R}^d)$ is said to be of *bounded variation*, and we write $u \in BV(\Omega; \mathbb{R}^d)$, if its distributional derivative Du is a finite matrix–valued Radon measure.

Given $u \in BV(\Omega; \mathbb{R}^d)$ the *jump set* of u, or *singular set*, is defined as the set of points $x \in \Omega$ which are not Lebesgue points for u. It is well known that $S(u)$ is $N-1$ rectifiable, i.e.,

$$S(u) = \bigcup_{n=1}^{\infty} K_n \cup E,$$

where $\mathcal{H}^{N-1}(E) = 0$ and K_n is a compact subset of a C^1 hypersurface. It can be shown that for $\mathcal{H}^{N-1}$ a.e. $x \in S(u)$ there exist a unit vector $\nu_u(x) \in S^{N-1}$, normal to $S(u)$ at

x, and two vectors $u^-(x),\ u^+(x) \in \mathbb{R}^d$ (the traces of u on $S(u)$ at the point x) such that

$$\lim_{\varepsilon\to 0}\frac{1}{\varepsilon^N}\int_{\{y\in Q(x_0,\varepsilon):\,(y-x)\cdot\nu_u(x)>0\}}|u(y)-u^+(x)|^{N/(N-1)}dy=0$$

and

$$\lim_{\varepsilon\to 0}\frac{1}{\varepsilon^N}\int_{\{y\in Q(x_0,\varepsilon):\,(y-x)\cdot\nu_u(x)<0\}}|u(y)-u^-(x)|^{N/(N-1)}dy=0.$$

We denote the *jump of* u *across* $S(u)$ by $[u] := u^+ - u^-$. The distributional derivative Du may be decomposed as

$$Du = \nabla u\,\mathcal{L}^N\lfloor\Omega + [u]\otimes\nu_u\,\mathcal{H}^{N-1}\lfloor S(u) + C(u), \tag{3.1}$$

where ∇u is the density of the absolutely continuous part of Du with respect to $\mathcal{L}^N$ and $C(u)$ is the Cantor part of Du.

The space $SBV(\Omega;\mathbb{R}^d)$ of *special functions of bounded variation*, introduced by De Giorgi, and Ambrosio (1988), is the space of all functions $u \in BV(\Omega;\mathbb{R}^d)$ such that

$$Du = \nabla u\,\mathcal{L}^N\lfloor\Omega + [u]\otimes\nu_u\,\mathcal{H}^{N-1}\lfloor S(u).$$

We refer to Ambrosio et al. (2000) for more details on SBV functions.

We now introduce the space of *functions of bounded Hessian*

$$\begin{aligned}BH(\Omega;\mathbb{R}) :&= \left\{u\in W^{1,1}(\Omega;\mathbb{R}) : D^2u \text{ is a finite Radon measure}\right\}\\ &= \left\{u\in W^{1,1}(\Omega;\mathbb{R}) : \nabla u\in BV(\Omega;\mathbb{R}^N)\right\},\end{aligned}$$

where D^2u denotes the distributional Hessian of u.
In view of (3.1) now applied to ∇u in place of u, we may decompose D^2u as

$$D^2u = \nabla^2u\,\mathcal{L}^N\lfloor\Omega + [\nabla u]\otimes\nu_{\nabla u}\,\mathcal{H}^{N-1}\lfloor S(\nabla u) + C(\nabla u),$$

where $\nabla^2 u$ is the density of the absolutely continuous part of D^2u with respect to $\mathcal{L}^N$, i.e., $\nabla^2u = \nabla(\nabla u)$, $[\nabla u] := (\nabla u)^+ - (\nabla u)^-$, and $C(\nabla u)$ is the Cantor part of D^2u.

Since D^2u is a symmetric distribution, we then have that

$$\nabla^2u(x)\in\mathbb{R}^{N\times N}_{\text{sym}},\quad (\nabla u)^+ - (\nabla u)^- = a\,\nu_{\nabla u} \tag{3.2}$$

for some $a \in L^1_{\mathcal{H}^{N-1}\lfloor S(\nabla u)}(\Omega;\mathbb{R})$. The latter equality results immediately from the fact that $\left((\nabla u)^+ - (\nabla u)^-\right)\otimes\nu_{\nabla u}$ must be a symmetric rank-one matrix. Note also that if u is in $W^{2,1}(\Omega;\mathbb{R})$ then $D^2u = \nabla^2u\,\mathcal{L}^N\lfloor\Omega$ with

$$\nabla^2 u = \left(\frac{\partial^2 u}{\partial x_i\partial x_j}\right),\quad i,j = 1,\cdots,N.$$

The space $SBH(\Omega;\mathbb{R})$ of *special functions of bounded Hessian*, cf. [Temam (1983); Demengel (1984, 1989); Carriero et al. (1992)], is the space of all functions of bounded Hessian such that $\nabla u \in SBV(\Omega;\mathbb{R}^N)$, i.e., $u\in BH(\Omega;\mathbb{R})$ and

$$D^2u = \nabla^2u\,\mathcal{L}^N\lfloor\Omega + [\nabla u]\otimes\nu_{\nabla u}\,\mathcal{H}^{N-1}\lfloor S(\nabla u).$$

Finally we define the space $SBV^2(\Omega;\mathbb{R})$, cf. Carriero et al. (1996), as

$$SBV^2(\Omega;\mathbb{R}) := \{u \in SBV(\Omega;\mathbb{R}) : \nabla u \in SBV(\Omega;\mathbb{R}^N)\}.$$

Clearly $SBH(\Omega;\mathbb{R}) \subset SBV^2(\Omega;\mathbb{R})$, and if $u \in SBH(\Omega;\mathbb{R})$ then $Du = \nabla u$, while if $u \in SBV^2(\Omega;\mathbb{R})$ then Du is given by (3.1) and, again writing $\nabla^2 u = \nabla(\nabla u)$,

$$D\nabla u = \nabla^2 u\,\mathcal{L}^N \lfloor \Omega + [\nabla u] \otimes \nu_{\nabla u}\,\mathcal{H}^{N-1} \lfloor S(\nabla u).$$

As shown in Theorem 3.2 below, if $u \in SBV^2(\Omega;\mathbb{R})$ then the symmetry of $\nabla^2 u$ is no longer guaranteed.

3.2 The SBV^2 setting

In this section we introduce the concept of second-order structured deformations in the framework of SBV^2 functions and we prove the appropriate Approximation Theorem.

Definition 3.1. The space of second-order structured deformations (in the SBV^2 setting) consists of triples (g, G, Σ) where

$$g \in SBV^2(\Omega;\mathbb{R}^N), \quad G \in SBV(\Omega;\mathbb{R}^{N\times N}), \quad \Sigma \in L^1(\Omega;\mathbb{R}^{N\times N\times N}).$$

Note that no symmetry requirement is placed on Σ. The main result of this section is the following Approximation Theorem.

Theorem 3.2. *Let (g, G, Σ) be a second-order structured deformation (in the SBV^2 setting). Then there exists a sequence $n \mapsto f_n \in SBV^2(\Omega;\mathbb{R}^N)$ such that*

$$f_n \to g \text{ in } L^1(\Omega;\mathbb{R}^N), \quad \nabla f_n \to G \text{ in } L^1(\Omega;\mathbb{R}^{N\times N}), \quad \nabla^2 f_n = \Sigma.$$

Choksi and Fonseca (1997) proved the Approximation Theorem for first-order structured deformations (in the SBV setting) by means of the following result of Alberti (1991):

Theorem 3.3. *Let Ω be an open subset of $\mathbb{R}^N$ and let $F \in L^1(\Omega;\mathbb{R}^N)$. Then there exists $u \in SBV(\Omega;\mathbb{R})$ and a constant $C > 0$, depending only on N, such that*

$$Du = F\,\mathcal{L}^N \lfloor \Omega + [u]\nu_u\,\mathcal{H}^{N-1} \lfloor S(u),$$

and

$$\int_\Omega |u|\,dx + \int_{S(u)\cap\Omega} |[u]|\,d\mathcal{H}^{N-1} \le C \int_\Omega |F|\,dx.$$

Here $\nu_u(x) \in S^{N-1}$ is the unit normal to the jump set $S(u)$ at x.

Analoguosly, we prove the Approximation Theorem 3.2 by means of Theorem 3.4, below, which states that any function $\Xi \in L^1(\Omega;\mathbb{R}^{N\times N})$, not necessarily symmetric valued, is the the density with respect to the N-dimensional Lebesgue measure of the distributional Hessian of a function in $SBV^2(\Omega;\mathbb{R})$.

Theorem 3.4. *Let Ω be an open subset of $\mathbb{R}^N$ and let $\Xi \in L^1\left(\Omega; \mathbb{R}^{N\times N}\right)$. Then there exist $u \in SBV^2(\Omega;\mathbb{R})$ and a constant $C > 0$, depending only on N, such that*

$$\nabla^2 u(x) = \Xi(x) \ \textit{for}\ \mathcal{L}^N \ \textit{a.e.}\ x \in \Omega,$$

and

$$\int_\Omega (|u| + |\nabla u|)\, dx + \int_{(S(u)\cup S(\nabla u))\cap\Omega} (|[u]| + |[\nabla u]|)\, d\mathcal{H}^{N-1} \leq C \int_\Omega |\Xi|\, dx.$$

PROOF. The proof is straightforward and it may be obtained by applying Theorem 3.3 twice. Indeed, for every $i = 1, \ldots, N$, define

$$F_i := \sum_{j=1}^{N} \Xi_{ij}\mathbf{e}_j.$$

Since $F_i \in L^1\left(\Omega;\mathbb{R}^N\right)$, by Theorem 3.3 there exist $v_i \in SBV(\Omega;\mathbb{R})$ and a constant $C = C(N) > 0$ such that

$$Dv_i = F_i\, \mathcal{L}^N \lfloor \Omega + [v_i]\nu_{v_i}\, \mathcal{H}^{N-1}\lfloor S(v_i),$$

with

$$\int_\Omega |v_i|\, dx + \int_{S(v_i)\cap\Omega} |[v_i]|\, d\mathcal{H}^{N-1} \leq C \int_\Omega |F_i|\, dx.$$

Set $v := \sum_{i=1}^N v_i\mathbf{e}_i$. Clearly $v \in SBV\left(\Omega;\mathbb{R}^N\right)$, and we have

$$Dv = \Xi\, \mathcal{L}^N \lfloor \Omega + [v]\otimes \nu_v\, \mathcal{H}^{N-1}\lfloor S(v), \tag{3.3}$$

with

$$\int_\Omega |v|\, dx + \int_{S(v)\cap\Omega} |[v]|\, d\mathcal{H}^{N-1} \leq C \int_\Omega |\Xi|\, dx. \tag{3.4}$$

Now, since $v \in L^1\left(\Omega;\mathbb{R}^N\right)$, and invoking again Theorem 3.3, there exists a function $u \in SBV(\Omega;\mathbb{R})$ such that

$$Du = v\, \mathcal{L}^N \lfloor \Omega + [u]\otimes \nu_u\, \mathcal{H}^{N-1}\lfloor S(u), \tag{3.5}$$

and

$$\int_\Omega |u|\, dx + \int_{S(u)\cap\Omega} |[u]|\, d\mathcal{H}^{N-1} \leq C \int_\Omega |v|\, dx. \tag{3.6}$$

Therefore we conclude that $\nabla u = v\, \mathcal{L}^N$ a.e. $x \in \Omega$, thus $u \in SBV^2(\Omega;\mathbb{R})$, and (3.3)-(3.6) assert that u fulfills the statement of the theorem. □

We are now ready to prove the Approximation Theorem 3.2.

PROOF OF THEOREM 3.2 For every $i = 1, \ldots, N$ define

$$G^i := \sum_{j=1}^{N} G_{ij}\mathbf{e}_j, \qquad \Sigma^i := \sum_{j,k,=1}^{N} \Sigma_{ijk}\mathbf{e}_j \otimes \mathbf{e}_k.$$

Since $\Sigma^i \in L^1(\Omega;\mathbb{R}^{N\times N})$ by Theorem 3.4 there exist $\sigma^i \in SBV^2(\Omega;\mathbb{R})$ such that

$$\nabla^2\sigma^i = \Sigma^i.$$

Let $n \mapsto \tilde{u}^i_n$ be a sequence of continuous piecewise affine functions such that

$$\nabla\tilde{u}^i_n \to G^i - \nabla\sigma^i \quad \text{in } L^1(\Omega;\mathbb{R}^N),$$

and let $n \mapsto \bar{u}^i_n$ be a sequence of piecewise constant functions such that

$$\|\bar{u}^i_n - (g_i - \sigma^i - \tilde{u}^i_n)\|_{L^1} \le \frac{1}{n}.$$

Set $u^i_n := \sigma^i + \tilde{u}^i_n + \bar{u}^i_n$ and $u_n := \sum_{i=1}^N u^i_n \mathbf{e}_i$. Clearly $u_n \in SBV^2(\Omega,\mathbb{R}^N)$, moreover taking into account that $\nabla\bar{u}^i_n = 0$ and that $\nabla^2\tilde{u}^i_n = 0$, $\mathcal{L}^N$-a.e. we find

$$\begin{aligned}
\|u_n - g\|_{L^1} &= \sum_{i=1}^N \|u^i_n - g_i\|_{L^1} = \sum_{i=1}^N \|\bar{u}^i_n - (g_i - \sigma^i - \tilde{u}^i_n)\|_{L^1} \le \frac{N}{n},\\
\|\nabla u_n - G\|_{L^1} &= \sum_{i=1}^N \|\nabla u^i_n - G^i\|_{L^1} = \sum_{i=1}^N \|\nabla\sigma^i - \nabla\tilde{u}^i_n - G^i\|_{L^1} \to 0,
\end{aligned}$$

and

$$(\nabla^2 u_n)_{ijk} = (\nabla^2 u^i_n)_{jk} = (\nabla^2\sigma^i)_{jk} = (\Sigma^i)_{jk} = \Sigma_{ijk},$$

$\mathcal{L}^N$-a.e.. □

3.3 The *SBH* setting

Here we restrict our attention to deformations belonging to SBH, which is a subset of SBV^2. As a consequence the Hessian matrix $\nabla^2 f$ of a function $f \in SBH(\Omega;\mathbb{R})$, is symmetric valued. Roughly speaking, a second-order structured deformation in the SBH framework is a second-order structured deformation (g, G, Σ) in the SBV^2 setting in which g has no jump-discontinuities and $G = \nabla g$.

Definition 3.5. The space of second-order structured deformations (in the SBH setting) consists of pairs (g,Σ) where

$$g \in SBH(\Omega;\mathbb{R}^N), \quad \Sigma \in L^1(\Omega;\mathbb{S}^{N\times N\times N}),$$

where $\mathbb{S}^{N\times N\times N} = \{\Lambda \in \mathbb{R}^{N\times N\times N} : \Lambda_{ijk} = \Lambda_{ikj}, \forall i,j,k = 1,\dots,N\}$.

The following Approximation Theorem is then valid.

Theorem 3.6. *Let (g,Σ) be a second-order structured deformation (in the SBH setting). Then there exist a sequence $n \mapsto u_n \in SBH(\Omega;\mathbb{R}^N)$ such that*

$$u_n \to g, \ \ in\ W^{1,1}(\Omega;\mathbb{R}^N), \quad \nabla^2 u_n = \Sigma.$$

Again, to prove the theorem above we shall use an appropriate generalization of Alberti's result. Indeed, in the next section we shall prove the following

Theorem 3.7. *Let Ω be an open subset of $\mathbb{R}^N$ and let $\Xi \in L^1\left(\Omega;\mathbb{R}^{N\times N}_{\text{sym}}\right)$. Then there exists $u \in BH(\Omega;\mathbb{R})$ and a constant $C>0$ depending only on N such that*

$$D^2u = \Xi\,\mathcal{L}^N\lfloor\Omega + [\nabla u]\otimes\nu_{\nabla u}\,\mathcal{H}^{N-1}\lfloor S(\nabla u),$$

and

$$\int_\Omega (|u|+|\nabla u|)\,dx + \int_{S(\nabla u)\cap\Omega} |[\nabla u]|\,d\mathcal{H}^{N-1} \le C\int_\Omega |\Xi|\,dx.$$

With this theorem in hand the proof of the Approximation Theorem 3.6 easily follows.

Proof of Theorem 3.6. For every $i=1,\dots,N$ define

$$\Sigma^i := \sum_{j,k=1}^N \Sigma_{ijk}\mathbf{e}_j\otimes\mathbf{e}_k.$$

Since $\Sigma^i \in L^1(\Omega;\mathbb{R}^{N\times N}_{\text{sym}})$ by Theorem 3.7 there exist $\sigma^i \in SBH(\Omega;\mathbb{R})$ such that

$$\nabla^2\sigma^i = \Sigma^i.$$

Let $n\mapsto \tilde{u}^i_n$ be a sequence of continuous piecewise affine functions such that

$$\tilde{u}^i_n \to g^i - \sigma^i, \qquad \text{in } W^{1,1}(\Omega;\mathbb{R}).$$

Set $u^i_n := \sigma^i + \tilde{u}^i_n$ and $u_n := \sum_{i=1}^N u^i_n\mathbf{e}_i$. Clearly $u_n \in SBH(\Omega;\mathbb{R}^N)$, $u_n \to g$ in $W^{1,1}(\Omega;\mathbb{R}^N)$ and $\nabla^2 u^i_n = \nabla^2\sigma^i$, $\mathcal{L}^N$-a.e.. □

Proof of Theorem 3.7. Guided by Alberti's work [Alberti (1991)], the proof of Theorem 3.7 will be a direct consequence of a Lusin-type result for second-order derivatives. Specifically, we have

Lemma 3.8. *Let $\Xi \in L^1\left(\mathbb{R}^N;\mathbb{R}^{N\times N}_{\text{sym}}\right)$ and let $\eta>0$. Then there exist $u\in SBH\left(\mathbb{R}^N;\mathbb{R}\right)$ and a constant $C>0$, depending only on N, such that*

$$\int_{\mathbb{R}^N}\left(|u|+|\nabla u|+\left|\Xi-\nabla^2 u\right|\right)dx \le \eta, \tag{3.7}$$

$$\int_{S(\nabla u)} |[\nabla u]|\,d\mathcal{H}^{N-1} \le C\int_{\mathbb{R}^N}|\Xi|\,dx. \tag{3.8}$$

Assuming that the lemma holds, we now prove Theorem 3.7.

Proof of Theorem 3.7. We proceed as in [Alberti (1991)] to construct by recursively two sequences $n\mapsto u_n \subset SBH\left(\mathbb{R}^N;\mathbb{R}\right)$ and $n\mapsto \Xi_n \subset L^1\left(\mathbb{R}^N;\mathbb{R}^{N\times N}_{\text{sym}}\right)$ as follows:

set $u_0 := 0$ and $\Xi_0 := \Xi$, and for given $n \in N$ suppose that $u_{n-1} \in SBH\left(\mathbb{R}^N;\mathbb{R}\right)$ and $\Xi_{n-1} \in L^1\left(\mathbb{R}^N;\mathbb{R}^{N\times N}_{\text{sym}}\right)$ have been selected. By Lemma 3.8 there exists $u_n \in SBH\left(\mathbb{R}^N;\mathbb{R}\right)$ such that

$$\int_{\mathbb{R}^N} \left(|u_n| + |\nabla u_n| + \left|\Xi_{n-1} - \nabla^2 u_n\right|\right) dx \le \frac{1}{2^n}\int_{\mathbb{R}^N} |\Xi_{n-1}|\, dx,$$

$$\int_{S(\nabla u_n)} |[\nabla u_n]|\, d\mathcal{H}^{N-1} \le C\int_{\mathbb{R}^N} |\Xi_{n-1}|\, dx.$$

Set $\Xi_n := \Xi_{n-1} - \nabla^2 u_n$. We claim that the function

$$u := \sum_{n=1}^{\infty} u_n$$

satisfies the requirements. Indeed, note that

$$\int_{\mathbb{R}^N} |\Xi_n|\, dx \le \frac{1}{2^{\alpha_n}}\int_{\mathbb{R}^N} |\Xi|\, dx \quad \text{for all} \quad n \in,$$

where $\alpha_1 := 1$, $\alpha_n := \alpha_{n-1} + n$ for $n \ge 2$. Thus

$$\begin{aligned}\int_{\mathbb{R}^N} \left(|u| + |\nabla u|\right) dx &\le \sum_{n=1}^{\infty}\int_{\mathbb{R}^N} \left(|u_n| + |\nabla u_n|\right) dx\\ &\le \sum_{n=1}^{\infty}\frac{1}{2^n}\int_{\mathbb{R}^N} |\Xi_{n-1}|\, dx\\ &\le \int_{\mathbb{R}^N} |\Xi|\, dx,\end{aligned}$$

and, similarly,

$$\int_{S(\nabla u)} |[\nabla u]|\, d\mathcal{H}^{N-1} \le C\sum_{n=1}^{\infty}\int_{\mathbb{R}^N} |\Xi_{n-1}|\, dx \le C\int_{\mathbb{R}^N} |\Xi|\, dx.$$

Basic properties of $SBH\left(\mathbb{R}^N;\mathbb{R}\right)$ and the definitions of $n \mapsto u_n$ and $n \mapsto \Xi_n$ imply directly that $\nabla^2 u(x) = \Xi(x)$ for $\mathcal{L}^N$ a.e. $x \in \mathbb{R}^N$, and this concludes the proof. □

We now prove Lemma 3.8 in the two dimensional case $N = 2$, where the main ideas may be easily introduced with the help of figures.

Proof of Lemma 3.8, N=2. Fix $h > 0$, $i, j \in \mathbb{Z}$, and denote by $A^h_{i,j}$ the lattice points

$$A^h_{i,j} := ih\mathbf{e_1} + jh\mathbf{e_2} = (ih, jh) \ \text{ with } i, j \in \mathbb{Z},$$

$L^h_{i,j}$ is the closed triangle whose vertices are the points $A^h_{i,j}$, $A^h_{i+1,j}$ and $A^h_{i,j+1}$, and $U^h_{i,j}$ denotes the closed triangle whose vertices are $A^h_{i,j}$, $A^h_{i-1,j}$ and $A^h_{i,j-1}$. Clearly the union of all triangles $L^h_{i,j}$ and $U^h_{i,j}$ forms a triangulation τ^h of $\mathbb{R}^2$ (see Figure 2). Let

$$\mathbf{t} := \frac{1}{\sqrt{2}}(\mathbf{e_2} - \mathbf{e_1}) = \frac{(-1,1)}{\sqrt{2}}$$

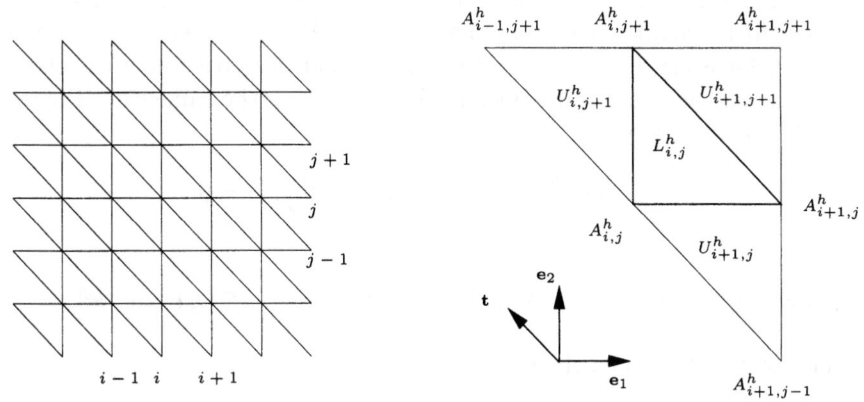

Figure 2. Triangulation for $N = 2$

where $\{\mathbf{e_1}, \mathbf{e_2}\}$ denotes an orthonormal basis of $\mathbb{R}^2$.

STEP 1. Assume first that $\Xi \in C_c\left(\mathbb{R}^2; \mathbb{R}^{2\times 2}_{\text{sym}}\right)$. Let $\ell > 0$ be so large that $\Xi(x) = 0$ outside the square $(-\ell/2, \ell/2)^2$. Fix $\eta > 0$ and let $h > 0$ be sufficiently small such that

$$|z - z'| < 2h \Rightarrow |\Xi(z) - \Xi(z')| < \frac{\eta}{\ell^2}. \tag{3.9}$$

Consider a triangulation τ^h as described above, and define

$$F^h_{A^h_{i,j},A^h_{i+1,j}} := \frac{1}{h^2}\left(\int_{L^h_{i,j}\cup U^h_{i+1,j}} \Xi\, dx\right)\mathbf{e_1}\cdot\mathbf{e_1} = \frac{1}{h^2}\int_{L^h_{i,j}\cup U^h_{i+1,j}} \Xi_{11}\, dx,$$

$$F^h_{A^h_{i,j},A^h_{i,j+1}} := \frac{1}{h^2}\left(\int_{L^h_{i,j}\cup U^h_{i,j+1}} \Xi\, dx\right)\mathbf{e_2}\cdot\mathbf{e_2} = \frac{1}{h^2}\int_{L^h_{i,j}\cup U^h_{i,j+1}} \Xi_{22}\, dx,$$

$$\begin{aligned} F^h_{A^h_{i,j+1},A^h_{i+1,j}} &:= \frac{1}{h^2}\left(\int_{L^h_{i,j}\cup U^h_{i+1,j+1}} \Xi\, dx\right)\mathbf{t}\cdot\mathbf{t} \\ &= \frac{1}{h^2}\int_{L^h_{i,j}\cup U^h_{i+1,j+1}} \left(\frac{\Xi_{11}+\Xi_{22}}{2} - \Xi_{12}\right) dx. \end{aligned}$$

If the function Ξ was the Hessian of a function v, i.e., $\Xi = \nabla^2 v$, then $F^h_{A^h_{i,j},A^h_{i+1,j}}$ would be the average over $L^h_{i,j} \cup U^h_{i+1,j}$ of the second derivative of v calculated along the direction of the segment $\overrightarrow{A^h_{i,j}A^h_{i+1,j}}$, i.e., $\partial^2 v/\partial x^2$ (note that the two triangles have as a common side the segment $\overrightarrow{A^h_{i,j}A^h_{i+1,j}}$). A similar interpretation holds for $F^h_{A^h_{i,j},A^h_{i,j+1}}$ and $F^h_{A^h_{i,j+1},A^h_{i+1,j}}$. We now define a function u which is continuous, piecewise quadratic (u is quadratic on any triangle $L^h_{i,j}$ and $U^h_{i,j}$), and whose Hessian is close to Ξ in L^1 norm.

On each triangle $L^h_{i,j}$ and $U^h_{i,j}$ the function u shall have the form

$$u(x, y) = \frac{1}{2}a_1x^2 + a_2xy + \frac{1}{2}a_3y^2 + a_4x + a_5y + a_6,$$

where the constants $a_1, \dots, a_6$ may vary from triangle to triangle. Since we want the function u to be small, continuous, and with its Hessian close to Ξ, we require that $u(A^h_{i,j}) = 0$ for every $i, j \in \mathbb{Z}$, and that the second derivative of u calculated along a side of a triangle be equal to the mean value of Ξ along the same side. Precisely, in the triangle $L^h_{i,j}$ we prescribe

$$u(A^h_{i,j}) = u(A^h_{i+1,j}) = u(A^h_{i,j+1}) = 0 \tag{3.10}$$

and

$$\nabla^2 u\,\mathbf{e_1} \cdot \mathbf{e_1} = F^h_{A^h_{i,j},A^h_{i+1,j}}, \quad \nabla^2 u\,\mathbf{e_2} \cdot \mathbf{e_2} = F^h_{A^h_{i,j},A^h_{i,j+1}} \quad \nabla^2 u\,\mathbf{t} \cdot \mathbf{t} = F^h_{A^h_{i,j+1},A^h_{i+1,j}}.$$

A simple computation then shows that

$$\begin{aligned} u\,|_{L^h_{i,j}}\,(x,y) = \frac{1}{2} a_1^{i,j}(x-ih)(x-(i+1)h) + a_2^{i,j}(x-ih)(y-jh) \\ + \frac{1}{2} a_3^{i,j}(y-jh)(y-(j+1)h), \end{aligned} \tag{3.11}$$

where the constants a_i, $i = 1, \dots, 6$, are given by

$$\begin{cases} a_1^{i,j} := F^h_{A^h_{i,j},A^h_{i+1,j}}, \\ a_2^{i,j} := \frac{1}{2}\left(a_1^{i,j} + a_3^{i,j}\right) - F^h_{A^h_{i,j+1},A^h_{i+1,j}}, \\ a_3^{i,j} := F^h_{A^h_{i,j},A^h_{i,j+1}}. \end{cases}$$

Clearly the function u is continuous, since it is continuous on each triangle and its values on the boundary of each triangle are assigned via (3.10) and by fixing the second order derivative of u along the sides. Moreover, using (3.11) we have

$$\int_{L^h_{i,j}} |u|\,dx \le Ch^2 |L^h_{i,j}|(|a_1^{i,j}| + |a_2^{i,j}| + |a_3^{i,j}|) \le Ch^2 \int_{B^h_{i,j}} |\Xi|\,dx,$$

where $B^h_{i,j} := L^h_{i,j} \cup U^h_{i,j+1} \cup U^h_{i+1,j+1} \cup U^h_{i+1,j}$. Similarly, we find

$$\int_{L^h_{i,j}} |\nabla u|\,dx \le Ch \int_{B^h_{i,j}} |\Xi|\,dx.$$

Since

$$\begin{aligned} (\nabla^2 u)_{12} - \Xi_{12} &= \left(\frac{\Xi_{11} + \Xi_{22}}{2} - \Xi_{12}\right) - \left(\frac{(\nabla^2 u)_{11} + (\nabla^2 u)_{22}}{2} - (\nabla^2 u)_{12}\right) \\ &\quad + \frac{(\nabla^2 u)_{11} - \Xi_{11}}{2} + \frac{(\nabla^2 u)_{22} - \Xi_{22}}{2} \\ &= \Xi\mathbf{t} \cdot \mathbf{t} - \nabla^2 u\,\mathbf{t} \cdot \mathbf{t} + \frac{(\nabla^2 u)_{11} - \Xi_{11}}{2} + \frac{(\nabla^2 u)_{22} - \Xi_{22}}{2} \\ &= \Xi\mathbf{t} \cdot \mathbf{t} - F^h_{A^h_{i,j+1},A^h_{i+1,j}} + \frac{F^h_{A^h_{i,j},A^h_{i+1,j}} - \Xi_{11}}{2} + \frac{F^h_{A^h_{i,j},A^h_{i,j+1}} - \Xi_{22}}{2}, \end{aligned}$$

by virtue of (3.9) we have

$$\begin{aligned}\int_{L^h_{i,j}} |\Xi - \nabla^2 u|\, dx &= \int_{L^h_{i,j}} \left(|\Xi_{11} - (\nabla^2 u)_{11}| + |\Xi_{22} - (\nabla^2 u)_{22}| + 2|\Xi_{12} - (\nabla^2 u)_{12}|\right) dx \\ &\leq C \int_{L^h_{i,j}} \left(\left|\Xi \mathbf{e_1} \cdot \mathbf{e_1} - F^h_{A^h_{i,j}, A^h_{i+1,j}}\right| + \left|\Xi \mathbf{e_2} \cdot \mathbf{e_2} - F^h_{A^h_{i,j}, A^h_{i,j+1}}\right| \right. \\ &\qquad \left. + \left|\Xi \mathbf{t} \cdot \mathbf{t} - F^h_{A^h_{i,j+1}, A^h_{i+1,j}}\right| \right) dx \\ &\leq C \frac{\eta}{\ell^2} |L^h_{i,j}|.\end{aligned}$$

Clearly the estimates above hold also for the triangles U_{ij}. Summing over all the triangles of τ^h contained in $(-\ell, \ell)^2$ yields

$$\int_{\mathbb{R}^2} |\Xi - \nabla^2 u|\, dx \leq C\eta,$$

where we have used the fact that for h sufficiently small both u and Ξ vanish outside $(-\ell, \ell)^2$ and hence (3.7) is established.

To obtain (3.8) notice that $S(\nabla u)$ is contained in the union of the sides of the triangles, and thus

$$\int_{S(\nabla u)} |[\nabla u]|\, d\mathcal{H}^{N-1} \leq Ch^2 \sum_{i,j} (|a_1^{i,j}| + |a_2^{i,j}| + |a_3^{i,j}|) \leq C \int_{\mathbb{R}^N} |\Xi|\, dx.$$

This concludes Step 1.
STEP 2. Given $\Xi \in L^1\left(\mathbb{R}^N; \mathbb{R}^{N\times N}_{\text{sym}}\right)$ and $\eta > 0$, by density we may find $\Xi_\eta \in C_c\left(\mathbb{R}^N; \mathbb{R}^{N\times N}_{\text{sym}}\right)$ such that

$$\int_{\mathbb{R}^N} |\Xi - \Xi_\eta|\, dx \leq \eta \int_{\mathbb{R}^N} |\Xi|\, dx. \tag{3.12}$$

By Step 1 applied to Ξ_η, there exists $u \in SBH\left(\mathbb{R}^N\right)$ such that

$$\int_{\mathbb{R}^N} |u| + |\nabla u| + \left|\Xi_\eta - \nabla^2 u\right|\, dx \leq \eta,$$

$$\int_{S(\nabla u)} |[\nabla u]|\, d\mathcal{H}^{N-1} \leq C \int_{\mathbb{R}^N} |\Xi_\eta|\, dx,$$

which, together with (3.12), yields

$$\int_{\mathbb{R}^N} |u| + |\nabla u| + \left|\Xi - \nabla^2 u\right|\, dx \leq \eta \left(1 + \int_{\mathbb{R}^N} |\Xi|\, dx\right),$$

$$\int_{S(\nabla u)} |[\nabla u]|\, d\mathcal{H}^{N-1} \leq C\,(1+\eta) \int_{\mathbb{R}^N} |\Xi|\, dx$$

and the proof is complete. □

4 Energetics

The purpouse of this short section is to discuss the energetics of second-order structured deformations. Let us start with the classical setting. By Theorem 2.7 a second-order structured deformation (κ, g, G, Σ) from Ω may be determined (in the sense of Definition 4) by simple deformations (κ_n, f_n). The energy associated to a simple deformation can be taken to be

$$\begin{aligned} E(\kappa, f) &:= \int_{\Omega\setminus\kappa} W(x, f, \nabla f, \nabla^2 f)\, dx + \int_{\kappa\cup S(f)} \psi(x, [f], \nu)\, d\mathcal{H}^N \\ &\quad + \int_{\kappa\cup S(\nabla f)} \eta(x, [\nabla f], \nu)\, d\mathcal{H}^N, \end{aligned} \tag{4.1}$$

where the first term is the bulk energy of the material in the configuration Ω, the second and third terms take into account the contribution of the surface energy due to slips and separation and to interfaces between two phases of material, respectively. In (4.1) ν represents the normal to $\kappa \cup S(f)$or $\kappa \cup S(\nabla f)$ and $[\cdot]$ denotes the jump. By means of this energy, in analogy with the work of Choksi and Fonseca (1997), we define the energy of a second-order structured deformation as the energetically most economical way to attain it from simple deformations, i.e.,

$$I(\kappa, g, G, \Sigma) := \inf\{\liminf E(\kappa_n, f_n) : (\kappa_n, f_n) \text{ determines } (\kappa, g, G, \Sigma)\}, \tag{4.2}$$

where again the word "determines" is specified in Definition 4.

Two difficulties arise when we try to handle the energies above: one is due to the fact that piecewise smooth functions have no good compactness property and the other is due to the fact that the energy does not simply depend on the deformation but also on the set κ. The first difficulty is overcome by introducing the space $SBV^2(\Omega)$, cf. [Carriero et al. (1996)]. Using this function space and the corresponding notion of structured deformation we easily surmount also the second difficulty mentioned earlier by rewriting the energy of a simple deformation as

$$\begin{aligned} E(f) &= \int_{\Omega} W(x, f, \nabla f, \nabla^2 f)\, d\mathcal{L}^N + \int_{S(f)} \psi(x, [f], \nu_f)\, d\mathcal{H}^{N-1} \\ &\quad + \int_{S(\nabla f)\setminus S(f)} \eta(x, [\nabla f], \nu_{\nabla f})\, d\mathcal{H}^{N-1}. \end{aligned} \tag{4.3}$$

With this energy and a relaxed version of the Approximation Theorem 3.2 for second-order structured deformations in the SBV^2 setting, we define the energy of a second-order structured deformation as

$$I(g, G, \Sigma) = \inf\{\liminf_{n\to\infty} E(f_n) : f_n \in \mathrm{SBV}^2(\Omega), \text{ and } n \mapsto f_n \text{ determines } (g, G, \Sigma)\}.$$

Once a precise meaning to the word "determines" has been given, the goal is to find an explicit representation of I in terms of W, ψ and η. This will require the use and development of fine geometric measure theory and analysis techniques recently introduced in relaxation theory.

The word "determines" may take different meanings which, of course, should be compatible with the Approximation Theorem 3.2. Trying to follow the classical setting we may say that f_n determines (g, G, Σ) if $f_n \to g$, $\nabla f_n \to G$ and $\nabla^2 f_n \to \Sigma$ in L^1. This approximation though, for many problems, might be too strong since it prevents oscillations of the gradient and the Hessian of f_n. A much weaker statement could be: f_n determines (g, G, Σ) if $f_n \to g$ in L^1 while $\nabla f_n \overset{*}{\rightharpoonup} G$, and $\nabla^2 f_n \overset{*}{\rightharpoonup} \Sigma$ in the sense of measures. Of course other possibilities are possible. It is clear that by changing the meaning of the word "determines", the energy associated to a simple deformation will change accordingly. In other words, the energy associated to a second-order structured deformation depends on the regularity of the determining sequences; this property was already pointed out by Choksi and Fonseca (1997) and more recently was discussed by Del Piero (2001) and in [Del Piero, Lecture 6 in this volume].

Ongoing work addresses the relaxation problem described in (4.3) in the space SBH. A function $f \in SBH$ has no jump discontinuities, i.e., $\mathcal{H}^{N-1}(S(f)) = 0$, hence the second integral on the right hand side of (4.3) is equal to zero. This fact immediately points out that second-order structured deformation in the SBH setting are not appropriate in problems where slipping and fracture play a dominant role. Again also in the SBH setting different meaning could be given to the word "determines".

Even in the "simpler" SBH setting the characterization of the energy I associated to a second-order structured deformation in terms of the potentials W, ψ and η turns out to be a challenging problem, cf. [Fonseca et al. (2002), Fonseca et al. (2003)]. The main difficulty comes from the fact that the bulk energy W in (4.3) depends not only on the Hessian of f but also on the gradient.

Bibliography

G. Alberti. A Lusin type theorem for gradients. *J. Funct. Anal.*, 100:110–118, 1991.

L. Ambrosio, N. Fusco, and D. Pallara. *Functions of Bounded Variation and Free Discontinuity Problems.* Mathematical Monographs, Oxford University Press, 2000.

M. Carriero, A. Leaci, and F. Tomarelli. Special bounded Hessian and elastic-plastic plate. *Rend. Accad. Naz. Sci. XL Mem. Mat.*, 16:223–258, 1992.

M. Carriero, A. Leaci, and F. Tomarelli. A second order model in image segmentation: Blake & Zisserman functional. In *Progr. Nonlinear Differential Equations Appl., Birkhäuser* 25:57–72, 1996.

R. Choksi, and I. Fonseca. Bulk and interfacial energy densities for structured deformations of continua. *Arch. Rational Mech. Anal.*, 138:37–103, 1997.

E. De Giorgi, and L. Ambrosio. Un nuovo tipo di funzionale del calcolo delle variazioni. *Atti Accad. Naz. Lincei Rend. Cl. Sci. Fis. Mat. Natur.*, 82:199–210, 1988.

G. Del Piero. The energy of a one-dimensional structured deformation. *Math. and Mech. of Solids*, 6:387–408, 2001.

G. Del Piero, and D.R. Owen. Structured deformations of continua. *Arch. Rational Mech. Anal.*, 124:99–155, 1993.

G. Del Piero, and D.R. Owen. Integral-gradient formulae for structured deformations. *Arch. Rational Mech. Anal.*, 131:121-138, 1995.

F. Demengel. Fonctions á hessien borné. *Ann. Inst. Fourier, (Grenoble)*, 34:155-190, 1984.

F. Demengel. Compactness theorems for spaces of functions with bounded derivatives and applications to limit analysis problems in plasticity. *Arch. Rational Mech. Anal.*, 105:123–161, 1989.

I. Fonseca, G. Leoni, J. Maly, and R. Paroni. A note on Meyers' theorem in $W^{k,1}$. *Trans. Amer. Math. Soc.*, 354:3722-3751, 2002.

I. Fonseca, G. Leoni, and R. Paroni. On lower semicontinuity in BH^p and 2-quasiconvexification. *Calc. Var. Partial Differential Equations*, 17:283–309, , 2003.

I. Fonseca, G. Leoni, and R. Paroni. On Hessian matrices in the space BH. To appear in *Commun. Contemp. Math.*

D.R. Owen, and R. Paroni. Second-order structured deformations. *Arch. Rational Mech. Anal.*, 155:215–235, 2000.

W. Noll, and E. Virga. Fit regions and functions of bounded variation. *Arch. Rational Mech. Anal.*, 102:1–22, 1988.

T. Temam. *Problémes mathématiques en plasticité.* Gauthier-Villars, Paris, 1983.

CRYSTALLINE PLASTICITY AND STRUCTURED DEFORMATIONS

L. DESERI(*)

1. Two level shears and single slip in single crystals

1.1. **Experimental background.** The connection between the geometrical changes at small length scales in single crystals and their macroscopic response has been the subject of extensive experimental studies. In (Deseri and Owen (2000)), experimental evidence that points to a connection between hardening behavior of single crystals undergoing single slip and changes in the separation of active slip-bands has been collected. Although the relation between the structure and separation of both slip-bands and of slip-lines, on the one hand, and the strain hardening of crystals, on the other hand, is still not well understood (Kubin (1993), pp. 145-146), the existence of such a relation was already established by the year 1950 (Hill (1950), p. 6). Experimental evidence is cited in (Barrett (1952)), p. 349 for two basic phenomena: (i) in crystals that deform without appreciable hardening, such as lead, further deformation due to slip continues on existing slip-lines, and (ii) in crystals that deform with appreciable hardening, such as aluminum, further deformation due to slip entails the formation of new slip-lines (Crussard (1945), p. 291; Brown (1952) p. 468). Moreover, in aluminum crystals, a significant number of active slip-lines become inactive as deformation progresses (Yamaguchi (1993), pp. 301-305; Crussard (1945), p. 290). In spite of the fact that the average separation of *all* slip-lines, i.e., active and inactive together, decreases with deformation in aluminum crystals (Crussard (1945), p. 291), there is evidence that the separation of *active* slip-bands increases with deformation in the f.c.c. alloy Cu_3Au (Salama et al. (1971), *et al,* 1971, p. 402).

The reader will notice that the experimental evidence cited above pertains in part to the distribution of slip-lines and in part to the distribution of slip-bands. Although some authors carefully maintain a distinction between slip-lines and slip-bands (Neuhäuser,

---

(*) The contents of these notes is based on the papers Deseri and Owen I (2002), Deseri and Owen (2000), Deseri and Owen II (2002), written with D. R. Owen, Department of Mathematical Sciences, Carnegie-Mellon University, 15213-3890 Pittsburgh PA-USA.

The author wishes also to thank the Center for Nonlinear Analysis of Carnegie-Mellon University, 15213-3890 Pittsburgh PA-USA for the partial financial support during his visits in 1999 and 2000, and the Department of Mathematics of the University of Kentucky, Lexington KY-USA, together with Prof. C.-S. Man for the support provided during the Spring semester 2002 while he was visiting professor.
Special thanks go to the author's colleagues Prof. A. Tralli, Dr. R. Rizzoni and Dr. E. Benvenuti for their support and help at the Engineering Dept. of the University of Ferrara.
The author also acknowledges his advisor, Prof. G. Del Piero, for his constant and substantial support and encouragement through the years.
It is and it has been a pleasure to collaborate with Prof. D. R. Owen through the past five years: D. R. Owen is also acknowledged for his great patience and support.

1983, p. 323), others appear to use these terms interchangeably. The *active slip-bands* for the crystal are here modelled as surfaces between which deformation procedes smoothly and across which tangential discontinuities in displacement occur. This conception of active slip-bands agrees with the descriptions in (Hill (1950)), p. 6, and in (Brown (1952)), pp. 434, 436. The assumption that active slip-bands have zero thickness requires that the active slip-lines within an active slip-band are not distinguished one from another.

1.2. **Geometrical measures of slip-band separation and of slip.** A specific class of non-classical, structured deformations called *two-level shears* is considered to model single slip in single crystals (Choksi et al. (1999), Del Piero and Owen (2000)). Two-level shears provide a simple tool for describing geometrically the main aspects of single slip in single crystals. If the unit vector $\mathbf{m}$ denotes the direction of shear and the unit vector $\mathbf{n}$ denotes the normal to the plane of slip, a two-level shear is specified by giving two real numbers μ and γ, along with the following two mappings g and G :

$$\begin{aligned} g(x,y,z) \quad &: \quad = (x+\mu z)\mathbf{m} + y\mathbf{a} + z\mathbf{n} \\ \mathbf{a} \quad &: \quad = \mathbf{n}\times\mathbf{m} \\ G(x,y,z) \quad &: \quad = I + \gamma\mathbf{m}\otimes\mathbf{n}. \end{aligned} \tag{1.1}$$

Here the disarrangement tensor M takes the form:

$$M(x,y,z) := \nabla g(x,y,z) - G(x,y,z) = (\mu-\gamma)\mathbf{m}\otimes\mathbf{n}. \tag{1.2}$$

A determining sequence for the given two-level shear (see Choksi et al. (1999), Del Piero and Owen (2000)) can be chosen such that for each n the jump set $\Gamma(f_n)$ is a family of planes, each with normal $\nu = \mathbf{n}$ in the z direction, $[f_n]$ is in the x-direction, and ∇f_n is of the same form as G:

$$\nabla f_n = I + (\nabla f_n)_{xz}\mathbf{m}\otimes\mathbf{n}. \tag{1.3}$$

Each plane in the family $\Gamma(f_n)$ may be interpreted as an individual active slip-band in the crystal, and each tangential jump $[f_n]$ as the slip experienced by the crystal across that slip-band. This interpretation and eqns (1.1) and (1.2) permit the number γ to be identified with the *shear without slip* and the number $\mu-\gamma$ with the *shear due to slip.* Moreover, each piecewise smooth mapping f_n may be thought of a mesolevel view of smooth lattice deformations and discontinuous slips associated with the two-level shear. Henceforth, for the sake of simplicity, only determining sequences $n \mapsto f_n$ with the features described above will be considered. For these determining sequences, the identification relation for M (Del Piero and Owen (1995)) permits eqn (1.2) to be written in the scalar form

$$\mu-\gamma = \lim_{r\to 0}\lim_{n\to\infty} s_n(x,y,z;r), \tag{1.4}$$

where the dimensionless quantity

$$s_n(x,y,z;r) := (vol\,\mathcal{B}(x,y,z;r))^{-1} \int_{\Gamma(f_n)\cap\mathcal{B}(x,y,z;r)} [f_n]_x\, dA \tag{1.5}$$

is the *amount of tangential jump per unit volume* within the ball $\mathcal{B}(x,y,z;r)$, and where $[f_n]_x$ is the x-component of the jump $[f_n]$. Because the numbers μ and γ are part of the specification of the two-level shear, the limit $\lim_{r\to 0}\lim_{n\to\infty} s_n(x,y,z;r) = \mu - \gamma$ is independent of the choice of determining sequence for the given two-level shear, even though each term $s_n(x,y,z;r)$ need not be.

An analog of relations (1.4) and (1.5) holds for the shear without slip γ. In particular,

$$\gamma = \lim_{r\to 0}\lim_{n\to\infty} \gamma_n(x,y,z;r), \tag{1.6}$$

where the dimensionless quantity

$$\gamma_n(x,y,z;r) = (vol\,\mathcal{B}(x,y,z;r))^{-1} \int\limits_{\mathcal{B}(x,y,z;r)} (\nabla f_n)_{xz}\, dV \tag{1.7}$$

is the average lattice shear within the ball $\mathcal{B}(x,y,z;r)$. Again, the limit γ is independent of the choice of determining sequence, whereas each term $\gamma_n(x,y,z;r)$ need not be.

It is useful to specify for each index n a length scale in terms of which we may view the deformation f_n for various determining sequences: such a scale may be identified through a positive number c_n, the *cell-size of the crystal lattice* in the level of magnification for the index n. In this scheme, the cell size c_n decreases with increasing n.

An averaged dimensionless measure of the distance between (active) slip planes, relative to the lattice cell size, within a given ball can be defined for every given determining sequence $n \mapsto f_n$ and for every sequence $n \mapsto c_n$. In particular, this measure can be expressed through the family $(r,n) \mapsto d_n(x,y,z;r)$ as follows:

$$d_n(x,y,z;r) := \frac{1}{c_n}\frac{vol\,\mathcal{B}(x,y,z;r)}{area(\Gamma(f_n)\cap\mathcal{B}(x,y,z;r))}. \tag{1.8}$$

The number $d_n(x,y,z;r)$ will be called the *average relative separation of active slip-bands for the index n.*

By eqns (1.8) and (1.5) one may define a dimensionless averaged measure of jumps in the given ball $\mathcal{B}(x,y,z;r)$, for the given sequences $n \mapsto f_n$, $n \mapsto c_n$, as follows:

$$j_n(x,y,z;r) := d_n(x,y,z;r)\, s_n(x,y,z;r). \tag{1.9}$$

By (1.5) and (1.8), the following formula holds:

$$j_n(x,y,z;r) = \frac{\int\limits_{\Gamma(f_n)\cap\mathcal{B}(x,y,z;r)} ([f_n]_x / c_n) dA}{area(\Gamma(f_n)\cap\mathcal{B}(x,y,z;r))}, \tag{1.10}$$

so that $j_n(x,y,z;r)$ is the average with respect to area of non-dimensional tangential jumps. Because the non-dimensionalization of the jumps in eqn (1.10) is taken with respect to the cell-size of the crystal lattice, $j_n(x,y,z;r)$ may be called the *average (possibly fractional) number of cells spanned by the tangential jumps for the index n* within the given ball.

As was pointed out in eqn (1.4), the tangential jump per unit volume $s_n(x,y,z;r)$ approaches the shear due to slip $\mu - \gamma$ as $n \to \infty$ and $r \to 0$, no matter what the choice of determining sequence. However, the condition that $n \mapsto f_n$ is a determining sequence for the two-level shear does not, in itself, guarantee that the relative separation of active slip-bands $d_n(x,y,z;r)$ has a limit. Both the the average separation of active slip-bands within the ball $\frac{vol\,\mathcal{B}(x,y,z;r)}{area(\Gamma(f_n)\cap\mathcal{B}(x,y,z;r))}$ and the cell-size of the crystal lattice c_n now are assumed to tend to zero in such a way that $d_n(x,y,z;r)$ does have a limit $d(x,y,z)$, which will be called the *relative separation of active slip bands for the given determining sequence.* Eqn (1.4) implies that $j_n(x,y,z;r)$ then has the limit $(\mu-\gamma)d(x,y,z)$, i.e.

$$\lim_{r\to 0}\lim_{n\to\infty} j_n(x,y,z;r) = (\mu-\gamma)d(x,y,z), \tag{1.11}$$

so that the product $(\mu-\gamma)d(x,y,z)$ represents an *average (possibly fractional) number of cells spanned by the tangential jumps for the given determining sequence.* It is important to remember that this product depends on the given two-level shear through μ and γ, as well as on the given determining sequence through $d(x,y,z)$.

2. Energetics for two-level shears and slip in single crystals

2.1. **The free energy.** A formula for the Helmholtz free energy associated with a two-level shear was obtained in (Deseri and Owen (2000)) and the procedure used there will be recalled in this section. The idea proposed and analyzed in (Choksi and Fonseca (1997)) for determining the form of the Helmholtz free energy of arbitrary structured deformations has been followed in (Deseri and Owen (2000)), although the steps in the implementation of this idea differ in significant ways from those in that analysis. A two-level shear, a sequence $n \mapsto c_n$ of cell-sizes, sequences $n \mapsto d_n(x, y, z; r)$ and $n \mapsto s_n(x, y, z; r)$ associated with a determining sequence $n \mapsto f_n$ satisfying the assumptions stated in Section 1, are considered at the outset. For each index n and positive number r, a number $\mathcal{H}_{n,r}$ measuring the Helmholtz free energy per unit volume of the crystal undergoing the piecewise smooth deformation f_n is also assigned. An additive decomposition of $\mathcal{H}_{n,r}$ is then assumed in the form

$$\mathcal{H}_{n,r} = \mathcal{H}^d_{n,r} + \mathcal{H}^s_{n,r},$$

where $\mathcal{H}^d_{n,r}$ measures the *free energy density due to lattice distortion,* and $\mathcal{H}^s_{n,r}$ measures the *free energy density due to tangential jumps across slip-bands.* The field $\gamma_n(\cdot, \cdot, \cdot; r)$ given below eqn (1.7) in the previous lecture measures the deformation of the lattice away from slip-bands, and $\mathcal{H}^d_{n,r}$ is assumed to depend upon this measure. In particular, one can state this assumption as follows:

$$\mathcal{H}^d_{n,r} = \tilde{\varphi}(\gamma_n(x, y, z; r)), \tag{2.1}$$

where $\tilde{\varphi}$ is assumed to be continuous and convex: $\tilde{\varphi}$ is then a constitutive function determined by the slip-free portions of the crystal.

Different considerations have to be made for the Helmholtz free energy $\mathcal{H}^s_{n,r}$ per unit volume due to tangential jumps across slip-bands. This part of the free energy should reflect the fact that a single tangential jump of amount $[f_n]_x = kc_n$, with k an integer and c_n the cell-size for the index n, cannot be detected geometrically and, hence, should not change the free energy. By equation (1.10) such a jump gives the value $j_n(x, y, z; r) = k$, i.e., the number of cells spanned by this jump is k, so that $\mathcal{H}^s_{n,r}$ should be unchanged in jumps $j_n(x, y, z; r) = k$. This leads to the following assumption:

$$\mathcal{H}^s_{n,r} = \tilde{\psi}(j_n(x, y, z; r)), \tag{2.2}$$

where $\tilde{\psi}$ is continuous and *periodic* of period 1. The assumed oscillatory nature of the constitutive function $\tilde{\psi}$ reflects the tacit physical assumption that mesolevel control of relative tangential lattice displacements across a slip-band would result in recoverable work being performed without dissipation. The dissipation that arises when such mesolevel control lapses is discussed in Subsection 2.4 (see eqn (2.40)).

The assumptions made on the determining sequence $n \mapsto f_n$ and the continuity of the constitutive functions $\tilde{\varphi}$ and $\tilde{\psi}$ imply that $\mathcal{H}_{n,r}$ has a limit as n tends to infinity and as r tends to zero :

$$\begin{aligned} \mathcal{H}(x, y, z) \quad : &= \lim_{r \to 0} \lim_{n \to \infty} \mathcal{H}_{n,r} \\ &= \tilde{\varphi}(\gamma) + \tilde{\psi}\left((\mu - \gamma) d(x, y, z)\right). \end{aligned} \tag{2.3}$$

The number $\mathcal{H}(x,y,z)$ is the *Helmholtz free energy per unit volume at the point* (x,y,z), and this relation tells us that the $\mathcal{H}(x,y,z)$ is determined by the shear without slip γ, the shear due to slip $\mu-\gamma$, and the relative separation of active slip-bands $d(x,y,z)$. The product $(\mu-\gamma)d(x,y,z)$ measures the average number of lattice cells traversed during the shear, and it is worth noting that γ and $\mu-\gamma$ depend only on the given two-level shear, whereas $d(x,y,z)$ depends also on the given determining sequence $n\mapsto f_n$ for the two-level shear. Because of this, eqn (2.3) implies that the free energy density need not be determined by the two-level shear alone. In spite of the general lack of experimental observations on the separation of active slip-bands in single crystals, the reference (Salama, *et al*, 1971) suggests that relative separation $d_n(x,y,z;r)$ may be regarded in a particular crystal as an increasing function of the amount of tangential jump per unit volume $s_n(x,y,z;r)$:

$$d_n(x,y,z;r)=\hat{d}\left(s_n(x,y,z;r)\right). \tag{2.4}$$

If the *constitutive function* $\hat{d}$ is continuous, then the double limit in both sides of this relation yields

$$d(x,y,z)=\hat{d}(\mu-\gamma), \tag{2.5}$$

and eqns (2.3) and (2.5) yield the following constitutive formula for the free energy density $\mathcal{H}(x,y,z)$:

$$\mathcal{H}(x,y,z)=\tilde{\varphi}(\gamma)+\tilde{\psi}\left((\mu-\gamma)\hat{d}(\mu-\gamma)\right). \tag{2.6}$$

The second term on the right hand side of (2.6) is $\mathcal{H}^s$, the Helmholtz free energy due to slip, which can be conveniently expressed as follows

$$\mathcal{H}^s=\frac{1}{\hat{d}(0)}\psi_1((\mu-\gamma)\hat{d}(\mu-\gamma)). \tag{2.7}$$

Here ψ_1 is a periodic function of period 1, and $\hat{d}(0):=d_0$ is the reference value of the relative separation of active slip-bands at $s=0$, as a normalization constant. According to Hill (1950), d_0 is of the order 10^4 in many crystals.

It also is convenient to employ a function $\mathfrak{h}$ defined in terms of the ratio between the relative separation of active slip-bands $d=\hat{d}(\mu-\gamma)$ and its reference value $d_0=\hat{d}(0)$ through the relation

$$\frac{\hat{d}(\mu-\gamma)}{\hat{d}(0)}=\frac{1}{\mu-\gamma}\int_0^{\mu-\gamma}\mathfrak{h}(r)dr. \tag{2.8}$$

This relation implies that $\mathfrak{h}(s)=\frac{d}{ds}\left(\frac{s\hat{d}(s)}{\hat{d}(0)}\right)$, i.e. $\mathfrak{h}$ represents the rate of change of averaged number of cells traversed in slip with respect to the shear due to slip, normalized by $d_0=\hat{d}(0)$ (Deseri and Owen II (2002)). By virtue of relation (2.8) the Helmholtz free energy due to slip can be rewritten as

$$\mathcal{H}^s=p\psi_1(\frac{1}{p}\int_0^{\mu-\gamma}\mathfrak{h}(r)\,dr), \tag{2.9}$$

where $p := \frac{1}{\hat{d}(0)}$ represents the shear due to tangential jumps across one lattice cell when $s = 0$. Thus, because of the order of magnitude of d_0, the shear p is of the order 10^{-4} in many crystals.

By relations (2.9) and (2.3), the Helmholtz free energy density $\mathcal{H} =: \Psi(\mu, \gamma)$ for a two-level shear then is given by

$$\Psi(\mu, \gamma) = \varphi(\gamma) + p\,\psi_1(\frac{1}{p}\int_0^{\mu-\gamma} \mathfrak{h}(r)\,dr), \tag{2.10}$$

with the slight notational change of replacement of $\tilde{\varphi}$ by φ. The function $\mathfrak{h}$ defined in (2.8) is assumed to be a non-decreasing function of s for $s \geq 0$. Because $\mathfrak{h}(s) = \frac{\hat{d}(s)}{\hat{d}(0)} + s\frac{\hat{d}'(s)}{\hat{d}(0)}$, there holds $\mathfrak{h}(0) = 1$ and $\mathfrak{h}(s) \geq \mathfrak{h}(0) = 1$. Since slip bands are separated by at least one lattice cell, one may assume that $\hat{d}(s) \geq 1$ for all $s \geq 0$. Moreover, $\mathfrak{h}$ is a constant function if and only if $\hat{d}$ is a constant function, and the assumed monotonicity of $\mathfrak{h}$ is equivalent to monotonicity of $\hat{d}$. If the separation of active slip bands does not vary with the amount of slip, then $\hat{d}(s) = \hat{d}(0)$, $\mathfrak{h}(s) = \mathfrak{h}(0) = 1$ for all s, and eqn.s (2.7), (2.10) and the definition $p = \hat{d}(0)^{-1}$ yield

$$\Psi(\mu, \gamma) = \varphi(\gamma) + p\psi_1(\frac{\mu - \gamma}{p}). \tag{2.11}$$

The mapping $s \mapsto \psi(s) := p\psi_1(\frac{s}{p})$ is a periodic function with period p (which is of the order 10^{-4}). Thus, when the separation of active slip bands does not vary, the Helmholtz free energy due to slip is a periodic function of the shear due to slip whose period is the reciprocal of the (constant) relative separation of active slip bands. The formula (2.11) was the starting point of the analysis of hysteresis, dissipation, and yielding in the study (Choksi et al. (1999)).

2.2. **Modelling G. I. Taylor's tensile tests.** The loading of a crystal in a soft-device is described in the sequel (see Deseri and Owen II (2002), Sect. 3); in particular, the modelling of tensile tests described by Sir G. I. Taylor in 1927 (see e.g. Taylor (1958), p.212) is considered. A single crystal in its reference configuration is assumed to occupy a cylindrical region $\mathcal{C}_R := \Omega_R \times (0, \ell_R)$ in the three dimensional euclidean space, with $\ell_R > 0$ the height of the crystal and with $\Omega_R \subset \mathbb{R}^2$ a bounded region representing its constant cross section. The common direction of the generators of $\mathcal{C}_R$ is given by the unit vector $\mathbf{e}_3$. If Ω_R^+ and Ω_R^- are the top and bottom faces of the crystal, with outer normals $\mathbf{e}_3$ and $-\mathbf{e}_3$, respectively, then the boundary $\partial\mathcal{C}_R$ of $\mathcal{C}_R$ is given by the formula $\partial\mathcal{C}_R = \Omega_R^+ \cup \Omega_R^- \cup (\partial\Omega_R \times (0, \ell_R))$.

Taylor assumed as reference parameters for experimental observations of crystals the angle θ_0 between a slip plane with unit normal $\mathbf{n}$ and the top face of the crystal, i.e., the angle between the vectors $\mathbf{n}$ and $\mathbf{e}_3$, and the angle η_0 between the direction of slip $\mathbf{m}$ and the direction of greatest slope in the slip plane. If $\mathbf{d}$ denotes a unit vector in this direction, then $\cos\eta_0 = \mathbf{d} \cdot \mathbf{n}$ and $\mathbf{e}_3 = \sin\theta_0\,\mathbf{d} + \cos\theta_0\,\mathbf{n} = \sin\theta_0(-\sin\eta_0\,\mathbf{n} \times \mathbf{m} + \cos\eta_0\,\mathbf{m}) + \cos\theta_0\,\mathbf{n}$. It turns out that the vector $\mathbf{e}_1 := \mathbf{n} \times \mathbf{d}$ is parallel both to the slip plane and the cross-section Ω_R. A right-handed orthonormal basis can be defined in Ω_R by putting $\mathbf{e}_2 := \mathbf{e}_3 \times \mathbf{e}_1$, and the following relationships hold:

$$\mathbf{e}_1 = \cos\eta_0 \; \mathbf{n} \times \mathbf{m} + \sin\eta_0 \; \mathbf{m} \tag{2.12}$$

$$\mathbf{e}_2 = \cos\theta_0 \, (\sin\eta_0 \; \mathbf{n} \times \mathbf{m} - \cos\eta_0 \; \mathbf{m}) + \sin\theta_0 \, \mathbf{n}. \tag{2.13}$$

The crystal is assumed to undergo a two-level shear from the reference configuration that is specified by giving: (a) two scalars μ and γ, (b) the two fixed and mutually orthogonal vectors $\mathbf{n}$, $\mathbf{m}$ introduced above, and a homogeneous deformation from $\mathcal{C}_R$ whose gradient is given by

$$\mathbf{F} = \mathbf{I} + \mu \, \mathbf{m} \otimes \mathbf{n}, \tag{2.14}$$

and (c) the rotation $\mathbf{R}$ such that $\mathbf{RFe}_3 = \varepsilon \mathbf{e}_3$,with ε the elongation of a fiber in the direction $\mathbf{e}_3$, $\varepsilon := |\mathbf{RFe}_3| = |\mathbf{Fe}_3|$, and (d) the tensor

$$\mathbf{G} = \mathbf{I} + \gamma \, \mathbf{m} \otimes \mathbf{n}. \tag{2.15}$$

From now on, μ will be assumed to be non-negative, so that the vector $\mathbf{m}$ points in the direction of shearing.

Taylor's tensile experiments may be modelled by assuming that total tensile loads $\mathbf{l}$ and $-\mathbf{l}$ are applied to the top and bottom faces Ω_R^+ and Ω_R^- in the form of dead loads. In particular:

$$\mathbf{l} = f \; area(\Omega_R) \, \mathbf{e}_3 = b \, area(\Omega) \, \frac{\mathbf{RFe}_3}{\varepsilon}, \tag{2.16}$$

where b and $area(\Omega)$ represent the magnitude of the applied force per unit area and the area in the deformed configuration, respectively, while f is the magnitude of the applied force per unit area in the reference configuration. In addition, it is assumed that the lateral surface of the crystal is traction free.

If $\partial\mathcal{C} = \Omega^+ \cup \Omega^- \cup \mathcal{L}$ denotes the boundary of the crystal in the current configuration, the previous assumptions on the applied loads and the relation $b = \frac{f}{|(\mathbf{RF})^*\mathbf{e}_3|}$ (with $\mathbf{A}^* := (\det \mathbf{A})\mathbf{A}^{-T}$ denoting the adjugate of an invertible tensor $\mathbf{A}$) then yield on $\mathcal{L}$

$$\mathbf{T}(\mathbf{RF})^*\mathbf{e}_1 = \mathbf{T}(\mathbf{RF})^*\mathbf{e}_2 = 0 \tag{2.17}$$

and on $\Omega^+ \cup \Omega^-$

$$\mathbf{T}\frac{(\pm(\mathbf{RF})^*\mathbf{e}_3)}{|(\mathbf{RF})^*\mathbf{e}_3|} = \pm\frac{f}{|(\mathbf{RF})^*\mathbf{e}_3|}\frac{\mathbf{RFe}_3}{\varepsilon}. \tag{2.18}$$

Here $\mathbf{T}$ is the Cauchy stress and the "+" and "-" signs are taken on Ω^+ and Ω^-, respectively. It is easy to check that if we take for $\mathbf{T}$ the constant tensor field

$$\mathbf{T} = \frac{f}{\varepsilon}\mathbf{RFe}_3 \otimes \mathbf{RFe}_3 = \varepsilon \, f \, \mathbf{e}_3 \otimes \mathbf{e}_3, \tag{2.19}$$

then $\mathbf{T}$ satisfies (2.17) and (2.18) as well as the balance of forces and moments. It is worth noting that the initial normals $\pm \, \mathbf{e}_3$ to the top and bottom faces of the crystal do not remain as such as the deformation progresses. Indeed, if $\pm \, \mathbf{e}$ denotes, respectively, the (non-unit) normals to such faces in the deformed configuration, the following relations do hold: $\pm \, \mathbf{e} = \pm \, (\mathbf{RF})^*\mathbf{e}_3 \neq \pm \, \mathbf{e}_3$.

By the expression for the vector $\mathbf{e}_3$ in terms of $\mathbf{m}$ and $\mathbf{n}$, the elongation ε defined above can be evaluated explicitly as: $\varepsilon = (1 + 2\mu \sin\theta_0 \cos\theta_0 \cos\eta_0 + \mu^2 \cos^2\theta_0)^{\frac{1}{2}}$. The scalar $\tau_{res} := \mathbf{T}\frac{(\mathbf{RF})^*\mathbf{n}}{|(\mathbf{RF})^*\mathbf{n}|} \cdot \frac{(\mathbf{RF})\mathbf{m}}{|(\mathbf{RF})\mathbf{m}|}$, which represents the shear component of the Cauchy stress on the slip plane resolved in the direction of slip, is then easily calculated:

$$\begin{aligned} \tau_{res} &= f\cos\theta_0 \frac{\mu + \tan\theta_0 \cos\eta_0}{\left(1 + \tan^2\theta_0 \sin^2\eta_0 + (\mu + \tan\theta_0\cos\eta_0)^2\right)^{\frac{1}{2}}} \\ &= : \hat{\tau}_f(\mu). \end{aligned} \tag{2.20}$$

This expression agrees with the one obtained by Taylor (Taylor, 1958, p.212, (6)). The function $\hat{\tau}_f$ depends only on the orientation of the crystal in the loading device and on the amount of load, and we call $\hat{\tau}_f$ the *load-orientation function* corresponding to f. The graph of $\hat{\tau}_f$ will be called the *load-orientation curve* corresponding to f.

2.3. **Gibbs free energy and metastable equilibrium.** For a body undergoing a two-level shear with f prescribed as in the previous subsection, the Gibbs free energy per unit volume can be obtained by subtracting from the Helmholtz free energy in (2.10) the work done per unit volume by the tractions, i.e.,

$$\mathcal{E}(\mu,\gamma;f) := \Psi(\mu,\gamma) - w(\mu;f), \tag{2.21}$$

where, by virtue of the boundary conditions for Taylor's soft device,

$$w(\mu;f) = \frac{1}{\ell_R\, area\Omega_R}\left(\int_{\Omega_R^+} \mathbf{S}\mathbf{e}_3\cdot\mathbf{u}\,dA - \int_{\Omega_R^-}\mathbf{S}\mathbf{e}_3\cdot\mathbf{u}\,dA\right) = \int_0^\mu \hat{\tau}_f(\zeta)\,d\zeta, \tag{2.22}$$

with $\mathbf{S}$ the Piola-Kirchhoff stress tensor (the dependence of $\mathbf{S}$ on f is determined by the formula (2.19) and by the formula $\mathbf{S} = \mathbf{T}(\mathbf{RF})^*$), and with $\mathbf{u}$ a displacement field satisfying $grad\,\mathbf{u} = \mathbf{RF} - \mathbf{I}$. The right-most member of (2.22) has been obtained using (2.14), (2.19), and (2.20).

Two-level shears for prescribed f that are local minimizers of the Gibbs free energy function $(\mu,\gamma) \mapsto \mathcal{E}(\mu,\gamma;f)$ are the goal of the subsequent analysis. A pair (μ,γ) is a *local minimizer for* $\mathcal{E}$ only if the partial derivatives of $\mathcal{E}$ with respect to μ and with respect to γ both vanish, i.e., only if both of the following equations hold:

$$\mathfrak{h}(\mu-\gamma)\psi_1'(\frac{1}{p}\int_0^{\mu-\gamma}\mathfrak{h}(r)\,dr) - \hat{\tau}_f(\mu) = 0, \tag{2.23}$$

$$\varphi'(\gamma) - \mathfrak{h}(\mu-\gamma)\psi_1'(\frac{1}{p}\int_0^{\mu-\gamma}\mathfrak{h}(r)\,dr) = 0, \tag{2.24}$$

where the prime $'$ denotes differentiation with respect to γ or with respect to $\mu-\gamma$. From such stationary points (μ,γ), local minimizers (μ,γ) for $\mathcal{E}$ for a given f are obtained by imposing the condition that the Hessian of the Gibbs free energy be positive definite. It is easy to show that positive definiteness of the Hessian is equivalent to the positivity of the determinant of the Hessian, i.e.,

$$\varphi''(\gamma)\frac{\partial^2\Psi}{\partial\mu^2}(\mu,\gamma) - \frac{d}{d\mu}\hat{\tau}_f(\mu)\left(\varphi''(\gamma) + \frac{\partial^2\Psi}{\partial\mu^2}(\mu,\gamma)\right) > 0. \tag{2.25}$$

For a given f and for a given stationary point (μ, γ) for $\mathcal{E}$ a scalar variable τ with the dimension of a stress may be defined as follows:

$$\tau := \varphi'(\gamma). \tag{2.26}$$

Equations (2.20), (2.23), and (2.24) imply that

$$\tau = \tau_{res}.$$

From now on a pair (μ, τ) will be called a *stationary point* corresponding to $f \geq 0$ if the corresponding pair $(\mu, (\varphi')^{-1}(\tau))$ is a stationary point for $\mathcal{E}$. By (2.20), (2.23), (2.24), (2.26), and by the earlier assumption $\mu \geq 0$, all of the stationary points lie in the first quadrant of the $\mu - \tau$ plane.

For each number $f \geq 0$, a stationary point (μ, τ) is called *metastable* if the inequality (2.25) holds with $\gamma = (\varphi')^{-1}(\tau)$. When the relation (2.25) holds with ">" replaced by "=", then the number f is called a *threshold load*, and the (non-metastable) stationary point (μ, τ) is called a *threshold point*.

It is worth noting that the collection of threshold points and loads is determined by the functions φ, ψ_1, and $\mathfrak{h}$ in the formula (2.10) for the Helmholtz free energy Ψ and by the geometry of the crystal through the formula (2.20) for the load-orientation function $\hat{\tau}_f$.

The assumptions (i) and (ii) made in Section 4.1 of (Choksi et al. (1999)) here will be made on the functions φ and $s \mapsto \psi_1(\frac{s}{p})$. In addition, the function $\mathfrak{h}$ is assumed here to be continuously differentiable.

In contrast to the dependence of threshold loads and threshold points on the functions φ, ψ_1, and $\mathfrak{h}$ in (2.10) as well as on the function $\hat{\tau}_f$ in relation (2.20), the parametric curve defined through the relations

$$\mu = \bar{\mu}(s) := s + (\varphi')^{-1}\left(\mathfrak{h}(s)\psi_1'(\frac{1}{p}\int_0^s \mathfrak{h}(r)\,dr)\right) \tag{2.27}$$

$$\tau = \bar{\tau}(s) := \mathfrak{h}(s)\psi_1'(\frac{1}{p}\int_0^s \mathfrak{h}(r)\,dr), \tag{2.28}$$

depends only upon φ, ψ_1, and $\mathfrak{h}$. If s is allowed to be any non-negative number, then this curve traces out all of the stationary points (μ, τ) for $f \geq 0$ as well as other points in the fourth and possibly in the third quadrants in the μ-τ plane. The curve described by (2.27) and (2.28) with $s \geq 0$ will be called the *constitutive curve* for the crystal for the class of two-level shears under consideration.

Equations (2.23), (2.24), (2.27), and (2.28) ensure that the stationary points (μ, τ) for a given f form the intersection of the constitutive curve and the load orientation curve and provide, in conjunction with (2.20), a formula for the applied force per unit reference area f as a function of the shear due to slip s:

$$f = \bar{f}(s) := \frac{\bar{\tau}(s)}{\hat{\tau}_1(\bar{\mu}(s))} = \frac{\bar{\tau}(s)}{\hat{\tau}_1\left(s + (\varphi')^{-1}(\bar{\tau}(s)\right)}. \tag{2.29}$$

Henceforth, in referring to $s = \mu - \gamma$ we will use the shorter term *slip* in place of *shear due to slip*.

Differentiation of equation (2.10) twice with respect to μ and differentiation of equations (2.27), and (2.28) with respect to s, together with the positivity of φ'', yield the following form of the metastability inequality (2.25):

$$\bar{\tau}'(s) > \frac{d}{d\mu}\hat{\tau}_{\bar{f}(s)}(\mu)\mid_{\mu=\bar{\mu}(s)} \bar{\mu}'(s). \tag{2.30}$$

When the inequality (2.30) holds, the non-negative number s will be called a *metastable slip*; when equality (in place of ">") holds in (2.30), s will be called a *threshold slip*: loss of metastability is then encountered at the threshold slips. For threshold slips we may write

$$\frac{\bar{\tau}'(s)}{\bar{\mu}'(s)} = \frac{d}{d\mu}\hat{\tau}_{\bar{f}(s)}(\mu)\mid_{\mu=\bar{\mu}(s)} \tag{2.31}$$

which expresses a tangency condition: threshold slips correspond to points $(\bar{\mu}(s), \bar{\tau}(s))$ in the μ-τ plane at which the constitutive curve (2.27),(2.28) and the load-orientation curve (2.20) for the load $\bar{f}(s)$ are tangent. The terminology introduced above (2.27) and (2.28) along with the equivalence of (2.25) and (2.31) justify our using of the term *threshold points* for these points of tangency.

In order to describe the constitutive curve and, in particular, the metastable points, *fundamental intervals* of slip are singled out and defined as follows: $I_j := [s^o_{2j}, \mu^o_{2j+1}]$, $j = 0, 1, 2, \cdots$, with s^o_{2j} and μ^o_{2j+1} distinguished values of slip to be defined, and the set $\{(\bar{\mu}(s), \bar{\tau}(s)) \mid s \in I_j\}$ will be called the $j+1^{st}$ *fundamental branch* of the constitutive curve. For each $j = 0, 1, 2, \cdots$, both the left endpoint $s^o_{2j} := \mu^o_{2j}$ and the right endpoint μ^o_{2j+1} of I_j are obtained from the collection $\{\mu^o_k \mid k = 0, 1, 2, \cdots\}$ of slips defined as the solutions of the equation $\int_0^{\mu^o_k} \mathfrak{h}(r)\, dr = \frac{kp}{2}$. It is also worth noting that $\bar{\tau}(s^0_{2j}) = \bar{\tau}(\mu^0_{2j+1}) = 0$.

A general analysis of the form of the constitutive curve (2.27),(2.28) is provided in (Deseri and Owen II (2002), Sect.s 4, 5 and 6), particularly in relation to thresholds and other distinguished points in any *fundamental branch* of the curve itself. The metastable points on the constitutive locus are described through the definition of a *fundamental interval* of slip. In spite of the rather complicated shape of the constitutive curve, an intrinsic energetic relation can be established. For every pair of points $A := (\bar{\mu}(s^a), \bar{\tau}(s^a))$ and $B := (\bar{\mu}(s^b), \bar{\tau}(s^b))$ on the constitutive curve, with $0 \leq s^a \leq s^b$, the difference in free energy is given by

$$\begin{aligned}\Psi_B - \Psi_A &= \int_{\Gamma^B_A} (\frac{\partial \Psi}{\partial \mu} d\mu + \frac{\partial \Psi}{\partial \gamma}\frac{d\gamma}{d\tau} d\tau) = \int_{\Gamma^B_A} \varphi'((\varphi')^{-1}(\tau)) d\mu \\ &= \int_{\Gamma^B_A} \tau d\mu,\end{aligned} \tag{2.32}$$

with $\Gamma^B_A := \{(\bar{\mu}(s), \bar{\tau}(s)) \mid s \in [s^a, s^b]\}$ the segment of the constitutive curve connecting A and B. This relation follows from the fact that every point (μ, τ) on the constitutive curve corresponds to a stationary point $(\mu, \gamma) = (\mu, (\varphi')^{-1}(\tau))$ for the total energy $\mathcal{E}$, so that $\frac{\partial \mathcal{E}}{\partial \mu} = \frac{\partial \Psi}{\partial \mu} - \hat{\tau}_f(\mu) = 0$ and $\frac{\partial \mathcal{E}}{\partial \gamma} = \frac{\partial \Psi}{\partial \gamma} = 0$, as in (2.23) and (2.24), with Ψ given by (2.10). Because the last term of (2.32) is the work done by the resolved shear stress as the point $(\bar{\mu}(s), \bar{\tau}(s))$ smoothly follows Γ_A^B, the dissipation inequality

$$\Psi_B - \Psi_A \leq \int_{\Gamma_A^B} \tau d\mu \tag{2.33}$$

is satisfied with equality: this implies that smooth evolution following the constitutive curve occurs without dissipation.

2.4. **"Staircase" hardening.** In this subsection, the source of dissipation during loading processes of the crystal is identified. The loading of a crystal now may be analyzed for the particular case where every fundamental branch of the constitutive curve is a simple curve that has exactly one slip $s_{2j}^m \in (s_{2j}^o, \mu_{2j+1}^o) \subset I_j$ at which $\bar{\tau}'(s_{2j}^m) = 0$, and at most one slip $\tilde{s}_{2j}^m \in (s_{2j}^o, \mu_{2j+1}^o) \subset I_j$ satisfying $\bar{\mu}'(\tilde{s}_{2j}^m) = 0$. In addition, it is assumed that for each load-orientation curve and for each fundamental branch of the constitutive curve either the two loci cross exactly twice or they do not cross (the latter includes both the case where the loci are disjoint and the case where the loci are tangent). This special case for the constitutive curve is depicted in (Deseri and Owen II (2002), Figure 3). For future reference it is worth noting that the slips s_{2j}^m and $\tilde{s}_{2j}^m$ satisfy $s_{2j}^m < \tilde{s}_{2j}^m$ (see Deseri and Owen II (2002), Sect. 5).

Suppose now that the crystal initially is deformed in the trivial two-level shear $\mu = \gamma = 0$ under zero applied load and, subsequently, is subjected to a monotonically increasing applied load. The crystal is assumed to respond first by attaining two-level shears with slips $s = \mu - \gamma$ in the first fundamental interval $I_0 = [s_0^o, \mu_1^o] = [0, \mu_1^o]$. As the applied load is increased from 0 to the smallest threshold load $\bar{f}(s_0^t)$, it is consistent with metastability to expect that the crystal will deform in such a way that μ and τ will follow the metastable points $\{(\bar{\mu}(s), \bar{\tau}(s)) \mid 0 \leq s < s_0^t\}$. According to (2.32) and the sentence containing (2.33), for each $s^a, s^b \in [0, s_0^t)$, with $s^a < s^b$, one may write

$$\Psi\left(\bar{\mu}(s^b), (\varphi')^{-1}(\bar{\tau}(s^b))\right) - \Psi\left(\bar{\mu}(s^a), (\varphi')^{-1}(\bar{\tau}(s^a))\right) = \int_{s^a}^{s^b} \bar{\tau}(s)\bar{\mu}'(s)ds, \tag{2.34}$$

so that *the shearing associated with the increase in f from 0 to $\bar{f}(s_0^t)$ occurs without dissipation.* The threshold point $A := (\bar{\mu}(s_0^t), \bar{\tau}(s_0^t))$ in Figure 4 of (Deseri and Owen II (2002)), reached when f assumes the threshold value $\bar{f}(s_0^t)$, is *not* metastable. It is consistent with this lack of metastability to expect that the crystal under the load $\bar{f}(s_0^t)$ will deform so that μ and τ will achieve a different point B on the intersection of the constitutive curve and the load-orientation curve $\tau = \hat{\tau}_{\bar{f}(s_0^t)}(\mu)$ through the threshold point A. The quality of the equilibrium at B has then to be investigated. To this end, the set

$$\mathcal{I}_1 := \left\{(\bar{\mu}(s), \bar{\tau}(s)) \mid s \in I_1 \text{ and } \bar{\tau}(s) = \hat{\tau}_{\bar{f}(s_0^t)}(\bar{\mu}(s))\right\} \tag{2.35}$$

of points of intersection of the load-orientation curve for $\bar{f}(s_0^t)$ with the next fundamental branch $\{(\bar{\mu}(s), \bar{\tau}(s)) \mid s \in I_1\}$ of the constitutive curve is introduced. The interval I_1 determines the second fundamental branch of the constitutive curve and is given by $I_1 = [s_2^o, \mu_3^o]$. The assumptions made in the first paragraph of this subsection yield two possibilities: (1) $\mathcal{I}_1$ contains no crossing points of the load orientation curve and the second fundamental branch or (2) $\mathcal{I}_1$ contains exactly two crossing points. In the first case (1), the set $\mathcal{I}_1$ consists of at most one point, necessarily a threshold point and, therefore, not metastable. This case would mean that the crystal under the threshold load $\bar{f}(s_0^t)$ will not deform from the threshold slip $s_0^t \in I_0$ by means of a two-level shear with slip $s = \mu - \gamma \in I_1$. For the analysis of the case (2) the slip s_2^c may be highlighted to be the smaller of the two slips in the interval $(s_2^o, \mu_3^o) \subset I_1$ for which $B := (\bar{\mu}(s_2^c), \bar{\tau}(s_2^c)) \in \mathcal{I}_1$ is a crossing point, as depicted in the Figure 4 cited above. It is possible to show (Deseri and Owen II (2002), Sect. 5) that B is metastable and corresponds to a lower Gibbs free energy for the crystal than does the threshold point $A = (\bar{\mu}(s_0^t), \bar{\tau}(s_0^t))$. The fact that the load-orientation curve crosses the fundamental branch $\mathcal{I}_1$ at the slip $s_2^c > s_0^t$ implies that relation (2.30) is satisfied with $s = s_2^c$ and, therefore, that B is a metastable point.

In order to compare the Gibbs free energies at A and at B, the points $C := (\mu_1^o, 0)$ and $D := (s_2^o, 0)$ are introduced to denote the points at the end of the first fundamental branch $\mathcal{I}_0$ and at the beginning of the second fundamental branch $\mathcal{I}_1$, respectively (see Figure 4 in Deseri and Owen II (2002)). The change in Helmholtz free energy between B and A can then be written as follows:

$$\Psi_B - \Psi_A = (\Psi_B - \Psi_D) + (\Psi_D - \Psi_C) + (\Psi_C - \Psi_A), \tag{2.36}$$

and the work done on going from A to B may be evaluated by noting that

$$\int_{\Gamma_D^B \cup \bar{\Gamma}_C^D \cup \Gamma_A^C} \tau \, d\mu = \int_{\Gamma_D^B} \tau \, d\mu + \int_{\bar{\Gamma}_C^D} \tau \, d\mu + \int_{\Gamma_A^C} \tau \, d\mu,$$

where:

$\Gamma_D^B := \{(\bar{\mu}(s), \bar{\tau}(s)) \mid s \in [s_2^o, s_2^c]\}$,
$\bar{\Gamma}_C^D := \{(\mu, 0) \mid \mu \in [\mu_1^o, s_2^o]\}$,
$\Gamma_A^C := \{(\bar{\mu}(s), \bar{\tau}(s)) \mid s \in [s_0^t, \mu_1^o]\}$.

Relations (2.10), (2.21), the definition of the points μ_{2j+1}^o above and the periodicity of ψ_1 imply that

$$\Psi_D - \Psi_C = -p\psi_1(\frac{1}{2}), \; \mathcal{E}_C - \mathcal{E}_O = \Psi_C - \Psi_O = p\psi_1(\frac{1}{2}) \tag{2.37}$$

where O denotes the origin in the μ-τ plane. It is worth noting that $\tau = 0$ at each point of $\bar{\Gamma}_C^D$ and that relations (2.36)-(2.37) yield $\Psi_B - \Psi_A = \int_{\Gamma_D^B \cup \bar{\Gamma}_C^D \cup \Gamma_A^C} \tau \, d\mu - p\psi_1(\frac{1}{2}) = \int_{\Gamma_D^B \cup \bar{\Gamma}_C^D \cup \Gamma_A^C \cup \mathcal{L}_B^A} \tau \, d\mu + \int_{\mathcal{L}_A^B} \tau \, d\mu - p\psi_1(\frac{1}{2})$, where

$$\mathcal{L}_A^B := \left\{(\bar{\mu}(s), \hat{\tau}_{\bar{f}(s_0^t)}(\bar{\mu}(s)) \mid s \in [s_0^t, s_2^c]\right\}$$

is the segment of the load-orientation curve that connects the threshold point $A = (\bar{\mu}(s_0^t), \bar{\tau}(s_0^t))$ to the metastable crossing point $B = (\bar{\mu}(s_2^c), \bar{\tau}(s_2^c))$, and $\mathcal{L}_B^A$ is its reversal. Green's Theorem may now be applied to transform the line integral along the simple, closed, positively-oriented curve $\Gamma_D^B \cup \bar{\Gamma}_C^D \cup \Gamma_A^C \cup \mathcal{L}_B^A$ to obtain the formula

$$\Psi_B - \Psi_A = -area(\mathcal{A}_{ACDBA}) - p\psi_1(\frac{1}{2}) + \int_{\bar{\mu}(s_0^t)}^{\bar{\mu}(s_2^c)} \hat{\tau}_{\bar{f}(s_0^t)}(\mu) d\mu \tag{2.38}$$

with $\mathcal{A}_{ACDBA}$ denoting the interior of $\Gamma_D^B \cup \bar{\Gamma}_C^D \cup \Gamma_A^C \cup \mathcal{L}_B^A$. From (2.21) and (2.38) it follows that

$$\mathcal{E}_B - \mathcal{E}_A = -area(\mathcal{A}_{ACDBA}) - p\psi_1(\frac{1}{2}) < 0, \tag{2.39}$$

so that the Gibbs free energy at the threshold point A is greater than the Gibbs free energy at the metastable crossing point B. Because $\mathcal{E}_B < \mathcal{E}_A$, this situation may be interpreted to mean that it is energetically favorable for the crystal under the fixed load $\bar{f}(s_0^t)$ to deform from the two-level shear with $\mu = \bar{\mu}(s_0^t), \gamma = (\varphi')^{-1}(\bar{\tau}(s_0^t))$ to the two-level shear with $\mu = \bar{\mu}(s_2^c), \gamma = (\varphi')^{-1}(\bar{\tau}(s_2^c))$. The mechanism for this deformation may be described as a *material instability at the mesolevel*: the slip s experiences jumps in this deformation from the threshold value $s_0^t \in I_0$ to the value $s_2^c \in I_1$. Obviously, there are corresponding jumps in the macroshear μ, in the shear without slip γ, and in the free energy Ψ, and the relation (2.38) ensures that the change in free energy in this jump is less than the work done by the applied threshold load. Thus, the dissipation inequality

$$\Psi_B - \Psi_A \leq \int_{\bar{\mu}(s_0^t)}^{\bar{\mu}(s_2^c)} \hat{\tau}_{\bar{f}(s_0^t)}(\mu) d\mu \tag{2.40}$$

is satisfied with strict inequality, i.e., *the jump in slip $s_2^c - s_0^t$ is dissipative.* In order to complete the analysis of case (2), the second crossing point on the second fundamental branch $E := (\bar{\mu}(\tilde{s}_2^c), \bar{\tau}(\tilde{s}_2^c))$ can be considered, where $s_2^c < s_2^m < \tilde{s}_2^c \in I_1$. It is then not difficult to show that the following relation holds:

$$\mathcal{E}_E - \mathcal{E}_B = \Psi_E - \Psi_B - \int_{\mathcal{L}_B^E} \tau \, d\mu = area(int(\Gamma_B^E \cup \mathcal{L}_E^B)) > 0. \tag{2.41}$$

Between the two crossing points B and E in the fundamental branch $\mathcal{I}_1$, the point B corresponds to a lower Gibbs free energy for the crystal, so that, under the threshold load $\bar{f}(s_0^t)$, the dissipative jump in slip from s_0^t to s_2^c is energetically more favorable than that from s_0^t to $\tilde{s}_2^c$. Because s_2^c corresponds to a metastable crossing point that has lowest energy among the crossing points in that fundamental branch, the slip s_2^c is said to be *attainable* from s_0^t under the threshold load $\bar{f}(s_0^t)$, and, analogously, that the crossing point $B = (\bar{\mu}(s_2^c), \bar{\tau}(s_2^c))$ is said to be *attainable* from the threshold point $A = (\bar{\mu}(s_0^t), \bar{\tau}(s_0^t))$ under that load.

The monotonically decreasing energy profile of successive attainable points for a given threshold load permits a crystal to undergo further dissipative shearing from a given attainable point for that load. Nevertheless, the passage from one attainable point $(\bar{\mu}(s_{2j}^c), \bar{\tau}(s_{2j}^c)) \in \mathcal{I}_j$ to another $(\bar{\mu}(s_{2k}^c), \bar{\tau}(s_{2k}^c)) \in \mathcal{I}_k$ for a given threshold load entails overcoming an energy barrier, analogous to that calculated in (2.41), between the two crossing points $(\bar{\mu}(s_{2j}^c), \bar{\tau}(s_{2j}^c))$ and $(\bar{\mu}(\tilde{s}_{2j}^c), \bar{\tau}(\tilde{s}_{2j}^c))$ in $\mathcal{I}_j$ (see Figure 3, Deseri and Owen II (2002)). Depending upon the availability of energy for the crystal in the loading device, this and successive energy barriers may provide a limitation on the attainable slips.

The alternating shearing without dissipation along fundamental branches of the constitutive curve and shearing with dissipation along threshold load-orientation curves produces a rising "staircase locus" as depicted in Figure 5 in (Deseri and Owen II (2002)). Qualitatively similar staircase loci are observed in experiments on single crystals in soft loading devices, and their appearance is called the Portevin-le Chatelier effect (see e.g. Sharpe (1966), Bell (1968), Bell (1973)).

3. Structured deformations and the geometry of multiple slip in single crystals

3.1. **Introduction.** Counterparts of the kinematical decompositions postulated in the classical formulations of single crystal plasticity are derived in this section. These relations and their derivation within the framework of structured deformations were obtained in (Deseri and Owen I (2002)). The framework of classical crystal plasticity is described, for instance, in (Bassani (1993)). Although other theories have been developed to include strain gradient effects (see e.g. Gurtin (2002), and references cited therein) these effects are not addressed here.

Unlike the classical formulations, where a notion of (virtual) local intermediate configuration is required and some constitutive attributes are usually assigned at the outset to the kinematical objects, the formulation followed here is free of such notions. In particular, all the derivations here are purely geometrical. Structured deformations lead in fact to a rigorous identification at the outset of the part G of the macroscopic deformation ∇g that is due to smooth deformations at smaller length scales (Del Piero and Owen (1993), Del Piero (2004), Owen (2004)). The *disarrangement tensor* $M := F - G$, where $F := \nabla g$, is the part of the macroscopic deformation solely due to non-smooth deformations at smaller length scales. The further condition $\det F = \det G$ that defines an *invertible* structured deformation (Del Piero and Owen (1993)) permits one to specify that disarrangements occur with no volume change. *Crystallographic slip systems* are identified in the usual way by the diads $s^{(a)} \otimes m^{(a)}$ associated with the geometry of the crystal in a given reference configuration: $s^{(a)}$ is a unit vector in the direction of slip, and $m^{(a)}$ is a unit vector normal to the plane associated with $s^{(a)}$.

Central to the kinematics of multiple slip in classical crystalline plasticity are two assumptions (see Bassani (1993) and Asaro (1983)):

(1) the macroscopic deformation gradient F can be factored as $F = F^e F^p$, where F^e is the 'elastic" part and F^p is the 'plastic" part of F; indeed, F^p is meant to represent residual deformation after a material element is unloaded, and F^e is intended to represent stretching and rotation of the crystalline lattice;

(2) the rate of change of residual deformation relative to the lattice after slip is assumed to be given by

$$\dot{F}^p F^{p-1} = \sum_{a=1}^{A} \dot{\gamma}^{(a)} s^{(a)} \otimes m^{(a)} \tag{3.1}$$

in which $\dot{\gamma}^{(a)}$ is meant to represent the slipping rate of the ath-slip system.

The above-mentioned framework for crystalline plasticity has led to successful calculations of geometrical changes and forces in single crystals that provide a practical justification for the framework. Nevertheless, the two assumptions above contain a number of primitive quantities F^e, F^p, $\dot{\gamma}^{(a)}$ along with assumptions about their interrelation, leading to a variety of questions raised in (Deseri and Owen I (2002), Sect. 1).

One approach to a rational framework for the geometry of multiple slip is provided by the theory of structured deformations. In particular, a multiplicative decomposition $F = GM_r$ follows immediately from the additive decomposition (see (3.7) below) for the macroscopic deformation gradient $F = G+M = G(I+G^{-1}M)$. Here, the factors G and $M_r := I + G^{-1}M$ have precise meaning in terms of the lattice distortion and slips as provided by the identification relations for M and G (Del Piero and Owen (1993)), recalled in the next subsection. The form of the relation $F = GM_r$ and the fact that G is the deformation due to smooth changes at smaller length scales do suggest an analogy with the standard relation $F = F^eF^p$.

The analogy between F^eF^p and GM_r is further strengthened by comparison of the form of the relation (3.45) derived in the sequel (see also Deseri and Owen I (2002) eqn(6.8)):

$$\dot{M}_r M_r^{-1} = \sum_{a=1}^{A} \lambda^{(a)} s^{(a)} \otimes m^{(a)} \tag{3.2}$$

with the form of (3.1) assumed in classical treatments. This analogy has to be cautiously approached, because in $F = GM_r$ all of the quantities have been identified in a rigorous way and are purely geometrical, whereas in the relation $F = F^eF^p$ the quantities F^e and F^p are described informally, often in terms of the concepts of reversible loading and unloading as well as irreversible slips and, hence, are not purely geometrical. In (Havner (1992), pp. 35 - 38), a factorization $A^*\tilde{A}$ corresponding to F^eF^p is introduced without the attribution of constitutive properties to the factors A^* and $\tilde{A}$, and with the use of purely geometrical terms in the interpretation of the factors. However, the (local) intermediate configuration associated with $\tilde{A}$ (Havner (1992), pp. 35 - 36) and the incremental slips $d\tilde{\gamma}$ (Havner (1992), p.37, eqn (3.2)) from that configuration are not computed from definite mappings of one region of the crystal into another. In contrast, the present approach based on structured deformations provides actual deformations of the body (the piecewise smooth deformations f_n in (1.3) that approximate a given invertible structured deformation) in terms of which slips and lattice distortion can be computed through the identification relations (3.5), (3.6), (3.10) computed in the following section.

3.2. **Invertible Structured Deformations.** Specification of an invertible structured deformation requires giving a piecewise smooth, injective mapping g that represents the macroscopic deformation of a body, as well as giving a piecewise continuous tensor field G that satisfies the equation

$$\det G(X) = \det \nabla g(X) \tag{3.3}$$

at every point X in the reference configuration $\mathcal{A}$ (Del Piero and Owen (1993), Del Piero (2004)). Equation (3.3) asserts that the change in volume without disarrangements equals the macroscopic change in volume. Following Del Piero and Owen (1993), Del Piero (2004), a structured deformation (g, G) satisfying (3.3) at every point X is called an *invertible structured deformation*, and the set of invertible structured deformations (g, G) from a given reference configuration $\mathcal{A}$ is denoted by $InvStd(\mathcal{A})$. In the following, g and G are required to be smooth to avoid technical complications.

The Approximation Theorem (Del Piero and Owen (1993)), valid for any structured deformation, then assures that there exists a sequence $n \mapsto f_n$ of piecewise smooth, injective mappings (a *determining sequence* for g and G) satisfying

$$g = \lim_{n\to\infty} f_n \qquad \text{and} \qquad G = \lim_{n\to\infty} \nabla f_n, \tag{3.4}$$

with the precise sense of convergence not relevant to the present discussion (see Del Piero (2004) for a discussion of appropriate notions of convergence). Moreover, for every determining sequence, the following "identification relations" can be proven (Del Piero and Owen (1993), Del Piero and Owen (1995)):

$$M(X) := \nabla g(X) - G(X) = \lim_{r\to 0}\lim_{n\to\infty} \frac{\int_{\Gamma(f_n)\cap\mathcal{B}(X,r)} [f_n](Y)\otimes\nu(Y)\, dA_Y}{vol\mathcal{B}(X,r)}, \tag{3.5}$$

$$G(X) = \lim_{r\to 0}\lim_{n\to\infty} \frac{\int_{\mathcal{B}(X,r)} \nabla f_n(Y)\, dV_Y}{vol\mathcal{B}(X,r)}, \tag{3.6}$$

where $\mathcal{B}(X,r)$ denotes the open ball centered at X of radius r, $vol\mathcal{B}(X,r)$ is its volume, $\Gamma(f_n)$ is the set on which f_n has jumps, $[f_n](Y)$ is the jump of f_n at the point $Y \in \Gamma(f_n)$, and $\nu(Y)$ is a unit normal to $\Gamma(f_n)$ at Y. The relation (3.5) shows that $M(X)$ is a double limit of surface integrals of the jumps in approximating deformations, so that the tensor $M(X)$ is identifiable as the deformation due to disarrangements (Owen, 1995). In (3.6), $G(X)$ is computed as a double limit of volume averages of deformation gradients of approximating deformations, and we thus are justified calling $G(X)$ the *deformation without disarrangements.* In view of the identification relations (3.5) and (3.6), the simple formula

$$\nabla g(X) = G(X) + (\nabla g(X) - G(X)) = G(X) + M(X) \tag{3.7}$$

becomes a rigorously derived, *additive decomposition of the macroscopic deformation gradient into the deformation without disarrangements* $G(X)$, *reflecting smooth geometrical changes at a smaller length scale, and the deformation due to disarrangements* $M(X)$, *reflecting non-smooth geometrical changes at a smaller length scale.*

An immediate issue that can be clarified in the setting of (not necessarily invertible) structured deformations is that of multiplicative decompositions of the deformation gradient $F := \nabla g$. In fact, multiplicative decompositions can be obtained starting from the additive decomposition (Del Piero and Owen (1993)). In particular, (3.7) implies the following multiplicative decomposition of F :

$$F = GM_r, \tag{3.8}$$

after setting (Del Piero and Owen (1993))

$$M_r := I + G^{-1}M \tag{3.9}$$

in (3.7). Going beyond the analysis provided in (Del Piero and Owen (1993)), it can be noted that substitution of the identification relation (3.5) for M into the formula defining M_r provides the new identification relation:

$$M_r(X) = I + \lim_{r\to 0}\lim_{n\to\infty} \frac{\int_{\Gamma(f_n)\cap\mathcal{B}(X,r)} G(X)^{-1}[f_n](Y)\otimes\nu(Y)\,dA_Y}{vol\mathcal{B}(X,r)}. \tag{3.10}$$

This relation shows that $M_r(X)$ equals the identity I if there are no jumps in the approximating deformations f_n. Moreover, because the jump in position $[f_n](Y)$ represents a vector determined by points in the deformed configuration of the body, the factor $G(X)^{-1}[f_n](Y)$ in the identification relation (3.10) for $M_r(X)$ may be interpreted as a "pull-back"of the jump $[f_n](Y)$ by the deformation without disarrangements $G(X)$. Hence, $M_r(X)$ *measures the deformation due to disarrangements, pulling back the jumps via the deformation without disarrangements.*

It is important to remember that M and M_r are measures of the effect at the macrolevel of non-smooth deformations at a smaller length scale in the precise sense given in the identification relations (3.5) and (3.10). As such, these tensors are purely geometrical quantities and, in the present context, do not carry the attributes of reversibility or irreversibility. Only with the prescription of constitutive relations would reversibility and irreversibility become meaningful.

A useful identification relation for the quantity

$$tr(M_r(X) - I) = tr(M(X)G(X)^{-1}) = tr(G(X)^{-1}M(X))$$

can be derived from (3.9) and (3.10):

$$tr(M_r(X) - I) = \lim_{r\to 0}\lim_{n\to\infty} \frac{\int_{\Gamma(f_n)\cap\mathcal{B}(X,r)} G(X)^{-1}[f_n](Y)\cdot\nu(Y)\,dA_Y}{vol\mathcal{B}(X,r)}. \tag{3.11}$$

In the following the dependence upon X will be omitted, unless specifically required by the context.

According to the identification relation (3.11), the vanishing of $tr(MG^{-1})$ for a given structured deformation indicates that, on average, the deformations due to disarrangements arise through tangential jumps. This provides the motivation for introducing the following subspace of $Lin\mathcal{V}$:

$$\{G^{-T}\}^{\perp} := \{A \in Lin\mathcal{V}\ |tr(AG^{-1}) = A\cdot G^{-T} = 0\}. \tag{3.12}$$

If $\mathbb{P}^0$ is the (orthogonal) projector from $Lin\mathcal{V}$ onto $\{G^{-T}\}^{\perp}$, i.e.

$$\mathbb{P}^0 := \mathbb{I} - \frac{G^{-T}\otimes G^{-T}}{|\,G^{-T}\,|^2},$$

then, for a given disarrangement tensor M, the tensor $M^0 := \mathbb{P}^0[M]$ is the closest element of $\{G^{-T}\}^{\perp}$to M.

If an invertible structured deformation $(g,G) \in InvStd(\mathcal{A})$ and a point $X \in \mathcal{A}$ are given, $M(X) = \nabla g(X) - G(X)$ and the corresponding $M^0(X)$ are such that their mutual distance is given by the formula:

$$\left|M - M^0\right| = \frac{\left|M\cdot G^{-T}\right|}{\left|G^{-T}\right|} = \frac{\left|tr(MG^{-1})\right|}{\left|G^{-T}\right|}. \tag{3.13}$$

An estimate for this distance is the starting point for the subsequent approximations of disarrangements by-slips:

$$\left|M - M^0\right| \leq \left(2\left|G^{-T}\right| + |M|\left|G^{-T}\right|^{-1}\det G^{-1}\right)|M|^2. \tag{3.14}$$

This estimate provides a bound for the distance from M to the given subspace $\{G^{-T}\}^{\perp}$ that goes to zero with the square of the norm of M. This bound is valid even for large disarrangements, but it provides accurate approximations as disarrangements become small. The fact that the coefficient $2\left|G^{-T}\right| + |M|\left|G^{-T}\right|^{-1}\det G^{-1}$ of $|M|^2$ in (3.14) depends continuously upon G and M implies that *there exist positive numbers k^0 and η^0 such that for each $(g, G) \in InvStd(\mathcal{A})$ and $X \in \mathcal{A}$:*

$$\left|M(X) - M^0(X)\right| \leq k^0 \left|M(X)\right|^2 \tag{3.15}$$

whenever $|G(X) - I|$ and $|M(X)|$ are less than η^0. A proof of (3.14) is given in (Deseri and Owen I (2002), Sect. 3).

3.3. **Slip-approximations for deformations of single crystals.** Any description of the deformation due to disarrangements that can occur in a given crystal must take into account observed restrictions at sub-macroscopic length scales on the directions of the disarrangements and on the orientation of the disarrangement sites. These restrictions on sub-macroscopic geometrical changes are included within the present geometrical setting through the choice of a subspace of $\{G^{-T}\}^{\perp}$, a subspace that provides an exact description of the restrictions just mentioned. In this section an approximation to the deformation due to disarrangements M that lies in the identified subspace is given, and error-bounds that decrease with the square of the norm of M are provided.

For a crystal in the reference configuration $\mathcal{A}$ the crystallographic data required for the analysis of multiple slip consist of pairs of orthogonal unit vectors $s^{(a)}, m^{(a)}$, $a = 1, \cdots, A$, with A the number of potentially active slip systems. The vector $s^{(a)}$ is the direction of slip, and the vector $m^{(a)}$ is the normal to the slip plane for the ath slip-system (Asaro (1983)). These data and the deformation without disarrangements $G(X)$ determine the subspace of $Lin\mathcal{V}$

$$\mathcal{C}(G(X)) := Span\left\{G(X)s^{(a)} \otimes m^{(a)} \mid a = 1, \cdots, A\right\}. \tag{3.16}$$

In fact, $\mathcal{C}(G(X))$ is a subspace of $\{G(X)^{-T}\}^{\perp}$, because

$$\begin{aligned}(G(X)s^{(a)} \otimes m^{(a)}) \cdot G(X)^{-T} &= tr(G(X)s^{(a)} \otimes m^{(a)}G(X)^{-1}) \\ &= tr(s^{(a)} \otimes m^{(a)}) = 0.\end{aligned}$$

Henceforth, the symbol $M_{\mathcal{C}(G(X))}$ denotes the tensor in $\mathcal{C}(G(X))$ that is closest to the disarrangement tensor $M(X)$. It is worth noting that, when $\mathcal{C}(G(X))$ is a proper subspace of $\{G(X)^{-T}\}^{\perp}$, the tensor $M_{\mathcal{C}(G(X))}$ may be different from the tensor $M^0(X)$. Furthermore, the fact that $M_{\mathcal{C}(G(X))} \in \mathcal{C}(G(X)$, implies that there exists a list of scalars $\{\delta^{(a)}(X) : a = 1, ..., A\}$ such that:

$$M_{\mathcal{C}(G(X))} = \sum_{a=1}^{A} \delta^{(a)}(X) G(X) s^{(a)} \otimes m^{(a)}. \tag{3.17}$$

Moreover because $M^0 = \mathbb{P}^0[M]$, both $M_{\mathcal{C}(G(X))}$ and $M^0(X)$ are in $\{G(X)^{-T}\}^{\perp}$ and the following relationship holds:

$$\left|M(X) - M_{\mathcal{C}(G(X))}\right|^2 = \left|M(X) - M^0(X)\right|^2 + \left|M^0(X) - M_{\mathcal{C}(G(X))}\right|^2. \tag{3.18}$$

Unlike $\left|M - M^0\right|$ for which there exists the estimate (3.15), there appears to be no counterpart for the quantity $\left|M^0 - M_{\mathcal{C}(G)}\right|$ valid for arbitrary invertible structured deformations. The aim now is to provide a non-empty subset of $InvStd(\mathcal{A})$ where such an approximation could be valid.

To this end, a collection of invertible structured deformations for which the differences $\left|M^0(X) - M_{\mathcal{C}(G(X))}\right|$ do satisfy an estimate of the form (3.15) can be identified. Specifically, for given crystallographic data $s^{(a)}, m^{(a)}$, $a = 1, \cdots, A$, the symbol $Crys(\mathcal{A})$ will denote the set of all invertible structured deformations $(g, G) \in InvStd(\mathcal{A})$ such that for every $X \in \mathcal{A}$:

$$\left|M^0(X) - M_{\mathcal{C}(G(X))}\right| \leq k^0 \left|M(X)\right|^2 \tag{3.19}$$

whenever $|G(X) - I|$ and $|M(X)|$ are less than η^0; here $k^0 > 0$ and $\eta^0 > 0$ are the numbers identified above (3.15). It is worth noting that the set $Crys(\mathcal{A})$ just introduced is not empty: the set $Crys(\mathcal{A})$ contains indeed all structured deformations in which g is a homogeneous deformation, G is a constant, and $M = \nabla g - G$ is a constant of the form

$$M = \sum_{a=1}^{A} \beta^{(a)} G s^{(a)} \otimes m^{(a)} + \alpha G^{-T}, \tag{3.20}$$

where the only restriction on the numbers $\beta^{(a)}$ and α is imposed by the relation $\det(G + M) = \det G$. In this example, $M^0(X) = M_{\mathcal{C}(G(X))} = \sum_{a=1}^{A} \beta^{(a)} G s^{(a)} \otimes m^{(a)}$.

The main result on multiple slip obtained in (Deseri and Owen I (2002)) follows immediately from (3.15), (3.18), and (3.19) and gives an estimate for the difference between $M(X)$ and an appropriate combination of crystallographic diads $G(X)s^{(a)} \otimes m^{(a)}$: *there exist positive numbers k and η such that, for each $X \in \mathcal{A}$ and each invertible structured deformation $(g, G) \in Crys(\mathcal{A})$, there are numbers $\delta^{(a)}(X)$, $a \in \{1, \cdots, A\}$ for which*

$$\left|M(X) - M_{\mathcal{C}(G(X))}\right| \leq k \left|M(X)\right|^2 \tag{3.21}$$

whenever $|G(X) - I|$ and $|M(X)|$ are less than η, with $M_{\mathcal{C}(G(X))}$ of the form in (3.17).

Because of its representation formula , the tensor $M_{\mathcal{C}(G(X))}$ appearing in (3.17) is called a *slip-approximation of* $M(X)$ for the given single crystal. However, the numbers $\delta^{(a)}(X)$, $a \in \{1, \cdots, A\}$, in general, are not uniquely determined by this formula. If the set of diads $\left\{G(X)s^{(a)} \otimes m^{(a)} \mid a = 1, \cdots, A\right\}$ is a linearly independent subset of $Lin\mathcal{V}$, so that it is a basis of $\mathcal{C}(G(X))$, then $\delta^{(a)}(X)$, $a \in \{1, \cdots, A\}$ are uniquely determined. In any case, we call $\delta^{(a)}(X)$ the *shear in the ath slip-system* for the slip-approximation $M_{\mathcal{C}(G(X))}$ to $M(X)$. Even when this set of diads is linearly dependent, the inequality

(3.25) tells us that, *as the deformation without disarrangements* $M(X)$ *tends to zero, the slip-approximation* $M_{\mathcal{C}(G(X))}$ *converges quadratically to* $M(X)$.

It is worth noting that the approximation (3.21) for $M(X)$ yields immediately an approximation for the tensor $M_r(X)$ defined in (3.9) (see Deseri and Owen I (2002)).

A useful decomposition of the disarrangement tensor can now be considered in light of the relations developed above:

$$M = M_{\mathcal{C}(G)} + \left(M^0 - M_{\mathcal{C}(G)}\right) + \left(M - M^0\right),$$

where the second and the third terms on the right-hand side arise in (3.19) and (3.15), respectively. The identification relation (3.5) and the representation formula for $M_{\mathcal{C}(G)}$, (3.17), provide the following interpretation of the terms in the decomposition for M:

Term	**Interpretation**
$M_{\mathcal{C}(G)}$	crystallographic slip-like disarrangements
$M^0 - M_{\mathcal{C}(G)}$	non-crystallographic slip-like disarrangements
$M - M^0$	non-slip-like disarrangements.

3.4. **The case of f.c.c. crystals.** In the special case of face-centered cubic crystals, a simple analysis based on an observation of Cermelli (Cermelli, 2000) shows that the dimension of the crystallographic subspace $\mathcal{C}(G(X))$ defined in (3.16) is eight and, therefore, equals the dimension of the subspace $\left\{G(X)^{-T}\right\}^{\perp}$. Because $\mathcal{C}(G(X))$ is itself a subspace of $\left\{G(X)^{-T}\right\}^{\perp}$, we conclude that

$$\mathcal{C}(G(X)) = \left\{G(X)^{-T}\right\}^{\perp}, \tag{3.22}$$

so that the projections of $M(X)$ on the two subspaces are the same:

$$M^0(X) = M_{\mathcal{C}(G(X))}. \tag{3.23}$$

Consequently, the inequality (3.19) that defines the collection of structured deformations $Crys(\mathcal{A}) \subset InvStd(\mathcal{A})$ is satisfied for every invertible structured deformation, i.e.,

$$Crys(\mathcal{A}) = InvStd(\mathcal{A}). \tag{3.24}$$

Therefore, the main result on slip approximations in Section 3.3 can be restated in the case of f.c.c. crystals as: *there exist positive numbers* k *and* η *such that, for each* $X \in \mathcal{A}$ *and each* $(g, G) \in InvStd(\mathcal{A})$, *there are numbers* $\delta^{(a)}(X)$, $a \in \{1, \cdots, A\}$ *for which*

$$\left| M(X) - \sum_{a=1}^{A} \delta^{(a)}(X) G(X) s^{(a)} \otimes m^{(a)} \right| \leq k \left|M(X)\right|^2 \tag{3.25}$$

whenever $|G(X) - I|$ *and* $|M(X)|$ *are less than* η.

3.5. **Time-parameterized families of invertible structured deformations.** In order to follow the deformation of a body in time, we consider a time-interval $\mathcal{I}$ and assume that for each time $\tau \in \mathcal{I}$ there is given an invertible structured deformation $(g(\cdot,\tau), G(\cdot,\tau)) \in InvStd(\mathcal{A})$. We denote by $F(\cdot,\tau)$, $G(\cdot,\tau), M(\cdot,\tau)$ and $M_r(\cdot,\tau)$ the fields on $\mathcal{A}$ at time τ that correspond to F, G, M, M_r introduced in the Sect. 3.2. In particular, the fields at time τ obey additive and multiplicative decompositions as well as identification relations corresponding to that section.

In order to describe *structured deformations relative to the configuration* $\mathcal{A}(t)$ occupied by the body at time $t \in \mathcal{I}$, for each $x \in \mathcal{A}(t)$ and $\tau \in \mathcal{I}$ we define the following fields:

$$g_t(x,\tau) := g(X,\tau) \qquad \text{and} \qquad G_t(x,\tau) := G(X,\tau)G(X,t)^{-1}, \tag{3.26}$$

where $X \in \mathcal{A}$ is determined by the relation $x = g(X,t)$, so that $X = g(\cdot,t)^{-1}(x)$. It can be shown that $(g_t(\cdot,\tau), G_t(\cdot,\tau))$ is an invertible structured deformation from $\mathcal{A}(t)$. The fields $F_t(\cdot,\tau) := \nabla g_t(\cdot,\tau)$, $G_t(\cdot,\tau)$, and $M_t(\cdot,\tau)$ describe, respectively, the (macroscopic) deformation gradient, the deformation without disarrangements, and the deformation due to disarrangements at time τ, each relative to the configuration at time t. The additive decomposition of deformation gradient and the identification relation become in this context

$$F_t(x,\tau) = G_t(x,\tau) + M_t(x,\tau), \tag{3.27}$$

$$M_t(x,\tau) := F_t(x,\tau) - G_t(x,\tau) = \lim_{r\to 0}\lim_{n\to\infty} \frac{\int_{\Gamma_t(f_n(\cdot,\tau))\cap\mathcal{B}(x,r)}[f_n](y,\tau)\otimes\nu(y,\tau)\,dA_y}{vol\mathcal{B}(x,r)}, \tag{3.28}$$

with $n \mapsto f_n(\cdot,\tau)$ a determining sequence for the invertible structured deformation $(g_t(\cdot,\tau), G_t(\cdot,\tau))$, with $\Gamma_t(f_n(\cdot,\tau))$ the set of jump points of the nth-term $f_n(\cdot,\tau)$, with $\nu(y,\tau)$ a unit normal to $\Gamma_t(f_n(\cdot,\tau))$ at $y \in \Gamma_t(f_n(\cdot,\tau))\cap\mathcal{B}(x,r)$, and with $\mathcal{B}(x,r) \subset \mathcal{A}(t)$ the open ball of radius r centered at the point $x \in \mathcal{A}(t)$.

Appropriate rates of deformation relative to the configuration at time t are needed in the sequel. To this end, we note the relation

$$F_t(x,\tau) = F(X,\tau)F(X,t)^{-1}, \qquad X = g(\cdot,t)^{-1}(x) \tag{3.29}$$

among the relative deformation gradient and the deformation gradients at times τ and t. In addition to the smoothness tacit in the assumption $(g(\cdot,\tau), G(\cdot,\tau)) \in InvStd(\mathcal{A})$ for each $\tau \in \mathcal{I}$, it is assumed that $F(X,\cdot)$ and $G(X,\cdot)$ are differentiable. Relations(3.26), (3.28), and (3.27) then imply that $F_t(x,\cdot)$, $G_t(x,\cdot)$, and $M_t(x,\cdot)$ not only are differentiable but also satisfy the relation

$$\frac{\partial}{\partial\tau}F_t(x,\tau) = \frac{\partial}{\partial\tau}G_t(x,\tau) + \frac{\partial}{\partial\tau}M_t(x,\tau) \tag{3.30}$$

for every $t,\tau \in \mathcal{I}$. For each $X \in \mathcal{A}$ and $t \in \mathcal{I}$ relative rates of deformation $L(X,t)$, $L_G(X,t)$, and $L_M(X,t)$ can be defined through the following relations:

$$L(X,t) := \frac{\partial}{\partial\tau}F_t(g(X,t),\tau)\mid_{\tau=t}, \tag{3.31}$$

$$L_G(X,t) := \frac{\partial}{\partial\tau}G_t(g(X,t),\tau)\mid_{\tau=t}, \tag{3.32}$$

$$L_M(X,t) := \frac{\partial}{\partial \tau} M_t(g(X,t),\tau) \,|_{\tau=t} \,. \tag{3.33}$$

While $L(X,t)$ is the standard measure of rate of deformation relative to the configuration at time t (i.e. the spatial gradient of the velocity), the rates $L_G(X,t)$ and $L_M(X,t)$ are introduced here by analogy to the standard one. Relations (3.29), (2.15) and the identification relation for $G(X,t)$ motivate the following terminology for $L_G(X,t)$ and $L_M(X,t)$: $L_G(X,t)$ is called the *relative rate of deformation without disarrangements* and $L_M(X,t)$ is called the *relative rate of deformation due to disarrangements.* From (3.26), (3.29), and (3.30) the following formulae can be obtained:

$$L(X,t) = \dot{F}(X,t)F(X,t)^{-1}, \tag{3.34}$$

$$L_G(X,t) = \dot{G}(X,t)G(X,t)^{-1}, \tag{3.35}$$

and

$$L(X,t) = L_G(X,t) + L_M(X,t), \tag{3.36}$$

i.e., the relative rate of deformation is the sum of the relative rate of deformation without disarrangements and the relative rate of deformation due to disarrangements. It is worth noting that these rates of deformation are purely kinematical quantities that, in the present discussion without specified constitutive relations, carry no attributes of reversibility or irreversibility.

The relation $M_r(X,t) = I + G(X,t)^{-1}F(X,t)$ in (3.9) together with (3.34)-(3.36) yield the following relation:

$$\dot{M}_r(X,t)M_r(X,t)^{-1} = G(X,t)^{-1}L_M(X,t)G(X,t). \tag{3.37}$$

Besides the following obvious relations:

$$F_t(x,t) = G_t(x,t) = I, \quad M_t(x,t) = 0,$$

the equation

$$trL_M(X,t) = 0 \tag{3.38}$$

follows first by differentiating the relation:

$$\det G(X,t) = \det F(X,t), \tag{3.39}$$

which characterizes any one-parameter family of invertible structured deformations, to obtain

$$(\det G(X,t))G(X,t)^{-T} \cdot \dot{G}\,(X,t) = (\det F(X,t))F(X,t)^{-T} \cdot \dot{F}\,(X,t).$$

By using (3.39) again, and by computing the trace of both sides one gets (3.38). Thus, the relative rate of deformation due to disarrangements turns out to be traceless in a time-parametrized family of invertible structured deformations. This is exactly the case in metal plasticity, where a counterpart of (3.38) is assumed to be valid, such as $trL^p = 0$, where L^p is the so called plastic part of the velocity gradient.

3.5.1. **Slip approximations relative to the current configuration.** The geometrical changes that result at each time τ due to (relative) invertible structured deformations $(g_t(\cdot,\tau), G_t(\cdot,\tau)) \in InvStd(\mathcal{A}(t))$ from a point $x = g(X,t)$ in the current configuration $\mathcal{A}(t)$ are examined in this section. To this end, the approximations derived in Sect.3.3 turn out to be useful: here the slip approximation of the disarrangement tensor $M_t(x,\tau) = \nabla g_t(x,\tau) - G_t(x,\tau)$ is computed relative to the current configuration $\mathcal{A}(t)$, and the following replacements in the relations obtained in Sections 3.2-3.3 are needed:

Original item	Replacement
$\mathcal{A}$	$\mathcal{A}(t)$
$(g(\cdot), G(\cdot))$	$(g_t(\cdot,\tau), G_t(\cdot,\tau))$
$s^{(a)}$	$s^{(a)}(x,t) := G(X,t)s^{(a)}$
$m^{(a)}$	$m^{(a)}(x,t) := G(X,t)^{-T}m^{(a)}$
$\{G(X)^{-T}\}^{\perp}$	$\{G_t(x,\tau)^{-T}\}^{\perp}$
$\mathcal{C}(G(X))$	$\mathcal{C}(G_t(x,\tau))$
$Crys(\mathcal{A})$	$Crys_x(\mathcal{A}(t))$
$\delta^{(a)}(X)$	$\delta_t^{(a)}(x,\tau)$.

Let $t \in \mathcal{I}$ and $x \in \mathcal{A}(t)$ be given, and assume that there is an open interval $\mathcal{J}(t) \subset \mathcal{I}$ such that for every $\tau \in \mathcal{J}(t)$, $(g_t(\cdot,\tau), G_t(\cdot,\tau)) \in Crys_x(\mathcal{A}(t))$. This assumption implies that the inequality (3.19) has the following counterpart: *there are positive numbers $k(x,t)$ and $\eta(x,t)$ such that, for every $\tau \in \mathcal{J}(t)$, there exist numbers $\delta_t^{(a)}(x,\tau)$, $a \in \{1,\cdots,A\}$ for which*

$$\left|M_t(x,\tau) - M_{\mathcal{C}(G_t(x,\tau))}\right| \leq k(x,t)\,|M_t(x,\tau)|^2 \tag{3.40}$$

whenever $|G_t(x,\tau) - I|$ and $|M_t(x,\tau)|$ are less than $\eta(x,t)$, with $M_{\mathcal{C}(G_t(x,\tau))}$ represented through the formula:

$$M_{\mathcal{C}(G_t(x,\tau))} = \sum_{a=1}^{A} \delta_t^{(a)}(x,\tau) G_t(x,\tau) s^{(a)}(x,t) \otimes m^{(a)}(x,t) \tag{3.41}$$

We remark that, by (3.40) and the smoothness of $G_t(x,\cdot)$ and $M_t(x,\cdot)$, the numbers $|G_t(x,\tau) - I|$ and $|M_t(x,\tau)|$ tend to zero as τ tends to t, so that (3.40) is satisfied for $|\tau - t|$ sufficiently small. It is worth noting that the collection $Crys_x(\mathcal{A}(t))$ of invertible structured deformations for which the above approximation holds does depend upon x (as well as upon t) through the diads $s^{(a)}(x,t) \otimes m^{(a)}(x,t)$. For this reason, the dependence upon x appears explicitly in the symbol $Crys_x(\mathcal{A}(t))$.

The members of (3.40) may be divided by $|\tau - t|$, so that by invoking the smoothness of the family $(g(\cdot,\tau), G(\cdot,\tau)) \in InvStd(\mathcal{A})$ together with (3.39) and (3.33), and by letting $\tau \to t$, we conclude that the ratio:

$$Q_t(x,\tau) := \frac{M_{\mathcal{C}(G_t(x,\tau))}}{\tau - t} = \sum_{a=1}^{A} \delta_t^{(a)}(x,\tau) G_t(x,\tau) s^{(a)}(x,t) \otimes m^{(a)}(x,t)/(\tau - t)$$

has a limit as τ tends to t and

$$L_M(X,t) = \lim_{\tau \to t} Q_t(x,\tau). \tag{3.42}$$

In other words, the relative rate of deformation due to disarrangements for the given family of invertible structured deformations equals the limit of the quotient $Q_t(x,\tau)$ as τ tends to t. Thus, the error in the approximations $Q_t(x,\tau)$ to $M_t(x,\tau)/(\tau - t)$ vanishes in the limit as τ tends to t. It is worth noting that $L_M(X,t)$ is traceless, by (3.38), although the quotients $Q_t(x,\tau)$ need not be. Following the terminology in Section 1.3, the tensor $M_{\mathcal{C}(G_t(x,\tau))}$ is called a *slip-approximation to the relative disarrangement tensor* $M_t(x,\tau)$, and the numbers $\delta_t^{(a)}(x,\tau)$ are called the *relative shear in the* a^{th} *slip-system* for the slip-approximation.

An argument developed in (Deseri and Owen I (2002)) shows that the limit as τ tends to t of $Q_t(x,\tau)$ is an element of $\mathcal{C}(G(X,t))$. This property allows one to choose real numbers $\lambda^{(a)}(x,t)$, $a = 1, \cdots, A$, such that

$$\lim_{\tau \to t} \sum_{a=1}^{A} \frac{\delta_t^{(a)}(x,\tau)}{\tau - t} s^{(a)}(x,t) \otimes m^{(a)}(x,t) = \sum_{a=1}^{A} \lambda^{(a)}(x,t) s^{(a)}(x,t) \otimes m^{(a)}(x,t),$$

and (3.42) yields the following *slip-representation for the relative rate of deformation due to disarrangements*:

$$L_M(X,t) = \sum_{a=1}^{A} \lambda^{(a)}(x,t)\, s^{(a)}(x,t) \otimes m^{(a)}(x,t). \tag{3.43}$$

This is a purely kinematical relation and it is based on the geometry of structured deformations and on the derived representation (3.42) for "incremental" disarrangements, without reference to constitutive relations. Nevertheless, the slip-representation for the relative rate of deformation due to disarrangements is of the same form as the assumed formula for the "plastic part" of the velocity gradient in standard treatments of crystalline plasticity (Asaro (1983), eqn(2.10)). From the outset, these treatments interpret plastic deformation as dissipative and, hence, tacitly entail constitutive assumptions. In spite of this difference, each number $\lambda^{(a)}(x,t)$ in (3.43) corresponds to the number $\dot{\gamma}^{(a)}$ in (Asaro (1983)), called there the "slipping rate of the ath slip-system". In the present derivation of (3.43) above, the numbers $\lambda^{(a)}(x,t)$ arise through limits of sums of quotients in which the numerators contain the relative shears $\delta_t^{(a)}(x,\tau)$ for each slip system and the denominators are $\tau - t$. This fact actually supports the terminology chosen in standard treatments.

It should be noted that the diads $s^{(a)}(x,t) \otimes m^{(a)}(x,t)$, $a = 1, \cdots, A$, generally are linearly dependent, and this dependence implies that the numbers $\lambda^{(a)}(x,t)$ are not uniquely determined by (3.43). In the special case where the diads $s^{(a)}(x,t) \otimes m^{(a)}(x,t)$, $a = 1, \cdots, A$, form a linearly independent subset of $Lin\mathcal{V}$, each quotient $\frac{\delta_t^{(a)}(x,\tau)}{\tau - t}$ will have a limit as $\tau \to t$, and the numbers $\lambda^{(a)}(x,t)$, $a = 1, \cdots, A$, will be given by the formulas

$$\lambda^{(a)}(x,t) = \lim_{\tau \to t} \frac{\delta_t^{(a)}(x,\tau)}{\tau - t}, \qquad a = 1, \cdots, A. \tag{3.44}$$

Accordingly, the number $\lambda^{(a)}(x,t)$ is called the *relative shear rate for the ath slip-system* (even in the case where the diads are linearly dependent).

In a similar way, the formulas for $s^{(a)}(x,t)$ and $m^{(a)}(x,t)$ (in the table at the beginning of this section) and (3.37) yield a formula for $\dot{M}_r M_r^{-1}$:

$$\begin{aligned}\dot{M}_r(X,t)M_r(X,t)^{-1} &= G(X,t)^{-1}L_M(X,t)G(X,t) \\ &= \sum_{a=1}^{A}\lambda^{(a)}(x,t)s^{(a)}\otimes m^{(a)} \qquad (3.45)\end{aligned}$$

with $x=g(X,t)$. Keeping in mind the qualifications made earlier, it is worth pointing out the similarity of the derived formula for $\dot{M}_r M_r^{-1}$and the relation (3.1) assumed for $\dot{F}^p F^{p-1}$ in standard presentations, e.g., below (2.10) in (Asaro (1983)) and (2.8) in (Bassani (1993)).

The analysis above can also be carried out when the dependence upon τ of $M_t(x,\tau)$ and $G_t(x,\tau)$ is less smooth, provided only that there exist a sequence $n\mapsto\tau_n$ of times for which $M_t(x,\tau_n)/(\tau_n-t)$ has a limit, $M_t(x,\tau_n)$ tends to zero, and $G_t(x,\tau_n)$ is bounded as $\tau_n\to t$. This weaker smoothness hypothesis may be appropriate on time scales where slip occurs in small, discrete time-steps.

References

Asaro, R. J. (1983) Crystal plasticity, *J. Appl. Mechanics* **50**, 921-934.

Barrett,C.S. (1952) *Structure of Metals*, Second Edition, McGraw-Hill, New York, Toronto, London.

Bassani, J.L. (1993) Plastic flow of crystals. In *Advances in Applied Mechanics* **30,** ed. J. W. Hutchinson and T. Y. Wu, pp. 191-258.

Bell, J. F. (1968) *The Physics of Large Deformation of Crystalline Solids*, Springer Tracts in Natural Philosophy, 14, Springer, Berlin.

Bell, J. F. (1973) The Experimental Foundation of Solid Mechanics, in *Handbuch der Physik*, Vol. VIa/1, ed. C. Truesdell, pp 1-813, Springer, Berlin.

Brown, A.F. (1952) Surface effects in plastic deformation of metals. *Adv. in Phys.* **1**, 427-493.

Cermelli, P. (2001) Private communication.

Choksi, R. and Fonseca, I. (1997) Bulk and interfacial energy densities for structured deformations of continua. *Arch. Rational Mech. Anal.* **138**, 37-103.

Choksi, R., Del Piero, G., Fonseca, I. and Owen, D. R. (1999) Structured deformations as energy minimizers in models of fracture and hysteresis, Math. Mech. Solids, **4** (3), 321-356.

Crussard, M.C. (1945) Étude des glissements plastiques dans les cristaux d'aluminum. Rev. de Métallurgie **42**, 286-294. Crussard, M.C. (1945) Étude des glissements plastiques dans les cristaux d'aluminum. Rev. de Métallurgie **42**, 286-294.

Del Piero, G. and Owen, D.R. (1993) Structured deformations of continua, *Arch. Rational Mech. Anal.* **124**, 99-155.

Del Piero, G. and Owen, D.R. (1995) Integral-gradient formulae for structured deformations, *Arch. Rational Mech. Anal.* **131**, 121-138.

Del Piero, G. (2004) Lecture Notes, *CISM course on Multiscale Modelling in Continuum Mechanics and Sturctured Deformations*, forthcoming.

Del Piero, G. and Owen D. R. (2000) *Structured deformations*, XXII Scuola Estiva di Fisica Matematica, Quaderni dell'Istituto nazionale di Alta Matematica, INDAM.

Deseri, L. and Owen D. R. (2000) *Active slip band-separation and the energetics of slip in single crystals*, Int. J. Plasticity **16**, 1411-1418.

Deseri, L. and Owen D. R. (2002) Invertible structured deformations and the geometry of multiple slip in single crystals, *Int. J. Plasticity*, **18**, 833-849, 2002..

Deseri, L. and Owen, D. R. (2002) Energetics of two-level shears and hardening of single crystals, Math. Mech. Solids **7**, no. 2, 113-147.

Gurtin, M. E. (2002) A gradient theory of single-crystal plasticity that accounts for geometrically necessary dislocations, *J. Mech. Phys. Solids* **50**, 5-32.

Havner, K.S. (1992) *Finite Plastic Deformation of Crystalline Solids.* Cambridge Monographs in Mechanics and Applied Mathematics, Cambridge University Press, Cambridge.

Hill, R. (1950) *The Mathematical Theory of Plasticity*, Oxford University Press, London.

Kubin, L.P. (1993) Dislocation patterning. In *Materials Science and Technology, Vol. 6,* ed. H. Mughrabi, pp.138-187, VCH, Weinheim, New York, Basel, Cambridge.

Neuhäuser, H.(1983) Slip-line formation and collective dislocation motion. In *Dislocations in Solids*, Vol. 6, ed. F. R. N. Nabarro, pp. 319-440, North-Holland, Amsterdam, New York, Oxford.

Owen, D.R. (1995) Disarrangements in continua and the geometry of microstructure. In *Recent Advances in Elasticity, Viscoelasticity and Inelasticity*, ed. K. Rajagopal, 67-81.

Owen, D.R. (2004) Lecture Notes, *CISM course on Multiscale Modelling in Continuum Mechanics and Structured Deformations*, Springer Verlag.

Salama, A., Shaikh, F. and Roberts, J.M. (1971) Microstrain and electron micrographic slip line studies of ordered and disordered Cu_3Au. *Acta Metallurgica* **19**, 395-404.

Sharpe, W. N. (1966) The Portevin-le Chatelier effect in aluminium single crystals and polycrystals, *Journal of the Mechanics and Physics of Solids*, 14, 187-202.

Taylor, G. I. (1958) *The Scientific Papers of Sir Geoffrey Ingram Taylor*, vol I, Mechanics of Solids, ed. G. K. Batchelor, Cambridge University Press.

Yamaguchi, K. (1928) The slip-bands produced when crystals of aluminum are stretched. Part I *Sci. Papers Inst. Phys. Chem. Research* **6**, 289-317.

Dipartimento di Ingegneria, Università di Ferrara, Ferrara, Italia
E-mail address: `deseri@ing.unife.it`

ELASTICITY WITH DISARRANGEMENTS*

David R. Owen
Department of Mathematical Sciences, Carnegie Mellon, Pittsburgh, PA USA

Part I
Elasticity with Space-Like Disarrangements

1 Introduction

The subject of elasticity provides, among many other things, a continuum field theory for the dynamical evolution of bodies that undergo large deformations, that respond to changes in geometry through a stored energy function, and that experience internal dissipation in isothermal situations only during non-smooth processes. The central position of this field theory within mechanics has provided a starting point for many approaches to obtaining field theories that capture the effects of submacroscopic material structure or of submacroscopic geometrical changes on the macroscopic evolution of a body. The developments described here employ structured deformations and structured motions in order to formulate field theories for the dynamical evolution of bodies undergoing smooth geometrical changes at the macrolevel, while undergoing only piecewise smooth geometrical changes at submacroscopic levels. In keeping with elasticity, the new field equations should describe bodies that undergo large deformations and that store energy, while, in transcending elasticity in its standard form, the field equations should permit the body to experience internal dissipation during smooth dynamical processes and should provide a connection between the internal dissipation and the non-smooth geometrical changes ("disarrangements") experienced by the body.

This exposition describes two aspects of the emerging subject "elasticity with disarrangements." Part I provides a detailed summary of the field theory [4] under development by Luca Deseri, University of Ferrara, and myself. In this field theory, the effects of submacroscopic, space-like jumps in position ("space-like" disarrangements) are taken

* The author gratefully acknowledges support from the National Science Foundation, Division of Mathematical Sciences, under Grant #0102477, and thanks Amit Acharya and Morton Gurtin for valuable comments on the research described here and Gianpietro Del Piero for his careful reading and thoughtful comments on earlier versions of this article.

into account explicitly, while the effects of submacroscopic, "time-like" disarrangements —even though they may occur—are not. Accordingly, only one additional field G, the "deformation without disarrangements," need be adjoined to the macroscopic motion χ in order to describe by means of structured deformations the multiscale evolution of the body. Indeed, the difference $\nabla\chi - G$ captures explicitly the effects of space-like disarrangements. In Part II, an additional field $\dot{\chi}_\backslash$, the "velocity without disarrangements," is adjoined, and the difference $\dot{\chi} - \dot{\chi}_\backslash$ captures the effects of time-like disarrangements. In the context of this enriched kinematical setting, a hitherto unpublished set of field relations is developed in Part II. In either setting, the richer geometry provides refined measures of stress $S_\backslash$, the stress without disarrangements, and S_d, the stress due to disarrangements. Each is determined explicitly in terms of the Piola-Kirchoff stress S and the fields χ and G, and a simple consistency relation satisfied by $S_\backslash$ and S_d provides an additional field equation for elastic bodies undergoing disarrangements.

Each of the two approaches to elasticity with disarrangements presented here begins with an assumed expression for the power expended in terms of the fields described above, each obtains balance laws by means of assumed invariance properties of the power, and each identifies constitutively the portion of the stress power that is stored via the Helmholtz free energy and the portion of the stress power that can be dissipated during isothermal motions of the body (the internal dissipation). In addition, the requirements of frame-indifference and material symmetry are studied.

Due to the present, early stage in the development of elasticity with disarrangements, only one detailed example is presented, that of an "energetically nearsighted elastic body" whose free energy only is sensitive to submacroscopic geometrical changes through its dependence solely on the volume fraction associated with a structured deformation of the body. We show in Part I that there are two phases that can arise in any such elastic body, independent of the particular free energy response: a universal spherical phase and a universal elongated phase. Each universal phase easily is expressed in terms of the "golden mean" (the number $(\sqrt{5}-1)/2$), and the field relations are derived for each phase. The field relations for the spherical phase turn out to be those for an ideal gas that, according to the dissipation inequality, can undergo only non-expansive processes. The field relations for the elongated phase allow for expansive processes, provided that the direction of submacroscopic elongation and the stretching in the macroscopic flow are appropriately aligned.

2 Structured Deformations

For present purposes, we may specify a structured deformation from a region $\mathcal{A}$, contained in a Euclidean space $\mathcal{E}$ with translation space $\mathcal{V}$, by specifying two fields $g : \mathcal{A} \to \mathcal{E}$ and $G : \mathcal{A} \to \text{Lin}\,\mathcal{V}$ called the *macroscopic deformation* and the *deformation without disarrangements*, respectively. Here, we assume that the fields g and G are smooth, although discontinuites are permitted in the piecewise-smooth approximating deformations f_n in (2.2). This assumption rules out slip and separation at the macroscopic level while permitting such discontinuities at submacroscopic levels. Precise smoothness assumptions on these fields and on the region $\mathcal{A}$ will not be specified here, but sufficient smoothness requirements on the fields can be inferred from the context. (The expositions of Del Piero [1] and Paroni [2] in this volume provide detailed smoothness assumptions for the theory of structured deformations. The absence in the present development of the permanent disarrangement site κ corresponds to setting κ equal to the empty set $\emptyset$ in Del Piero's and Paroni's treatments.) Other than smoothness assumptions, we require that g be injective and that there exists a positive number m such that the inequalities

$$m < \det G(X) \leq \det \nabla g(X) \tag{2.1}$$

hold for all $X \in \mathcal{A}$.

The Approximation Theorem for structured deformations ([5], [1], [2]) assures that there is a sequence $n \mapsto f_n$ of piecewise smooth, injective mappings defined on $\mathcal{A}$ such that

$$g = \lim_{n\to\infty} f_n, \quad G = \lim_{n\to\infty} \nabla f_n, \tag{2.2}$$

with the limits taken in the sense of essentially uniform convergence (i.e., L^∞-convergence). The spatial derivatives ∇f_n are taken in the classical sense, and the limit $G = \lim_{n\to\infty} \nabla f_n$ of derivatives in $(2.2)_2$ need not equal the corresponding derivative ∇g of the macroscopic deformation $g = \lim_{n\to\infty} f_n$ (nor need G even be the gradient of some deformation). Specific quantitative information about the difference $M := \nabla g - G$ is provided below and justifies the terminology *deformation due to disarrangements* or *disarrangement tensor* for M.

2.1 An additive decomposition for ∇g

The algebraically obvious *additive decomposition*

$$\nabla g = G + M \tag{2.3}$$

for the macroscopic deformation gradient is given deeper significance by means of the following limit relation ([6], [1]) for M :

$$M(X) = \lim_{\delta\to 0} \lim_{n\to\infty} vol\mathcal{B}(X;\delta)^{-1} \int_{\Gamma(f_n)\cap\mathcal{B}(X;\delta)} [f_n](Y) \otimes \nu(Y) dA_Y. \tag{2.4}$$

In this relation, $n \mapsto f_n$ is any sequence of piecewise smooth, injective mappings that satisfies the limit relations (2.2). The symbol $\mathcal{B}(X;\delta) \subset \mathcal{E}$ denotes the ball of radius $\delta > 0$ centered at a point X in $\mathcal{A}$, and $\Gamma(f_n) \subset \mathcal{E}$, $[f_n](Y) \in \mathcal{V}$, and $\nu(Y) \in \mathcal{V}$ denote, respectively, the jump set of the piecewise smooth deformation f_n, the jump of f_n at a point $Y \in \Gamma(f_n)$, and the unit normal to the jump set $\Gamma(f_n)$ at the point Y. The integrand in (2.4) is the tensor product of $[f_n](Y)$ and $\nu(Y)$, both vectors in $\mathcal{V}$.

2.2 Factorizations, virgin configurations, and intermediate configurations

The definition of composition of two structured deformations ([5], [1]) reduces here to the formula:

$$(p, P) \circ (g, G) := (p \circ g, (P \circ g)G). \tag{2.5}$$

The symbol " $\circ$ " on the left-hand side denotes the composition of two structured deformations, while on the right-hand side it denotes the composition of two functions. In addition, $(P \circ g)G$ denotes the pointwise composition of the two tensor fields $P \circ g$ and G. This formula provides the following factorizations for a structured deformation (g, G):

$$(g, G) = (g, \nabla g) \circ (i, K) \tag{2.6}$$

$$(g, G) = (i, H) \circ (g, \nabla g), \tag{2.7}$$

where $i(X) := X$ for all $X \in \mathcal{E}$, $K := (\nabla g)^{-1}G$ and $H := (G(\nabla g)^{-1}) \circ g^{-1}$. The first factorization (2.6) represents the given structured deformation as the classical deformation $(g, \nabla g)$ *following* a "purely submacroscopic" structured deformation (i, K) that accomplishes all of the disarrangements associated with (g, G). Analogously, the second factorization (2.7) represents (g, G) as the same classical deformation *followed by* the purely submacroscopic structured deformation (i, H).

The factorization (2.6) provides a distinction between the body before and after it undergoes the purely submacroscopic deformation (i, K), a distinction that permits us to distinguish between the *reference configuration*, from which the classical deformation $(g, \nabla g)$ proceeds, and the *virgin configuration*, from which both (i, K) and (g, G) proceed. Similarly, we may distinguish by means of (2.7) between the *deformed configuration without disarrangements*, attained from the virgin configuration via the classical deformation $(g, \nabla g)$ alone, and the *deformed configuration*, attained from the deformed configuration without disarrangements via the purely submacroscopic deformation (i, H).

We note that the inequality (2.1) implies the relations

$$0 < \det K = \det H \leq 1 \tag{2.8}$$

and permits us to call $\det K = \det H = \det G \,/ \det \nabla g$ the *volume fraction* associated with the given structured deformation. Similarly, we may call $1 - \det K = 1 - \det H$ the *void fraction* associated with (g, G) . When the volume fraction is less than one or, equivalently, the void fraction is positive, the given structured deformation has created submacroscopic voids. Of particular interest in applications such as crystalline plasticity [3] are *invertible structured deformations* (g, G), i.e., structured deformations for which the volume fraction equals 1. The term "invertible" is appropriate, because the pair $(g^{-1}, G^{-1} \circ g^{-1})$ then is itself a structured deformation that is a two-sided inverse for (g, G) with respect to the composition in (2.5) and with (i, I) playing the role of the identity structured deformation (here, $Iv = v$ for all $v \in \mathcal{V}$). In this case, the purely submacroscopic factor (i, K) also is an invertible structured deformation with inverse $(i, K)^{-1} = (i, K^{-1})$, and we have the following factorization

$$(g, \nabla g) = (g, G) \circ (i, K)^{-1} \tag{2.9}$$

of the classical deformation $(g, \nabla g)$. For the structured deformation (g, G), the purely submacroscopic deformation (i, K) carried the virgin configuration into the reference configuration; consequently, its inverse $(i, K)^{-1}$ carries the reference configuration into the virgin configuration. Therefore, *the virgin configuration for the invertible structured deformation (g, G) plays the role of a (global) intermediate configuration for the classical* *deformation* $(g, \nabla g)$. Local intermediate configurations play an important role in descriptions using classical deformations alone of single and polycrystalline materials (see [7], [3] and references cited therein).

2.3 Motions via families of structured deformations; space-like disarrangements

A simple way of capturing the evolution in time of a body that undergoes structured deformations at each instant is to specify a positive number t_f and a pair of smooth mappings $\chi : \mathcal{A}\times(0,t_f) \to \mathcal{E}$ and $G : \mathcal{A}\times(0,t_f) \to \mathrm{Lin}\mathcal{V}$ such that the pair $(\chi(\cdot,t), G(\cdot,t))$ is a structured deformation for each $t \in (0,t_f)$. When the Approximation Theorem and the identification relation are invoked at each time t, the relations (2.2), (2.4) become:

$$\chi(\cdot,t) = \lim_{n\to\infty} \chi_n(\cdot,t), \quad G(\cdot,t) = \lim_{n\to\infty} \nabla\chi_n(\cdot,t) \tag{2.10}$$

and

$$M(X,t) = \lim_{\delta\to 0}\lim_{n\to\infty} vol\mathcal{B}(X;\delta)^{-1} \int_{\Gamma(\chi_n(\cdot,t))\cap\mathcal{B}(X;\delta)} [\chi_n(\cdot,t)](Y) \otimes \nu(Y) dA_Y. \tag{2.11}$$

In this context, the disarrangements associated with the approximating motions χ_n that are captured in the tensor field $M : \mathcal{A}\times(0,t_f) \to \mathrm{Lin}\mathcal{V}$ are space-like, because the jump set $\Gamma(\chi_n(\cdot,t))$ in (2.11) is a subset of the Euclidean space $\mathcal{E}$. Consequently, time-like jumps in χ_n are not captured by the family $t \mapsto (\chi(\cdot,t), G(\cdot,t))$ of structured deformations. With the present description of structured motions, the subject of Part I more accurately can be described as *elasticity with space-like disarrangements*.

The more complete treatment of structured motions described in [8] introduces not only a deformation without disarrangements G but also a "velocity without disarrangements" $\dot{\chi}_\backslash$ and permits both space-like and time-like jumps to be captured in two analogues of the identification relation (2.11). This setting forms the basis of the field theory developed in Part II, and the analogues of (2.11) are the identification relations (14.4) and (14.5).

3 Contact and Body Forces

3.1 Decompositions

Earlier studies of balance laws for bodies undergoing structured deformations ([8], Part Two, Section 1, and [9]) showed that the classical law of balance of forces in the reference configuration is equivalent to a "refined balance law" that may be written as:

$$\mathrm{div}(SK^*) + \mathrm{div}((\det K)S - SK^*) - S\nabla \det K + (\det K)b_{ref} = 0 \tag{3.1}$$

Here, $S : \mathcal{A}\times(0,t_f) \to Lin\mathcal{V}$ is the Piola-Kirchhoff stress field, $K := (\nabla\chi)^{-1}G$, b_{ref} is the body force per unit volume in the reference configuration, and $A^* := (\det A)A^{-T}$ for all invertible $A \in \mathrm{Lin}\mathcal{V}$. Moreover, identification relations derived in the earlier studies (see the Appendix in [4]) permit us to call

$$S_\backslash := SK^* = (\det K)SK^{-T} \tag{3.2}$$

the *stress without disarrangements*, to call $\mathrm{div}(SK^*)$ the *volume density of contact forces without disarrangements*, to call $\mathrm{div}((\det K)S - SK^*) - S\,\nabla \det K$ the *volume density of contact forces due to disarrangements*, and to call

$$S_d := (\det K)S - SK^* \tag{3.3}$$

the *stress due to disarrangements.* The availability through structured deformations of both a virgin configuration and a reference configuration permits one to view (3.1) as the law of balance of forces in the virgin configuration, a configuration differing from the reference configuration by the purely submacroscopic deformation (i, K) described in Section 2.2. The scalar field $\det K$ may be thought of as the volume fraction associated with the given time-parameterized family of structured deformations, and $1 - \det K$ represents the void fraction.

The considerations above lead us not only to the *decomposition*

$$F = G + M \tag{3.4}$$

of the macroscopic deformation gradient $F = \nabla\chi : \mathcal{A}\times(0, t_f) \to \mathrm{Lin}\mathcal{V}$ but also, upon adding relations (3.2) and (3.3), to the *decomposition of the stress*:

$$(\det K)S = S_\backslash + S_d. \tag{3.5}$$

The stress tensor $(\det K)S$ is an analogue of the "weighted Cauchy tensor" $(\det F)T$ discussed in [10], and equations (3.1) and (3.5) show that it is the weighted measure of stress $(\det K)S$ that readily decomposes into a part without disarrangements plus a part due to disarrangements.

3.2 Consistency relation

We note that the defining relations (3.2) and (3.3) for the stress without disarrangements $S_\backslash$ and the stress due to disarrangements S_d along with the decomposition (3.5) provide upon elimination of the Piola-Kirchhoff stress S a consistency relation between the stress due to disarrangements and the stress without disarrangements:

$$S_\backslash(K^T - I) = S_d. \tag{3.6}$$

The decomposition (3.5) of the weighted stress $(\det K)S$ into parts with and without disarrangements has the property that, given the field K, specification of S then determines both $S_\backslash$ and S_d. By way of contrast, the decomposition (3.4) of the macroscopic deformation gradient into parts with and without disarrangements only is restricted by the condition

$$m(t) < \det G(X,t) \le \det(G(X,t) + M(X,t))$$

where $m(t)$ is a positive number depending only upon time.

We refer to (3.6) as the *consistency relation.* It eventually will provide a supplementary constitutive relation that appears in the field relations in our theory. An equivalent form of the consistency relation,

$$S_{\backslash}M^T + S_d G^T + S_d M^T = 0, \tag{3.7}$$

follows from (3.6), after substitution of $G^T F^{-T}$ for K^T, and implies that

$$\mathrm{sk}(S_{\backslash}M^T) + \mathrm{sk}(S_d G^T) + \mathrm{sk}(S_d M^T) = 0, \tag{3.8}$$

where $\mathrm{sk}A := (A - A^T)/2$ denotes the skew part of $A \in \mathrm{Lin}\mathcal{V}$. This relation plays a role in the analysis of moment densities in Section 5.

4 Power Expended; Balance Laws

We postulate that in a time-parameterized family $t \mapsto (\chi(\cdot,t), G(\cdot,t))$ of structured deformations the power expended at time $t \in (0, t_f)$ on a subbody $\mathcal{S} \subset \mathcal{A}$ by its exterior is given by the classical formula

$$\begin{aligned} P(\mathcal{S},t) &= \int_{\mathrm{bdy}\mathcal{S}} S(X,t)\nu(X) \cdot \dot{\chi}(X,t)\, dA_X + \\ &\quad \int_{\mathcal{S}} b^*(X,t) \cdot \dot{\chi}(X,t)\, dV_X. \end{aligned} \tag{4.1}$$

Here, beyond the quantities S, $\dot{\chi}$, and b_{ref} introduced above, $\nu(X)$ denotes the outward unit normal at the point $X \in \mathrm{bdy}\mathcal{S}$, $b^* := b_{ref} - \rho_{ref}\ddot{\chi}$ is the total body force, and ρ_{ref} is the density in the reference configuration.

A standard argument [11],[12] based on invariance of the power expended under superposed rigid motions yields the classical laws of balance of linear and angular momentum:

$$\mathrm{div}S + b_{ref} = \rho_{ref}\ddot{\chi} \tag{4.2}$$

$$\mathrm{sk}(SF^T) = 0. \tag{4.3}$$

The definition (4.1) of the power expended and the balance law (4.2) yield by means of the divergence theorem and a standard product rule the following reduced expression for the power expended

$$P(\mathcal{S},t) = \int_{\mathcal{S}} S(X,t) \cdot \nabla\dot{\chi}(X,t)\, dV_X. \tag{4.4}$$

Our choice of the classical formula (4.1) for the power expended reflects a desire to retain as much as possible the structure of non-linear elasticity as a field theory. However, in spite of this classical approach, the formula $\nabla\dot{\chi} = (\nabla\chi)^{\cdot} = \dot{F}$ and the two basic decompositions (3.4) and (3.5) permit us now to decompose $(\det K)S \cdot \nabla\dot{\chi}$, the *density of stress-power in the virgin configuration*, in the following manner:

$$(\det K)S \cdot \nabla\dot{\chi} = S_{\backslash} \cdot \dot{G} + S_d \cdot \dot{M} + S_{\backslash} \cdot \dot{M} + S_d \cdot \dot{G}. \tag{4.5}$$

The contribution $S_\backslash \cdot \dot{G} + S_d \cdot \dot{M}$ to the stress-power has two terms, each of which is a product of "like" quantities (in the first term a stress *without* disarrangements and a rate of deformation *without* disarrangements, or, in the second term, corresponding quantities due to disarrangements). Because the remaining contribution $S_\backslash \cdot \dot{M} + S_d \cdot \dot{G}$ in (4.5) has two terms, each of which mixes factors with and without disarrangements, we refer to it as the *mixed (stress-) power.*

In a similar way, we may decompose the volume density of moments in the balance law (4.3):

$$(\det K)\mathrm{sk}(SF^T) = \mathrm{sk}(S_\backslash G^T) + \mathrm{sk}(S_d M^T) + \mathrm{sk}(S_\backslash M^T) + \mathrm{sk}(S_d G^T), \tag{4.6}$$

and individual terms on the right-hand side may be interpreted as particular moment densities, as described in the next section. By the consistency relation (3.8), the last three moment densities $\mathrm{sk}(S_d M^T)$, $\mathrm{sk}(S_\backslash M^T)$, and $\mathrm{sk}(S_d G^T)$ appearing on the right-hand side of (4.6) must add to zero. By the balance of angular momentum (4.3), by (4.6), and by (3.8), the first moment density $\mathrm{sk}(S_\backslash G^T)$ on the right-hand side of (4.6) must vanish.

5 Offset Moments

The volume densities of moments $\mathrm{sk}(S_\backslash G^T)$, $\mathrm{sk}(S_d M^T)$, $\mathrm{sk}(S_\backslash M^T)$, and $\mathrm{sk}(S_d G^T)$ arose in the formula (4.6), and in this section we identify the terms $\mathrm{sk}(S_d M^T)$ and $\mathrm{sk}(S_\backslash M^T)$ as volume densities of "offset moments."

The significance of the moment density $\mathrm{sk}(S_d M^T)$ is revealed by the following identification relation:

$$\mathrm{sk}(S_d(X,t)M^T(X,t)) = \lim_{\delta\to 0}\lim_{n\to\infty} \frac{\int_{\Gamma(\chi_n(\cdot,t))\cap\mathcal{B}(X;\delta)} S_d(X,t)\nu(Y) \times [\chi_n(\cdot,t)](Y)dA_Y}{\mathrm{vol}\mathcal{B}(\mathrm{X};\delta)}. \tag{5.1}$$

Relation (5.1) follows immediately, if we substitute the right-hand side of the identification relation (2.11) for $M(X,t)$ into the left-hand side of (5.1) and identify the resulting skew tensors in the integrand with their dual vectors, thereby obtaining the cross-product appearing on the right-hand side of (5.1).

The vector $S_d(X,t)\nu(Y)$ in the right-hand side of (5.1) is the traction due to disarrangements at the point Y on a disarrangement site in the virgin configuration, computed using the stress due to disarrangements at the *center* X of the ball $\mathcal{B}(X;\delta)$. The vector product $S_d(X,t)\nu(Y) \times [\chi_n(\cdot,t)](Y)$ is (minus) the moment per unit area produced by that traction acting against the *offset* $[\chi_n(\cdot,t)](Y)$ caused by disarrangements. An elementary instance of such moments would arise if a deck of cards, in equilibrium under a system of loads, is shifted near the middle card without changing either the shape of the individual cards or the applied loads. The moment arising from the change in geometry of the deck corresponds to the moment calculated on the right-hand side of (5.1). Consequently, we call $\text{sk}(S_d M^T)$ a *volume density of offset moments.* Of course, replacing S_d by $S_\backslash$ in (5.1) permits us also to call $\text{sk}(S_\backslash M^T)$ a *volume density of offset moments.*

Because G measures the deformation away from disarrangement sites, we interpret the moment density $\text{sk}(S_d G^T)$ in (4.6), arising even in motions involving disarrangements, as an analogue of the moment density $\text{sk}(SF^T)$ arising in the classical balance law. (Recall that in Section 4 we pointed out that the analogous density $\text{sk}(S_\backslash G^T)$ vanishes, because it is a scalar multiple of $\text{sk}(SF^T)$.) According to (3.8), the offset moment densities $\text{sk}(S_d M^T)$, $\text{sk}(S_\backslash M^T)$, and the moment density $\text{sk}(S_d G^T)$ in (4.6) add to zero. In Section 8 we obtain further restrictions on these moment densities by imposing the principle of material frame-indifference.

6 Dynamical Processes, Constitutive Classes, and the Dissipation Inequality

Given the mass density ρ_{ref} in the reference configuration, a dynamical process is specified here by giving a motion χ, the deformation without disarrangements G, the stress field S, and the volume density ψ of the Helmholtz free energy in the reference configuration. Of course, the stresses without and due to disarrangements $S_\backslash$ and S_d are determined by the Piola-Kirchhoff stress S, the motion χ, and the deformation without disarrangements G through the relations (3.2) and (3.3), and the body force b_{ref} also is determined by fields in our list from the balance of linear momentum (4.2). (We may guarantee that the balance of angular momentum (4.3) is satisfied on every dynamical process by imposing the condition $\text{sk}(SF^T) = 0$, but we refrain from doing so pending the discussion of frame-indifference.)

We use the concept of a constitutive class to specify the particular material that is to be considered. Here, following Gurtin [13], a *constitutive class* $\mathcal{C}$ simply is a collection of dynamical processes. A particular choice of constitutive class limits the dynamical processes that are to be considered. In practice, a constitutive class is specified by giving a list of *response functions*: the corresponding constitutive class is then the collection of dynamical processes that satisfy the relations on the fields χ, G, S, ψ provided by the response functions.

Another limitation on dynamical processes is provided by the second law of thermodynamics which, in the present context of isothermal processes, is the *dissipation inequality*:

$$\dot{\psi}(X,t) \leq S(X,t) \cdot \nabla\dot{\chi}(X,t), \tag{6.1}$$

asserting that *the rate of change of the density of the Helmholtz free energy does not exceed the stress power.* We denote by $\mathcal{D}$ the collection of all dynamical processes χ, G, S, ψ that satisfy the dissipation inequality. The dissipation inequality is imposed by means of the requirement

$$\mathcal{C} \subset \mathcal{D}. \tag{6.2}$$

In other words, every dynamical process for the given material must obey (6.1).

The dissipation inequality may be used to impose restrictions on the response functions that specify a constitutive class $\mathcal{C}$, as first described in the context of the Clausius-Duhem inequality by Coleman and Noll [14] and now widely followed in continuum thermodynamics. According to this procedure, one seeks necessary and sufficient conditions on the response functions that specify $\mathcal{C}$ in order that $\mathcal{C} \subset \mathcal{D}$. We indicate in the next section that, when the free energy and stresses depend only upon $F = \nabla\chi$ and G, the restrictions obtained from this procedure include the vanishing of the internal dissipation $S(X,t) \cdot \nabla\dot{\chi}(X,t) - \dot{\psi}(X,t)$ on dynamical processes in $\mathcal{C}$. We shall maintain the premise that it is useful to identify and study constitutive classes that admit a non-vanishing internal dissipation on a non-trivial class of dynamical processes. Consequently, we are led in the next section to specify a particular constitutive class $\mathcal{E}_d$ and to show directly the inclusion $\mathcal{E}_d \subset \mathcal{D}$, i.e., we impose sufficient conditions on the constitutive class in order that it be included in $\mathcal{D}$.

Although all of the fields in our description of a dynamical process are smooth, the present approach echoes the standard use of the second law of thermodynamics to limit the class of non-smooth processes that can occur in the presence of a shock wave (see, for example, [15]). In our context, the non-smoothness occurs at a submacroscopic level and is made explicit only through the piecewise-smooth motions χ_n that arise in the Approximation Theorem.

In spite of the present choice not to pursue the procedure of Coleman and Noll, the constitutive class $\mathcal{C}$ obtained via that procedure merits detailed study, because it admits the possibility that internal dissipation arises via small jumps between points on a constitutive manifold determined by the consistency relation. (See [16], [17], [3] for elementary examples.)

7 A Constitutive Class for Elastic Bodies Undergoing Disarrangements

The constitutive data that we employ temporarily for the specification of an elastic body undergoing disarrangements are the smooth response functions $(F,G) \mapsto \Psi(F,G)$, $(F,G) \mapsto S_\backslash(F,G)$, and $(F,G) \mapsto S_d(F,G)$ for the free energy, stress without disarrangements, and stress due to disarrangements, all defined on pairs of invertible tensors (F,G) satisfying the inequalities

$$0 < \det G \leq \det F. \tag{7.1}$$

An equivalent description of these response functions entails the specification of the mappings $(M,G) \mapsto \tilde{\Psi}(M,G) := \Psi(M+G,G)$, $(M,G) \mapsto \tilde{S}_\backslash(M,G) := S_\backslash(M+G,G)$, and $(M,G) \mapsto \tilde{S}_d(M,G) := S_d(M+G,G)$ defined on pairs of tensors (M,G) satisfying

$$0 < \det G \leq \det(M+G). \tag{7.2}$$

For future reference, we record here the relations

$$D_M\tilde{\Psi}(M,G) = D_F\Psi(M+G,G) \tag{7.3}$$

$$D_G\tilde{\Psi}(M,G) = D_F\Psi(M+G,G) + D_G\Psi(M+G,G). \tag{7.4}$$

We allow the free energy response function also to depend upon the material point X at which the free energy is to be computed, but we delay until Section 9 making explicit this dependence on X in the symbols $\Psi(F,G)$ and $\tilde{\Psi}(M,G)$.

The functions $\tilde{\Psi}$, $\tilde{S}_\backslash$, $\tilde{S}_d$ now permit us to define a provisional class $\mathcal{C}$ of dynamical processes satisfying the constitutive relations

$$\psi(X,t) = \tilde{\Psi}(M(X,t), G(X,t)), \tag{7.5}$$

$$S_\backslash(X,t) = \tilde{S}_\backslash(M(X,t), G(X,t)), \tag{7.6}$$

and

$$S_d(X,t) = \tilde{S}_d(M(X,t), G(X,t)) \tag{7.7}$$

for all X, t associated with a dynamical process. We now indicate how the requirement (6.2) imposed via the procedure of Coleman and Noll [14] leads to a constitutive class in which no internal dissipation occurs. (To simplify the relations below, we omit the argument (X,t) throughout.) We multiply both sides of the dissipation inequality (6.1) by $\det K$, use the constitutive relations (7.5), (7.6), (7.7) and the formula (4.5) for the stress power in the virgin configuration, and we conclude that the internal dissipation

$$\begin{aligned}
&\det K(S \cdot \nabla\dot{\chi} - \dot{\psi}) \\
= \; &(\tilde{S}_\backslash(M,G) + \tilde{S}_d(M,G) - (\det K) D_M \tilde{\Psi}(M,G)) \cdot \dot{M} \\
&+(\tilde{S}_\backslash(M,G) + \tilde{S}_d(M,G) - (\det K) D_G \tilde{\Psi}(M,G)) \cdot \dot{G}
\end{aligned} \tag{7.8}$$

is not negative on each dynamical process in $\mathcal{C}$. In spite of the restrictions that the consistency relation (3.6), the inequalities (7.2), and the constitutive relations (7.6), (7.7) together place on $\dot{M}$ and $\dot{G}$, we may reverse any dynamical process in $\mathcal{C}$ with respect to its time-evolution and obtain another dynamical process in $\mathcal{C}$.[1] Consequently, $\dot{M}$ and $\dot{G}$ may be replaced by $-\dot{M}$ and $-\dot{G}$ in (7.8), leaving all other quantities unchanged. Therefore, the internal dissipation as given in (7.8) must vanish for every dynamical process in $\mathcal{C}$, and the dissipation inequality (6.1) must be satisfied as an equality.

In order to obtain a theory that admits internal dissipation, we consider now a collection of dynamical processes different from $\mathcal{C}$: specifically, we consider the collection $\mathcal{E}_d$ of dynamical processes χ, G, S, ψ satisfying the *constitutive relations*

$$\begin{aligned}
\psi(X,t) &= \tilde{\Psi}(M(X,t), G(X,t)), && (7.9)\\
S_d(X,t) &= (\det K(X,t)) D_M \tilde{\Psi}(M(X,t), G(X,t)), && (7.10)\\
S_\backslash(X,t) &= (\det K(X,t)) D_G \tilde{\Psi}(M(X,t), G(X,t)), && (7.11)
\end{aligned}$$

and the *mixed power inequality*

$$0 \le S_\backslash(X,t) \cdot \dot{M}(X,t) + S_d(X,t) \cdot \dot{G}(X,t) \tag{7.12}$$

for all X, t. Relations (7.10) and (7.11) amount to assuming at the outset that the response functions for the stresses with and without disarrangements are determined in

[1]A rigorous proof of this assertion requires that smoothness conditions on dynamical processes be specified, e.g., χ, G, S, and ψ are twice continuously differentiable.

a particular way by the response function for the free energy. We note that the derivatives $\dot{M}(X,t)$ and $\dot{G}(X,t)$ on the right-hand side of (7.12) may *not* be varied arbitrarily while holding the pair $(M(X,t), G(X,t))$ and, hence, the pair $(S_\backslash(X,t), S_d(X,t))$ fixed. For example, replacing $\dot{M}(X,t)$ and $\dot{G}(X,t)$ by $-\dot{M}(X,t)$ and $-\dot{G}(X,t)$ without other change would, in general, cause the mixed power inequality (7.12) to be violated.

An important conclusion that can be drawn from the definition of the constitutive class $\mathcal{E}_d$ is that *the dissipation inequality is satisfied for every dynamical process in* $\mathcal{E}_d$ *i.e.,* $\mathcal{E}_d \subset \mathcal{D}$. Indeed, the constitutive relations (7.9), (7.10), (7.11), the mixed power inequality (7.12), and relations (3.4) and (3.5) tell us that

$$\begin{aligned}(\det K)\dot{\psi} &= (\det K)D_M\tilde{\Psi}\cdot\dot{M} + (\det K)D_G\tilde{\Psi}\cdot\dot{G}\\ &\leq (\det K)D_M\tilde{\Psi}\cdot\dot{M} + (\det K)D_G\tilde{\Psi}\cdot\dot{G} +\\ &\quad (\det K)D_M\tilde{\Psi}\cdot\dot{G} + (\det K)D_G\tilde{\Psi}\cdot\dot{M}\\ &= (\det K)S\cdot\dot{F},\end{aligned}$$

which is equivalent to the dissipation inequality (6.1).

Another significant observation is that *the consistency relation (3.6), through the constitutive relations (7.10) and (7.11), imposes a restriction on dynamical processes in* $\mathcal{E}_d$, namely, for every dynamical process χ, G, S, ψ in $\mathcal{E}_d$ there holds

$$D_G\tilde{\Psi}(M(X,t),G(X,t))(K(X,t)^T - I) = D_M\tilde{\Psi}(M(X,t),G(X,t)) \qquad (7.13)$$

for all (X,t). Consequently, the pairs $(M(X,t),G(X,t))$ available through dynamical processes in $\mathcal{E}_d$ lie in a submanifold of Lin$\mathcal{V}\times$Lin$\mathcal{V}$. In particular, for each (X,t), the pairs $(\dot{M}(X,t),\dot{G}(X,t))$ of time-derivatives available through dynamical processes in $\mathcal{E}_d$ lie in the tangent space of the submanifold at $(M(X,t),G(X,t))$ and, hence, cannot be arbitrary elements of Lin$\mathcal{V}\times$Lin$\mathcal{V}$. Similarly, *the mixed power inequality may be viewed as a restriction on the quantities* M, G, $\dot{M}$ *and* $\dot{G}$ *, or, equivalently, on* F, G, $\dot{F}$*and* $\dot{G}$, *that can arise for dynamical processes in the constitutive class* $\mathcal{E}_d$, and we shall discuss some of these restrictions in Section 10. Finally, for every *classical* dynamical process χ, $\nabla\chi$, S, ψ in $\mathcal{E}_d$, the consistency relation (7.13), and the implication - - if $M = F - G = 0$, then $K = I$ - - yield for all X,t:

$$D_M\tilde{\Psi}(0,\nabla\chi(X,t)) = 0, \qquad (7.14)$$

and, equivalently, by (7.3),

$$D_F\Psi(\nabla\chi(X,t),\nabla\chi(X,t)) = 0. \qquad (7.15)$$

8 Material Frame-Indifference

We consider here the transformation properties of the kinematical quantities associated with dynamical processes under changes of observer. These transformation properties can be obtained by replacing the motion χ and approximating motion χ_n from the Approximation Theorem by $(X,t) \mapsto r(\chi(X,t),t)$ and $(X,t) \mapsto r(\chi_n(X,t),t)$, respectively, where r denotes a rigid motion $(X,t) \mapsto x_0(t) + Q(t)(X - X_0)$ with $Q(t)$ a proper orthogonal tensor. From this observation and the fact that $G = \lim_{n\to\infty} \nabla\chi_n$, we obtain the transformation rules

$$\begin{array}{rcl} F & \to & QF \\ G & \to & QG \\ M & \to & QM \\ K & \to & K \\ \dot{F} & \to & Q\dot{F} + \dot{Q}F \\ \dot{G} & \to & Q\dot{G} + \dot{Q}G \\ \dot{M} & \to & Q\dot{M} + \dot{Q}M. \end{array}$$

In the present context of an elastic body undergoing disarrangements, we say that the response function $\tilde{\Psi}$ is it frame-indifferent if, for all proper orthogonal tensors Q and pairs (M,G) with $0 < \det G \le \det(M+G)$, there holds

$$\tilde{\Psi}(QM, QG) = \tilde{\Psi}(M,G), \tag{8.1}$$

or, equivalently,

$$\Psi(QF, QG) = \Psi(F,G) \tag{8.2}$$

for all proper orthogonal tensors Q and pairs (F,G) with $0 < \det G \le \det F$.

A useful characterization of this condition follows from the polar decompositions $F = R_F U_F$ and $G = R_G U_G$. Indeed, we may put $Q := R_G^T$ in (8.1) or (8.2), or $Q := R_F^T$ in (8.2) and use the relations $R_F^T = U_F^{-1}F^T$, $R_G^T = U_G^{-1}G^T$ to obtain for all pairs (M,G) with $0 < \det G \le \det(M+G)$, the representations

$$\begin{aligned} \tilde{\Psi}(M,G) &= \tilde{\Psi}(U_G^{-1}G^T M, U_G) = \overset{\leftrightarrow}{\Psi}(G^T M, C_G) && (8.3)\\ \Psi(F,G) &= \Psi(U_G^{-1}G^T F, U_G) = \widehat{\Psi}(G^T F, C_G) && (8.4)\\ \Psi(F,G) &= \Psi(U_F, U_F^{-1}F^T G) = \underline{\Psi}(C_F, F^T G), && (8.5) \end{aligned}$$

where $C_F := F^T F$ and $C_G := G^T G$ are the right Cauchy-Green tensors for F and G, respectively. Each one of these representations is both a necessary and a sufficient condition for the frame-indifference of the respective response function $\tilde{\Psi}$ or Ψ in the context of elastic bodies undergoing disarrangements.

A second characterization of the frame-indifference of the response function $\tilde{\Psi}$ follows by imposing (8.1) on smooth, time-parameterized families $t \mapsto Q(t)$ and $t \mapsto (M(t), G(t))$ and differentiating both sides of (8.1) with respect to t to conclude that

$$\begin{aligned} & D_M\tilde{\Psi}(QM,QG)\cdot[\dot{Q}M + Q\dot{M}] + D_G\tilde{\Psi}(QM,QG)\cdot[\dot{Q}G + Q\dot{G}] \\ = \; & D_M\tilde{\Psi}(M,G)\cdot\dot{M} + D_G\tilde{\Psi}(M,G)\cdot\dot{G}. \end{aligned} \tag{8.6}$$

Because the restriction (8.1) applies throughout the domain of $\tilde{\Psi}$, we may vary $\dot{Q}$, $\dot{G}$, and $\dot{M}$ independently (subject to the constraints $\mathrm{sym}(\dot{Q}Q^T) = 0$ and $0 < \det G \le \det(M+G)$ with "sym" denoting "symmetric part") to conclude from the smoothness of $\tilde{\Psi}$ and (8.1) that

$$D_M\tilde{\Psi}(QM,QG) = QD_M\tilde{\Psi}(M,G), \tag{8.7}$$

$$D_G\tilde{\Psi}(QM, QG) = QD_G\tilde{\Psi}(M, G), \tag{8.8}$$

and

$$\text{sk}(D_M\tilde{\Psi}(M, G)M^T + D_G\tilde{\Psi}(M, G)G^T) = 0 \tag{8.9}$$

for all proper orthogonal tensors Q and pairs (M, G) with $0 < \det G \leq \det(M + G)$. It is easy to verify that relations (8.7) - (8.9) imply that the response function $\tilde{\Psi}$ is frame-indifferent.

It is important to distinguish between, on the one hand, the smooth time-parameterized families $t \mapsto (M(t), G(t))$ used in establishing (8.6), and, on the other hand, the families

$$t \mapsto (M(X, t), G(X, t)) = (\nabla\chi(X, t) - G(X, t), G(X, t))$$

arising from dynamical processes in the constitutive class $\mathcal{E}_d$. In particular, the time derivatives of former pairs can be varied arbitrarily (when $\det G(t) < \det(M(t) + G(t))$), while those of the latter pairs cannot, as we observed near the end of Section 7.

We say that the mixed power $S_\backslash \cdot \dot{M} + S_d \cdot \dot{G} = \det K \left(D_G\tilde{\Psi} \cdot \dot{M} + D_M\tilde{\Psi} \cdot \dot{G}\right)$ is *frame-indifferent* if, for all smooth, time-parameterized families $t \mapsto Q(t)$ and for all families $t \mapsto (M(X, t), G(X, t))$ arising from dynamical processes in $\mathcal{E}_d$, there holds

$$\begin{aligned} & (\det K)D_G\tilde{\Psi}(M, G) \cdot \dot{M} + (\det K)D_M\tilde{\Psi}(M, G) \cdot \dot{G} \\ = \; & (\det K)D_G\tilde{\Psi}(QM, QG) \cdot (QM)^{\cdot} \\ & +(\det K)D_M\tilde{\Psi}(QM, QG) \cdot (QG)^{\cdot}. \end{aligned} \tag{8.10}$$

This condition amounts to the assertion that the mixed power is invariant under superpositions of rigid motions on dynamical processes. We now show that, *given the frame-indifference of the response function $\tilde{\Psi}$, the mixed power is frame-indifferent if and only if*

$$\text{sk}(D_G\tilde{\Psi}(M, G)M^T + D_M\tilde{\Psi}(M, G)G^T) = 0 \tag{8.11}$$

for all pairs (M, G) arising from dynamical processes in $\mathcal{E}_d$. In fact, expansion of the derivatives on the right-hand side of (8.10) tells us that (8.10) is equivalent to the relation

$$\begin{aligned} & (D_G\tilde{\Psi}(M, G) - Q^T D_G\tilde{\Psi}(QM, QG)) \cdot \dot{M} \\ & +(D_M\tilde{\Psi}(M, G) - Q^T D_M\tilde{\Psi}(QM, QG)) \cdot \dot{G} \\ = \; & (D_G\tilde{\Psi}(QM, QG)M^T + D_M\tilde{\Psi}(QM, QG)G^T) \cdot \dot{Q}. \end{aligned}$$

Given the frame-indifference of $\tilde{\Psi}$, we conclude from (8.7) and (8.8) that the previous relation is equivalent to

$$0 = (D_G\tilde{\Psi}(M, G)M^T + D_M\tilde{\Psi}(M, G)G^T) \cdot Q^T\dot{Q},$$

and the relation $\text{sym}(Q^T\dot{Q}) = 0$ along with the arbitrariness of $t \mapsto Q(t)$ provides the asserted characterization of frame-indifference of the mixed power.

Our main result on material frame-indifference is a generalization of a result of Noll [18] in classical elasticity: *if both the free energy response function and the mixed power are frame indifferent, then the balance of angular momentum (4.3) is satisfied for all dynamical processes in* $\mathcal{E}_d$. (Noll actually showed that the frame-indifference of the free energy response is equivalent to the law of balance of angular momentum in the classical context.) Indeed, if we add (8.11) to (8.9) we conclude that

$$\mathrm{sk}(D_M\tilde{\Psi}(M,G)F^T + D_G\tilde{\Psi}(M,G)F^T) = 0 \tag{8.12}$$

for all pairs (M, G) arising from dynamical processes in $\mathcal{E}_d$. For each dynamical process, the constitutive relations (7.10), (7.11), and the decomposition (3.5) may be applied to yield (4.3), the law of balance of angular momentum.

This main result permits us to impose the law of balance of angular momentum indirectly by requiring that both the free energy response function and the mixed power be frame-indifferent, *a requirement that we impose from now on through the relations (8.1) and (8.11).* In these considerations, it is important to remember that (8.11) is a restriction on dynamical processes, while (8.1) is a restriction on the free energy response function. Moreover, our main result and the imposition of (8.1) and (8.11) permit us not to list the law of balance of angular momentum among the field equations that we provide in Section 9.

We note from relations (7.10) and (7.11) that (8.11) implies

$$sk(S_\backslash M^T + S_d G^T) = 0 \tag{8.13}$$

on all dynamical processes in the constitutive class $\mathcal{E}_d$, and the consistency relation, through (3.8), then implies also that

$$\mathrm{sk}(S_d M^T) = 0. \tag{8.14}$$

Thus, it follows from frame-indifference as realized in (8.1) and (8.11) not only that the balance of angular momentum holds, but also that

(i) the offset moment density $\mathrm{sk}(S_\backslash M^T)$ and the moment density $\mathrm{sk}(S_d G^T)$ add to zero,

(ii) the offset moment density $\mathrm{sk}(S_d M^T)$ vanishes,

(iii) by the identification relation (5.1), the traction due to disarrangements $S_d\nu$ and the geometrical offset $[\chi_n]$ are colinear, on average, and

(iv) the moment density $\mathrm{sk}(S_\backslash G^T)$ vanishes.

Items (ii)and (iv) assert that each of the "pure" moment densities $\mathrm{sk}(S_d M^T)$ and $\mathrm{sk}(S_\backslash G^T)$ vanishes, so that the moments purely associated with disarrangements are self-equilibrating, as are those completely without disarrangements. By (i), "mixed" moment densities $\mathrm{sk}(S_\backslash M^T)$ and $\mathrm{sk}(S_d G^T)$ generally are not self-equilibrating, but together must produce no net moment.

It is useful to record the forms that relations (8.1) and (8.11) assume when the response function $(M, G) \mapsto \tilde{\Psi}(M, G)$ is replaced by $(F, G) \mapsto \Psi(F, G) = \tilde{\Psi}(F - G, G)$. Of course, (8.1) is replaced by (8.2), a restriction on the response function Ψ. In view of relations (3.4), (7.3), and (7.4), relation (8.11) is equivalent to

$$\mathrm{sk}(D_F\Psi(F,G)F^T + D_G\Psi(F,G)(F^T - G^T)) = 0, \tag{8.15}$$

a restriction on dynamical processes. Henceforth, when using F and G as the arguments of the free energy, we assume that (8.2) and (8.15) are satisfied, the first throughout the domain of Ψ and the second on all dynamical processes in $\mathcal{E}_d$.

9 Field Relations

Our analysis in Sections 7 and 8 has led us to the specification of one response function $(M, G) \mapsto \tilde{\Psi}(M, G)$ satisfying relations (8.1) and (8.11), the former throughout the domain of $\tilde{\Psi}$ and the latter on all dynamical processes in $\mathcal{E}_d$. Given such a response function $\tilde{\Psi}$, as well as the body force field b_{ref} and the density ρ_{ref} in the reference configuration, the conditions to be employed in deriving the field relations for elastic bodies undergoing disarrangements are restrictions on dynamical processes: the balance of linear momentum (4.2), the constitutive relations (7.9), (7.10), and (7.11), the mixed power inequality (7.12), and the consistency relation (7.13). As demonstrated in Section 8, the law of balance of angular momentum is a consequence of the assumptions of frame-indifference (8.1) and (8.11).

We now are in a position to record and derive from the conditions just cited the *field relations for an elastic body undergoing disarrangements*:

$$\operatorname{div}(D_M\tilde{\Psi} + D_G\tilde{\Psi}) + b_{ref} = \rho_{ref}\ddot{\chi}, \tag{9.1}$$

$$D_G\tilde{\Psi}(K^{-T} - I) + D_M\tilde{\Psi}K^{-T} = 0, \tag{9.2}$$

$$\operatorname{sk}(D_G\tilde{\Psi}M^T + D_M\tilde{\Psi}G^T) = 0 \tag{9.3}$$

$$D_G\tilde{\Psi} \cdot \dot{M} + D_M\tilde{\Psi} \cdot \dot{G} \geq 0. \tag{9.4}$$

The law of balance of linear momentum (9.1), the consistency relation (9.2), and the frame-indifference of the mixed power (9.3) amount to $3 + 6 + 3 = 12$ scalar equations for the unknowns χ and G, having a total of 12 scalar components. (That the consistency relation amounts to 6 equations, rather than 9, follows from the fact that when the original consistency relation (3.6) is rewritten in the form $S_\backslash K^T = S_\backslash + S_d$ and then multiplied by F^T, both sides are symmetric tensors.) The mixed power inequality (9.4) is a single scalar relation that further restricts the dynamical processes satisfying (9.1), (9.2), and (9.3). *We emphasize that, given the body force and density fields in the reference configuration, the field relations are restrictions on dynamical processes, while the relation (8.1) is a restriction on the response function* $\tilde{\Psi}$. We must also keep in mind that $\tilde{\Psi}$ and its derivatives depend not only upon $M(X,t)$ and $G(X,t)$ but also upon the material point X itself, and we make this dependence explicit when needed for clarity. For example, the first term on the left-hand side of the equation of balance of linear momentum (9.1) is the field

$$\begin{aligned}(X,t) \quad \mapsto \quad & \operatorname{div}_X[D_M\tilde{\Psi}(\nabla\chi(X,t) - G(X,t), G(X,t), X) \\ & + D_G\tilde{\Psi}(\nabla\chi(X,t) - G(X,t), G(X,t), X)].\end{aligned} \tag{9.5}$$

The field relations follow readily from the conditions cited above (9.1) - (9.4): the law of balance of linear momentum is a consequence of its counterpart (4.2), of the constitutive relations (7.10) and (7.11) for the stresses without and with disarrangements, and of the decomposition (3.5); the consistency relation (9.2) is the relation (7.13), rewritten with trivial algebraic changes; (9.3) is (8.11), and the mixed power inequality (9.4) is (7.12) with $S_\backslash$ and S_d replaced by the expressions in the formulas (7.10) and (7.11). Moreover, by the decomposition (3.5) and the constitutive relations (7.10), (7.11), one has the *stress relation*

$$S(X,t) = D_M\tilde{\Psi}(M(X,t), G(X,t), X) + D_G\tilde{\Psi}(M(X,t), G(X,t), X) \tag{9.6}$$

valid for all pairs (X, t) associated with dynamical processes in $\mathcal{E}_d$.

When χ and G determine a classical motion, i.e., $G = \nabla\chi$, then the relations $M = 0$, $K = I$, and $\det K = 1$ tell us that the balance of linear momentum (9.1) and the consistency relation (9.2) become

$$\mathrm{div}_X[D_G\tilde{\Psi}(0, \nabla\chi(X,t), X)] + b_{ref}(X,t) = \rho_{ref}(X)\ddot{\chi}(X,t), \tag{9.7}$$

$$D_M\tilde{\Psi}(0, \nabla\chi(X,t), X) = 0. \tag{9.8}$$

The remaining relations (9.3) and (9.4) are satisfied identically.

In some applications, it is easier to use the field relations when they are expressed in terms of the response function $(F, G) \mapsto \Psi(F, G, X) = \tilde{\Psi}(F - G, G, X)$. In this case, the response function Ψ is assumed to satisfy (8.2), and the field relations (9.1) - (9.4) become

$$\mathrm{div}(2D_F\Psi + D_G\Psi) + b_{ref} = \rho_{ref}\ddot{\chi} \tag{9.9}$$

$$D_F\Psi(2K^{-T} - I) + D_G\Psi(K^{-T} - I) = 0 \tag{9.10}$$

$$\mathrm{sk}((D_F\Psi + D_G\Psi)F^T - D_G\Psi G^T) = 0 \tag{9.11}$$

$$(D_F\Psi + D_G\Psi)\cdot\dot{F} - D_G\Psi\cdot\dot{G} \geq 0, \tag{9.12}$$

respectively, while the stress relation (9.6) becomes

$$S(X,t) = 2D_F\Psi(F(X,t), G(X,t), X) + D_G\Psi(F(X,t), G(X,t), X). \tag{9.13}$$

Corresponding to the expression (9.5), the first term on the left-hand side of the equation of balance of linear momentum (9.9) is the field

$$\begin{aligned}(X,t) \quad \mapsto \quad & \mathrm{div}_X[2D_F\Psi(\nabla\chi(X,t), G(X,t), X) \\ & + D_G\Psi(\nabla\chi(X,t), G(X,t), X)].\end{aligned} \tag{9.14}$$

In view of the significance of the purely submacroscopic factor (i, K) in (2.6), some applications become more accessible if one employs the response function

$$(F, K) \mapsto \overline{\Psi}(F, K, X) := \Psi(F, FK, X) \tag{9.15}$$

with domain the set of pairs (F, K) satisfying $0 < \det F$ and $0 < \det K \leq 1$. The relations

$$D_F\Psi(F, FK, X) = D_F\overline{\Psi}(F, K, X) - F^{-T}D_K\overline{\Psi}(F, K, X)K^T \tag{9.16}$$

and

$$D_G\Psi(F, FK, X) = F^{-T} D_K\overline{\Psi}(F, K, X) \tag{9.17}$$

permit us to express the field relations in terms of this choice of variables:

$$\operatorname{div}[2D_F\overline{\Psi} + F^{-T}D_K\overline{\Psi}(I - 2K^T)] + b_{ref} = \rho_{ref}\ddot{\chi} \tag{9.18}$$

$$D_F\overline{\Psi}(2K^{-T} - I) + F^{-T}D_K\overline{\Psi}K^T(\{K^{-T}\}^2 - 3K^{-T} + I) = 0 \tag{9.19}$$

$$\operatorname{sk}(D_F\overline{\Psi}F^T + F^{-T}D_K\overline{\Psi}(I - 2K^T)F^T) = 0 \tag{9.20}$$

$$(D_F\overline{\Psi} + F^{-T}D_K\overline{\Psi}(I - 2K^T)) \cdot \dot{F} - D_K\overline{\Psi} \cdot \dot{K} \geq 0. \tag{9.21}$$

In addition, the stress relation (9.13) becomes

$$\begin{aligned} S(X,t) &= 2D_F\overline{\Psi}(F(X,t), K(X,t), X) \\ &\quad + F^{-T}(X,t)D_K\overline{\Psi}(F(X,t), K(X,t), X)(I - 2K^T(X,t)). \end{aligned} \tag{9.22}$$

10 Internal Dissipation

The *internal dissipation* in the reference configuration for a dynamical process χ, G, S, ψ in $\mathcal{E}_d$ is defined to be the excess of the stress-power over the rate of change of free energy: $S \cdot \nabla\dot{\chi} - \dot{\psi} = S \cdot \dot{F} - \dot{\psi}$. Because the dissipation inequality (6.1) is satisfied for every dynamical process in $\mathcal{E}_d$, the internal dissipation is non-negative, and we consider from now on

$$\Upsilon := (\det K)(S \cdot \dot{F} - \dot{\psi}) \geq 0, \tag{10.1}$$

the *internal dissipation in the virgin configuration.* It follows immediately from (4.5), (7.10), and (7.11) that *the internal dissipation in the virgin configuration equals the mixed stress power:*

$$\begin{aligned} \Upsilon &= S_\backslash \cdot \dot{M} + S_d \cdot \dot{G} \\ &= (\det K)(D_G\tilde{\Psi} \cdot \dot{M} + D_M\tilde{\Psi} \cdot \dot{G}) \\ &= (\det K)[(D_F\Psi + D_G\Psi) \cdot \dot{F} - D_G\Psi \cdot \dot{G}] \geq 0 \end{aligned} \tag{10.2}$$

for each dynamical process χ, G, S, ψ in the constitutive class $\mathcal{E}_d$. Our aim in this section is to relate the internal dissipation to familiar quantities in the literature by investigating the relative contributions of the two terms $S_\backslash \cdot \dot{M}$ and $S_d \cdot \dot{G}$ in (10.2).

An elementary but useful consideration rests on re-examining the decomposition (4.5) of the stress-power in the virgin configuration in light of the constitutive relations (7.10) and (7.11). One obtains easily the relation

$$\begin{aligned} S \cdot \dot{F} &= (\det K)^{-1}(S_\backslash \cdot \dot{G} + S_d \cdot \dot{M} + S_\backslash \cdot \dot{M} + S_d \cdot \dot{G}) \\ &= D_G\tilde{\Psi} \cdot \dot{G} + D_M\tilde{\Psi} \cdot \dot{M} + (\det K)^{-1}\Upsilon \\ &= \dot{\psi} + (\det K)^{-1}\Upsilon, \end{aligned} \tag{10.3}$$

a *decomposition of the stress-power in the reference configuration into a non-dissipative part* $\dot{\psi}$ *and a dissipative part* $(\det K)^{-1}\Upsilon \geq 0$. Thus, by (10.2), *the dissipative part*

$(\det K)^{-1}\Upsilon$ *of the stress-power equals the mixed stress-power in the reference configuration.* Moreover, for classical dynamical processes χ, $\nabla\chi$, S, ψ in the constitutive class $\mathcal{E}_d$, the internal dissipation vanishes, because $\dot{M} = S_d = 0$.

For a given stress S and for given deformation rates $\dot{M}$ and $\dot{G}$, the relative magnitudes of the terms $S_\backslash \cdot \dot{M}$ and $S_d \cdot \dot{G}$ can be altered by adjusting $K = F^{-1}G$, because of the formulas $S_\backslash = SK^*$ and $S_d = (\det K)S - SK^*$. In particular, for K close to the identity I, we have $S_\backslash = S + O(K - I)$ and $S_d = O(K - I)$, and we expect that the term $S_\backslash \cdot \dot{M}$ dominates the term $S_d \cdot \dot{G}$ in the expression (10.2) for Υ as K tends to the identity I.

In order to understand this idea in more depth, it is enlightening to express the internal dissipation Υ in terms of the Cauchy stress T, the macroscopic deformation F and its time-derivative $\dot{F}$, and the deformation without disarrangements G and its derivative $\dot{G}$. In doing so, we employ (10.2) along with the the formulas $S_\backslash = SK^*$ and $S_d = (\det K)S - SK^*$, and we suppress the arguments X, t, and $\chi(X, t)$ for the sake of simplicity of notation. We record the result here, omitting its routine derivation:

$$\begin{aligned}(\det F)^{-1}\Upsilon &= TH^* \cdot (\dot{F}F^{-1} - \dot{G}G^{-1}) \\ &\quad + TH^* \cdot (\dot{G}G^{-1}(H - I)^2),\end{aligned} \tag{10.4}$$

where $H := GF^{-1}$ is the tensor field appearing in (2.7) and $H^* = (\det H)H^{-T}$. We note that the expression $TH^* \cdot (\dot{G}G^{-1}(H - I)^2)$ on the right-hand side of (10.4) is quadratic in $H - I$, while the first term $TH^* \cdot (\dot{F}F^{-1} - \dot{G}G^{-1})$ equals $T \cdot (\dot{F}F^{-1} - \dot{G}G^{-1})$ plus a term linear in $H - I$. In other words, we conclude from (10.4) that

$$\begin{aligned}(\det F)^{-1}\Upsilon &= TH^* \cdot (\dot{F}F^{-1} - \dot{G}G^{-1}) + O((H - I)^2) \\ &= T \cdot (\dot{F}F^{-1} - \dot{G}G^{-1}) + O(H - I).\end{aligned} \tag{10.5}$$

In order to relate the last formula for the internal dissipation to more familiar quantities, we note that the fields $L_G := \dot{G}G^{-1}$ and $L_M := \dot{F}F^{-1} - \dot{G}G^{-1}$, defined in an arbitrary structured motion, were identified in the study [7] (see also [3] in this volume) of multiple slip in single crystals as the *relative rate of deformation without disarrangements* and the *relative rate of deformation due to disarrangements*, respectively. (There the term "slip" replaced "disarrangement" because of the particular context.) Moreover, the factorization (2.7) implies that the tensor field $T_\backslash := TH^*$ in (10.4) and (10.5) is analogous to $S_\backslash = SK^*$ and may be called the *stress in the current configuration without disarrangements*, a configuration macroscopically identical to the current configuration but containing none of the disarrangements associated with χ and G. Accordingly, $T_\backslash$ represents a stress without disarrangements. (In view of (3.3), the tensor field $T_d := (\det H)T - T_\backslash$ is the analogue of S_d and represents a stress due to disarrangements.) Therefore, (10.5) may now be recast in the form

$$\begin{aligned}(\det F)^{-1}\Upsilon &= T_\backslash \cdot L_M + O((H - I)^2) \\ &= T \cdot L_M + O(H - I).\end{aligned} \tag{10.6}$$

The tensor $H - I$ measures the disarrangements from the current configuration without disarrangements to the current configuration, and the decomposition (10.6) tells us that *the quantities* $T \cdot L_M$ *and* $T_\backslash \cdot L_M$ *provide approximations to the internal dissipation to within, respectively, linear and quadratic terms in the disarrangements from the current configuration without disarrangements.* This result places in perspective with respect to the present theory the widespread identification of the internal dissipation as an expression of the form $T \cdot L_M$ (sometimes called "plastic power" in phenomenological theories of plasticity).

11 Energetically Nearsighted Elastic Bodies

The field relations obtained in Section 9, together with special properties of a body such as material uniformity and material symmetry, provide a setting in which to gain further understanding of the scope and range of applicability of elasticity with disarrangements. We provide in this section a preliminary step in this direction by considering elastic bodies that are "energetically nearsighted" in the sense that only the purely submacroscopic factors appearing in the factorizations (2.6) and (2.7) affect the free energy. Thus, submacroscopic slips or formation of voids would permit the body to change its free energy, while classical deformations would not, and we consider now elastic bodies for which the free energy response $\overline{\Psi}$ in (9.15) does not depend upon the macroscopic deformation F and, therefore, satisfies $D_F\overline{\Psi}(F,K,X) = 0$ for all triples F, K, X in the domain of $\overline{\Psi}$. Accordingly, the field relations (9.18) - (9.21) and the stress relation (9.22) take the special form

$$\mathrm{div}[F^{-T}\,\overline{\Psi}^{\prime}(I - 2K^T)] + b_{ref} = \rho_{ref}\ddot{\chi} \tag{11.1}$$

$$\overline{\Psi}^{\prime} K^T(\{K^{-T}\}^2 - 3K^{-T} + I) = 0 \tag{11.2}$$

$$\mathrm{sk}(F^{-T}\,\overline{\Psi}^{\prime}(I - 2K^T)F^T) = 0 \tag{11.3}$$

$$(F^{-T}\,\overline{\Psi}^{\prime}(I - 2K^T)) \cdot \dot{F} - \overline{\Psi}^{\prime} \cdot \dot{K} \geq 0. \tag{11.4}$$

$$S(X,t) = F^{-T}(X,t)\,\overline{\Psi}^{\prime}(F(X,t),K(X,t),X)(I - 2K^T(X,t)), \tag{11.5}$$

where we have written $\overline{\Psi}^{\prime}$ in place of $D_K\overline{\Psi}$ to simplify notation. In some of the considerations below, it is helpful to use the field $H = GF^{-1} = FKF^{-1}$ that arose in the factorization (2.7), and we note the relation

$$H^{-T} = F^{-T}K^{-T}F^T = (F^T)^{-1}K^{-T}F^T. \tag{11.6}$$

11.1 Universal phases and the golden mean

The consistency relation in the form (11.2) provides a restriction on the field K, and the form of this restriction depends in general upon the response function $\overline{\Psi}$. We observe, however, that there are solutions K of the consistency relation that do not depend upon $\overline{\Psi}$, because the expression $\{K^{-T}\}^2 - 3K^{-T} + I$ occurs multiplicatively there. Consequently, each tensor K with $0 < \det K \leq 1$ for which K^{-T} is a solution of the quadratic, tensor equation

$$X^2 - 3X + I = 0, \quad X \in \mathrm{Lin}\mathcal{V}, \tag{11.7}$$

determines a solution of the consistency relation (11.2). It is easy to see that K^{-T} is a solution of the quadratic equation (11.7) if and only K itself is a solution. In turn, this is equivalent to the assertion that the tensor $H = FKF^{-1}$ is a solution of (11.7). We call solutions K (with $0 < \det K \leq 1$) of the consistency relation *universal* if they are solutions of (11.7), because they do not depend upon the free energy response function of the energetically nearsighted elastic body. We also refer to the corresponding tensor $H = FKF^{-1}$as a universal solution of the consistency relation.

A necessary condition for H to be universal is the inclusion of the spectrum of H in the solution set $\left\{\frac{3+\sqrt{5}}{2}, \frac{3-\sqrt{5}}{2}\right\} = \{2+\gamma_0, 1-\gamma_0\}$ of the scalar quadratic equation $x^2 - 3x + 1 = 0$. Here, $\gamma_0 := \frac{\sqrt{5}-1}{2} \approx .618$ is the "golden mean," the positive number satisfying the relation $\frac{1}{x} = \frac{x}{1-x}$. By elementary linear algebra, H is universal if and only if it is diagonalizable over the reals with diagonal entries given up to permutations by one of the two triples: $(1-\gamma_0, 1-\gamma_0, 1-\gamma_0)$, $(1-\gamma_0, 1-\gamma_0, 2+\gamma_0)$. (Some algebraically valid possibilities for such triples are eliminated by the constraint $\det H \in (0, 1]$.)

Of course, the first triple $(1-\gamma_0, 1-\gamma_0, 1-\gamma_0)$ determines the tensor

$$H_{sph} := (1-\gamma_0)I \tag{11.8}$$

that, in turn, determines the purely submacroscopic structured deformation

$$(i, H_{sph}) = (i, (1-\gamma_0)I),$$

to follow a classical deformation $(\chi(\cdot, t), \nabla\chi(\cdot, t))$. A piecewise smooth approximation h_n for $(i, (1-\gamma_0)I)$ takes a body partitioned into congruent cubic cells of side $1/n$ and replaces each cell by one with the same center but now of side $(1-\gamma_0)/n$. The simultaneous shrinking of each cell creates voids, and the resulting structured deformation has volume fraction $\det H_{sph} = \det K_{sph} = (1-\gamma_0)^3 \approx .056$. This change of geometry will be referred to as the *(universal) spherical phase* of the energetically nearsighted elastic body.

For each choice of basis $(d_{(1)}, d_{(2)}, d_{(3)})$, with reciprocal basis $(d^{(1)}, d^{(2)}, d^{(3)})$ satisfying $d_{(i)} \cdot d^{(j)} = \delta^i_j$, the second triple $(1-\gamma_0, 1-\gamma_0, 2+\gamma_0)$ of diagonal entries determines a tensor

$$\begin{aligned} H_{long} \quad : &= (1-\gamma_0)d_{(1)} \otimes d^{(1)} + (1-\gamma_0)d_{(2)} \otimes d^{(2)} + (2+\gamma_0)d_{(3)} \otimes d^{(3)} \\ &= H_{sph} + (1+2\gamma_0)d_{(3)} \otimes d^{(3)} \end{aligned} \tag{11.9}$$

that, in turn, determines a purely submacroscopic deformation (i, H_{long}) that we refer to as the *(universal) elongated phase* of the body. A piecewise smooth approximation h_n for (i, H_{long}) takes each of the cubic cells of side $1/n$, with its edges now parallel to $d_{(1)}$, $d_{(2)}$, $d_{(3)}$, respectively, and stretches the "$d_{(3)}$" edge to the length $(2+\gamma_0)/n \approx 2.618/n$ while shrinking the other two edges to the length $(1-\gamma_0)/n \approx .382/n$. The simultaneous elongation of each cell creates voids, and the resulting structured deformation (i, H_{long}) has volume fraction

$$\det H_{long} = \det K_{long} = (1-\gamma_0)^2(2+\gamma_0) = (1-\gamma_0) \approx .382. \tag{11.10}$$

We have used in the last calculation the fact that $(1-\gamma_0)$ and $(2+\gamma_0)$ are reciprocals. Moreover, after elongation, the cells may have to be translated slightly in order to avoid interpenetration of neighboring cells, because the piecewise smooth approximations h_n are required to be injective. While the universal solution H_{sph} does not vary in space and time, the universal solution H_{long} may vary through dependence of the dyad $d_{(3)} \otimes d^{(3)}$ on X and t. In addition, the basis $(d_{(1)}, d_{(2)}, d_{(3)})$ need not be orthogonal, so that the approximating deformations h_n map unit cubes into possibly non-rectangular parallelepipeds.[2] For the class of energetically nearsighted elastic bodies considered in the next subsection, universal solutions H must be symmetric, and the basis $(d_{(1)}, d_{(2)}, d_{(3)})$ is taken to be orthonormal, so that $d^{(i)} = d_{(i)}$ for $i = 1, 2, 3$.

11.2 Field relations

We now specialize the discussion above to the case

$$\overline{\Psi}(F, K) = \bar{\psi}(\det K) = \bar{\psi}(\det H) \tag{11.11}$$

in which only the volume fraction $f := \det K = \det H$ produced by the purely submacroscopic deformations (i, H) and (i, K) affects the free energy. We shall restrict our attention to universal phases, so that the consistency relation need not be considered further, and the formula

$$\overline{\Psi}'(K) = f\bar{\psi}'(f)K^{-T} \tag{11.12}$$

along with the fact that $f \in \{1-\gamma_0, (1-\gamma_0)^3\}$ is a constant for each phase yields after some computations the following forms of the remaining field relations (11.1), (11.3), and (11.4), as well as the stress relation (11.5):

$$f\bar{\psi}'(f)\mathrm{div}((H^{-1} - 2I)F^{-T}) + b_{ref} = \rho_{ref}\ddot{\chi} \tag{11.13}$$

$$\mathrm{sk}H = 0 \tag{11.14}$$

$$\bar{\psi}'(f)[(H^{-1} - 2I) \cdot \dot{F}F^{-1} - I \cdot \dot{H}H^{-1}] \geq 0 \tag{11.15}$$

$$S = f\bar{\psi}'(f)(H^{-1} - 2I)F^{-T}. \tag{11.16}$$

We have used the frame-indifference of the mixed power (11.14) to replace H^{-T} by H^{-1} in the balance law (11.13) and in the mixed power inequality (11.15), and we have assumed that $\bar{\psi}'(f) \neq 0$ for $f \in \{1-\gamma_0, (1-\gamma_0)^3\}$. Because H is symmetric, we may now rewrite (11.9) as

$$H_{long} = H_{sph} + (1 + 2\gamma_0)d \otimes d, \tag{11.17}$$

with $d = d_{(3)} = d^{(3)}$ of unit magnitude. An easy calculation provides the formulas

$$H_{sph}^{-1} = (2+\gamma_0)I, \quad H_{long}^{-1} = (2+\gamma_0)I - (1+2\gamma_0)d \otimes d. \tag{11.18}$$

[2] If the basis is chosen from eigenvectors of the symmetric part of H, then the parallelepipeds can be formed from (possibly non-rectangular) cross-sections all perpendicular to one of the basis vectors.

Using the constancy of H_{sph}, we obtain the specific forms for the balance of linear momentum, the mixed power inequality, and the stress relation *in the spherical phase*

$$\gamma_0(1-\gamma_0)^3\bar{\psi}'((1-\gamma_0)^3)\mathrm{div}F^{-T}+b_{ref}=\rho_{ref}\ddot{\chi} \tag{11.19}$$

$$\bar{\psi}'((1-\gamma_0)^3)\mathrm{tr}(\dot{F}F^{-1})\geq 0 \tag{11.20}$$

$$S=\gamma_0(1-\gamma_0)^3\bar{\psi}'((1-\gamma_0)^3)F^{-T}. \tag{11.21}$$

The stress relation implies that the Cauchy stress in the spherical phase is given by

$$\begin{aligned} T &= SF^T/\det F=\rho SF^T/\rho\det F \\ &= C_{sph}\,\rho I \end{aligned} \tag{11.22}$$

where $C_{sph}:=\gamma_0(1-\gamma_0)^3\bar{\psi}'((1-\gamma_0)^3)/\rho_{ref}$ has the same sign as $\bar{\psi}'((1-\gamma_0)^3)$ and ρ denotes the density in the current configuration. Thus, in the spherical phase, the nearsighted elastic body experiences a hydrostatic stress that is proportional to the density in the current configuration. If $\bar{\psi}'((1-\gamma_0)^3)<0$, then the stress is a hydrostatic pressure, again proportional to the density, as in the case of an ideal gas. Consequently, *if* $\bar{\psi}'((1-\gamma_0)^3)<0$, *the equation of state of the energetically nearsighted elastic body in the spherical phase is that of an ideal gas undergoing isothermal dynamical processes.* Of course, the balance of linear momentum then takes the standard form in the current configuration for gas dynamics:

$$C_{sph}\,\mathrm{grad}\rho+b=\rho\dot{v} \tag{11.23}$$

where ρ and b now denote the density and body force in the current configuration, and v denotes the spatial velocity field. However, the mixed power inequality now requires that

$$\mathrm{div}v\leq 0, \tag{11.24}$$

which tells us that, *when* $\bar{\psi}'((1-\gamma_0)^3)<0$, *the spherical phase can arise only when the energetically nearsighted elastic body is not expanding.*

By employing relations (11.17), (11.18), and the formulas

$$\dot{H}_{long}=(1+2\gamma_0)(\dot{d}\otimes d+d\otimes\dot{d}),\quad d\cdot d=1$$

we obtain in a similar way specific forms for the balance of linear momentum, the mixed power inequality, and the stress relation *in the elongated phase*

$$(1-\gamma_0)\bar{\psi}'((1-\gamma_0))\mathrm{div}[(\gamma_0 I-(1+2\gamma_0)d\otimes d)F^{-T}]+b_{ref}=\rho_{ref}\ddot{\chi} \tag{11.25}$$

$$(1-\gamma_0)\bar{\psi}'((1-\gamma_0))[\gamma_0 I-(1+2\gamma_0)d\otimes d)]\cdot\dot{F}F^{-1}\geq 0 \tag{11.26}$$

$$S=(1-\gamma_0)\bar{\psi}'((1-\gamma_0))[\gamma_0 I-(1+2\gamma_0)d\otimes d]F^{-T}. \tag{11.27}$$

The stress relation implies that the Cauchy stress in the elongated phase is given by

$$T=C_{long}\rho[I-(3+\gamma_0)d\otimes d] \tag{11.28}$$

with $C_{long} := \gamma_0(1-\gamma_0)\bar{\psi}'((1-\gamma_0))/\rho_{ref}$ having the same sign as $\bar{\psi}'((1-\gamma_0))$, and the relations (11.23) and (11.24) become in the elongated phase:

$$C_{long}\,\mathrm{grad}\rho - C_{long}(3+\gamma_0)\mathrm{div}[\rho d\otimes d] + b = \rho\dot{v} \tag{11.29}$$

and

$$\mathrm{div}v \leq (3+\gamma_0)(\,\mathrm{grad}v)d\cdot d, \tag{11.30}$$

the latter when $\bar{\psi}'((1-\gamma_0)) < 0$. We conclude that the elongated phase, unlike the spherical phase, can persist even when div$v$ is positive, as long as the stretching (grad$v$)$d\cdot d$ in the direction of d is at least div$v/(3+\gamma_0)$. For example, suppose that d is chosen to be an eigenvector of the stretching tensor $D := \frac{1}{2}(\mathrm{grad}v + \mathrm{grad}v^T)$ such that d corresponds to the largest eigenvalue of D. As long as D has at least one non-negative eigenvalue, we then have

$$\mathrm{div}v = \mathrm{tr}D \leq 3Dd\cdot d = 3(\,\mathrm{grad}v)d\cdot d \leq (3+\gamma_0)(\,\mathrm{grad}v)d\cdot d, \tag{11.31}$$

and, for this choice of d, the dissipation inequality (11.30) is satisfied. Consequently, *when* $\bar{\psi}'((1-\gamma_0)) < 0$, *the body in the elongated phase may expand by means of an appropriate alignment of the direction of submacroscopic elongation d and the macroscopic flow.*

12 Material Symmetry

For each point X_0 in the region $\mathcal{A}$ undergoing a dynamical process, we consider the transformation properties of the kinematical quantities at that point under a change of virgin configuration determined by a given unimodular tensor $\mathbb{H}_0$. These transformation properties can be obtained by replacing the time-parameterized family of structured deformations (χ, G) by

$$\begin{aligned}(X,t) &\mapsto ((\chi,G)\circ(\xi_{\mathbb{H}_0},\mathbb{H}_0))(X,t) \qquad (12.1)\\ &= (\chi(\xi_{\mathbb{H}_0}(X,t),\,t),G(\xi_{\mathbb{H}_0}(X,t),\,t)\mathbb{H}_0), \qquad (12.2)\end{aligned}$$

where "$\circ$" denotes composition of structured deformations as defined by (12.2), and $\xi_{\mathbb{H}_0}$ denotes the homogeneous, time-independent deformation $(X,t)\mapsto X_0 + \mathbb{H}_0(X-X_0)$. From this observation we obtain the following *transformation rules*

$$\begin{array}{ccc} F & \rightarrow & F\mathbb{H}_0 \\ G & \rightarrow & G\mathbb{H}_0 \\ M & \rightarrow & M\mathbb{H}_0 \\ K & \rightarrow & \mathbb{H}_0^{-1}K\mathbb{H}_0 \end{array} \tag{12.3}$$

under change of virgin configuration. In this display, if a quantity on the left is evaluated at (X,t), the quantity on the right is evaluated at $(X_0 + \mathbb{H}_0(X-X_0),t)$.

We say that $\mathbb{H}_0$ is a *symmetry transformation* at X_0 *with respect to changes of virgin configuration* for the elastic body undergoing disarrangements if the response function $(F,G)\mapsto\Psi(F,G;X_0)$ satisfies

$$\Psi(F\mathbb{H}_0, G\mathbb{H}_0, X_0) = \Psi(F,G,X_0) \tag{12.4}$$

for all (F, G) with $0 < \det G \leq \det F$ or, equivalently, if the response function $(M, G) \mapsto \tilde{\Psi}(M, G, X_0)$ satisfies

$$\tilde{\Psi}(M\mathbb{H}_0, G\mathbb{H}_0, X_0) = \tilde{\Psi}(M, G, X_0) \tag{12.5}$$

for all (M, G) with $0 < \det G \leq \det(M + G)$. As in elasticity without disarrangements, the symmetry transformations at X_0 form a group $\mathcal{G}_{X_0}^{virgin}$.

In the special case of an isotropic solid, when $\mathcal{G}_{X_0}^{virgin}$ is the proper orthogonal group, it is easy to obtain necessary and sufficient conditions that (12.4) holds for all $\mathbb{H}_0 \in \mathcal{G}_{X_0}^{virgin}$. For an isotropic solid, we can choose $\mathbb{H}_0$ to be R_F^T or R_G^T, from the polar decompositions $F = V_F R_F$ and $G = V_G R_G$, to obtain

$$\begin{aligned} \Psi(F, G, X_0) &= \Psi(V_F R_F R_F^T, G R_F^T, X_0) = \Psi(V_F R_F R_F^T, G F^T V_F^{-1}, X_0) \\ &= \Psi(B_F^{1/2}, G F^T B_F^{-1/2}, X_0) = \hat{\Psi}(B_F, G F^T, X_0), \end{aligned} \tag{12.6}$$

or, alternatively,

$$\Psi(F, G, X_0) = \check{\Psi}(F G^T, B_G, X_0). \tag{12.7}$$

Here, $B_F = F F^T$ and $B_G = G G^T$ are the left Cauchy-Green tensors for F and G. The existence of a function $\hat{\Psi}$ such that (12.6) holds for all (F, G) with $0 < \det G \leq \det F$ (or, equivalently, of $\check{\Psi}$ such that (12.7) holds for all such pairs) is a necessary and sufficient condition that $\mathcal{G}_{X_0}^{virgin}$ be the proper orthogonal group. Similarly, the existence of a function $(M, G) \mapsto \Psi^{\#}(M, G, X_0)$ such that

$$\tilde{\Psi}(M, G, X_0) = \Psi^{\#}(M G^T, B_G, X_0) \tag{12.8}$$

for all (M, G) with $0 < \det G \leq \det(M + G)$ is both necessary and sufficient for $\mathcal{G}_{X_0}^{virgin}$ to be the proper orthogonal group.

When $\mathcal{G}_{X_0}^{virgin}$ is the proper unimodular group, we may put $\mathbb{H}_0 = (\det F)^{1/3} F^{-1}$ or $\mathbb{H}_0 = (\det G)^{1/3} G^{-1}$ to conclude that the Helmholtz free energy can be expressed as a function of either of the pairs $(\det F, H)$, $(\det G, H^{-1})$. In Section 11, we considered an elastic body such that the Helmholtz free energy depends on the volume fraction $f = \det H = \det K$, alone.

Alternatively, a notion of material symmetry may be formulated in terms of invariance of response to changes in *reference configuration.* For each point X_0 in the region $\mathcal{A}$ undergoing a dynamical process, we consider the transformation properties of the kinematical quantities at that point obtained first by factoring (χ, G) via the notion of composition introduced in (2.5),

$$(\chi, G) = (\chi, \nabla\chi) \circ (\pi, (\nabla\chi)^{-1} G). \tag{12.9}$$

Here, $\pi(X, t) = X$ for all X and t. The factor $(\chi, \nabla\chi)$ on the right-hand side of (12.9) is a family of classical deformations, while $(\pi, (\nabla\chi)^{-1} G)$ involves only purely submacroscopic deformations, because π leaves each point fixed. We next replace the expression on the right-hand side of (12.9) by

$$((\chi, \nabla\chi) \circ (\xi_{\mathbb{H}_0}, \mathbb{H}_0)) \circ (\pi, (\nabla\chi)^{-1} G), \tag{12.10}$$

where, as above, $\xi_{\mathbb{H}_0}$ denotes the homogeneous, time-independent deformation $(X,t) \mapsto X_0+\mathbb{H}_0(X-X_0)$. This replacement leaves the purely submacroscopic factor $(\pi, \nabla\chi^{-1}G)$ unchanged and changes only the classical factor $(\chi, \nabla\chi)$. From this replacement we obtain the following *transformation rules*

$$\begin{array}{ccl} F & \to & F\mathbb{H}_0 \\ G & \to & (F\mathbb{H}_0F^{-1})G \\ M & \to & (F\mathbb{H}_0F^{-1})M \\ K & \to & K \end{array} \tag{12.11}$$

under change of reference configuration. In this display, if a quantity on the left is evaluated at (X,t), the quantity on the right is evaluated at $(X_0 + \mathbb{H}_0(X - X_0), t)$.

For a pair $(\mathbb{H}_0, K_0)$, with $0 < \det K_0 \leq 1 = \det \mathbb{H}_0$, we say that $\mathbb{H}_0$ is a *K_0-symmetry transformation at X_0 with respect to changes of reference configuration* for the elastic body undergoing disarrangements if the response function $F \mapsto \Psi(F, FK_0, X_0)$ satisfies

$$\Psi(F\mathbb{H}_0, F\mathbb{H}_0K_0, X_0) = \Psi(F, FK_0, X_0) \tag{12.12}$$

for all tensors F with $0 < \det F$. Equivalently, we may use the definition in (9.15), $\overline{\Psi}(F,K,X_0) := \Psi(F, FK, X_0)$ for all pairs (F,K) with $0 < \det F$ and $0 < \det K \leq 1$, to write (12.12) in the simpler form

$$\overline{\Psi}(F\mathbb{H}_0, K_0, X_0) = \overline{\Psi}(F, K_0, X_0) \tag{12.13}$$

for all tensors F with $0 < \det F$. We denote by $\mathcal{G}^{ref}_{X_0,K_0}$ the group formed by the K_0-symmetry transformations at X_0.

The symmetry group $\mathcal{G}^{ref}_{X_0,K_0}$ defined through (12.12) or (12.13) corresponds to the usual symmetry group of an elastic body undergoing only classical deformations, because the influence of disarrangements is removed by fixing the value of $K = F^{-1}G$ at K_0. We remark that there is an obvious notion of invariance dual to (12.13):

$$\overline{\Psi}(F, K\mathbb{P}_0, X_0) = \overline{\Psi}(F, K, X_0) \tag{12.14}$$

for all tensors F and K such that $0 < \det F$ and $0 < \det K \leq 1$, with $\mathbb{P}_0$ a given unimodular tensor. This invariance arises by replacing the right-hand side of the factorization (12.9) by

$$(\chi, \nabla\chi) \circ ((\pi, (\nabla\chi)^{-1}G) \circ (\pi, \mathbb{P}_0)). \tag{12.15}$$

The purely submacroscopic factor $(\pi, \mathbb{P}_0)$ alters the given one $(\pi, (\nabla\chi)^{-1}G)$ without changing the classical factor $(\chi, \nabla\chi)$. The resulting symmetry group $\mathcal{G}^{submac}_{X_0}$ corresponds to a notion introduced by Šilhavý and Kratochvíl [19], in the context of Noll's new theory of simple materials, and adapted by Bertram [20] to formulate and solve problems in the plasticity of materials undergoing large deformations. This symmetry group is obtained by means of non-classical changes $(\pi, \mathbb{P}_0)$ in configuration, as distinct from the groups $\mathcal{G}^{ref}_{X_0,K_0}$ and $\mathcal{G}^{virgin}_{X_0}$, obtained via the classical changes $(\xi_{\mathbb{H}_0}, \mathbb{H}_0)$.

The assumptions about material behavior made in [19], [20], when translated into the present framework, take the form of an additional, special requirement: for all (X,t) encountered in a dynamical process, there holds

$$K(X,t) \in \mathcal{G}^{submac}_{X}. \tag{12.16}$$

This requirement will simplify substantially the field relations (9.18) - (9.21), because (12.16) and (12.14) imply, for example, the relations

$$D_F\overline{\Psi}(F, K(X,t), X) = D_F\overline{\Psi}(F, I, X) \tag{12.17}$$

and

$$D_K\overline{\Psi}(F, K(X,t), X)K(X,t)^T = D_K\overline{\Psi}(F, I, X) \tag{12.18}$$

for all F, X, and t. In fact, these two relations render trivial the dependence upon $K(X,t)$ of the fields $(X,t) \mapsto D_F\overline{\Psi}(F(X,t), K(X,t), X)$ and $(X,t) \mapsto D_K\overline{\Psi}(F(X,t), K(X,t), X)$ and, thereby, simplify considerably the field relations (9.18) - (9.21).

Part II
Elasticity with Space- and Time-Like Disarrangements

13 Introduction

The approach to elasticity with disarrangements set forth in [4] and treated in Part I takes submacroscopic geometrical changes into account, at a fixed but arbitrary time t, by means of a sequence $n \mapsto \chi_n(\cdot, t)$ of piecewise smooth functions that approximate $\chi(\cdot, t)$. No consideration is given to relating the approximations $\chi_n(\cdot, t)$ and $\chi_n(\cdot, \tau)$ corresponding to two different times t and τ, nor is the possible evolution in time of the approximations taken into account explicitly in the field theory of Part I. For example, one cannot distinguish in the context of that field theory between the macroscopic velocity $\dot{\chi}$ and the velocity of submacroscopic pieces of the body.

Part II of this work is devoted to the formulation of a field theory for elastic bodies that takes into account not only space-like but also "time-like" disarrangements. This formulation rests on a refined notion of structured motion [8] and on a derivation of balance laws similar to that given in [22], starting from a postulate of invariance of a non-classical form of the power under classical and under certain non-classical superposed rigid motions. When constitutive relations are introduced in a manner analogous to that in Part I, the resulting field relations that we obtain in Part II differ from those in Part I in three main respects:

- there is a supplementary balance equation for linear momentum;
- there is a new formula for the internal dissipation in which the "driving power" replaces the mixed power, and the assumption that the "driving power" is frame-indifferent replaces the corresponding assumption on the mixed power in Part I;
- there is a different symmetry restriction on stresses imposed by balance of angular momentum or, alternatively, by material frame-indifference.

The supplementary balance equation for linear momentum mentioned in the first bullet governs the evolution of the "velocity without disarrangements." Regarding the second bullet, positivity of the driving power and, hence, of the internal dissipation, places two restrictions on the motion and forces: the free energy cannot have a local minimum with respect to submacroscopic geometrical changes, and submacroscopic geometrical incompatibility, expressed in terms of stretchings, must be present. In connection with the last bullet, for the field theory in Part II it is a stress without disarrangements, rather than the Cauchy stress itself, that turns out to be symmetric (see Section 20.1 for a detailed comparison of the two theories).

In spite of differences between the field theories in Parts I and II, the reader will note that the overall structure of the field equations, including the methods and assumptions employed in their derivation, have much in common. This significant overlap points to a common foundation for the support of future developments.

14 Structured Motions Revisited

Here in Part II we refine the notion of structured motion by specifying not only the fields $\chi : \mathcal{A}\times(0,t_f) \to \mathcal{E}$ and $G : \mathcal{A}\times(0,t_f) \to \mathrm{Lin}\mathcal{V}$ used in Part I, but also a vector field $\dot{\chi}_\backslash : \mathcal{A}\times(0,t_f) \to \mathcal{V}$ called the *velocity without disarrangements.* (The notation w instead of $\dot{\chi}_\backslash$ was used when structured motions were introduced in [8], Part Two; in addition, the "disarrangement site" $\kappa^{(4)}$ in [8], Part Two, here is assumed to be the empty set. This assumption rules out in the present study discontinuities in the fields χ, G , and $\dot{\chi}_\backslash$, although discontinuites are permitted in the piecewise smooth approximating motions χ_n introduced below (14.1).)

While precise smoothness assumptions on the fields χ, G , and $\dot{\chi}_\backslash$ and on the region $\mathcal{A}$ will not be specified here, sufficient smoothness requirements on the fields can be inferred from the context. Other than smoothness requirements, the only conditions imposed on the fields χ, G, and $\dot{\chi}_\backslash$ are the injectivity of $X \mapsto \chi(X,t)$ for every $t \in (0,t_f)$, as well as the existence of a positive number m such that the inequalities

$$m < \det G(X,t) \leq \det \nabla\chi(X,t) \tag{14.1}$$

hold for all $(X,t) \in \mathcal{A}\times(0,t_f)$. The Approximation Theorem for structured motions ([8], Section 6.5) assures that there is a sequence $n \mapsto \chi_n$ of piecewise smooth motions defined on $\mathcal{A}\times(0,t_f)$ with the same injectivity property as χ such that

$$\chi = \lim_{n\to\infty} \chi_n, \quad G = \lim_{n\to\infty} \nabla\chi_n, \quad \dot{\chi}_\backslash = \lim_{n\to\infty} \dot{\chi}_n, \tag{14.2}$$

with the limits taken in the sense of L^∞-convergence on $\mathcal{A} \times (0,t_f)$. We note that in (2.10), which is the analogue of (14.2), the limits are taken in the sense of L^∞-convergence on $\mathcal{A}$, because there the time was held fixed at a given but arbitrary value. The spatial derivatives $\nabla\chi_n$ and the time derivatives $\dot{\chi}_n$ are taken in the classical sense, and the limits G and $\dot{\chi}_\backslash$ of derivatives in $(14.2)_{2,3}$ need not equal the corresponding derivatives $\nabla\chi$ and $\dot{\chi}$ of the macroscopic motion (nor need G even be the gradient of some motion). Specific quantitative information about the differences $M := \nabla\chi - G$ and $\dot{\chi}_d := \dot{\chi} - \dot{\chi}_\backslash$ are provided in the next section and justify the terminology *deformation due to disarrangements* for M and *velocity due to disarrangements* for $\dot{\chi}_d$.

14.1 Decompositions and identification relations

The algebraically trivial *additive decompositions*

$$\nabla\chi = G + M, \quad \dot{\chi} = \dot{\chi}_\backslash + \dot{\chi}_d \tag{14.3}$$

for the macroscopic deformation gradient and velocity are given deeper significance by means of the limit relations (14.2) for G and $\dot{\chi}_\backslash$ and by means of the following *identifi-*

cation relations obtained in [8], Part Two, Section 6.6, for M and $\dot{\chi}_d$:

$$M(X,t) = \lim_{\delta \to 0} \lim_{n \to \infty} vol\mathcal{B}(X,t;\delta)^{-1} \int_{\Gamma(\chi_n)\cap\mathcal{B}(X,t;\delta)} [\chi_n](Y,\tau) \otimes \nu^{(3)}(Y,\tau) dA_{(Y,\tau)} \quad (14.4)$$

and

$$\dot{\chi}_d(X,t) = \lim_{\delta \to 0} \lim_{n \to \infty} vol\mathcal{B}(X,t;\delta)^{-1} \int_{\Gamma(\chi_n)\cap\mathcal{B}(X,t;\delta)} \nu^{(1)}(Y,\tau)[\chi_n](Y,\tau) dA_{(Y,\tau)}. \quad (14.5)$$

In these relations, $n \mapsto \chi_n$ is an arbitrary sequence of piecewise smooth motions that satisfies the limit relations (14.2). The symbol $\mathcal{B}(X,t;\delta) \subset \mathcal{E} \times \mathbb{R}$ denotes the space-time ball of radius $\delta > 0$ centered at a point (X,t) in $\mathcal{A}\times(0,t_f)$, and $\Gamma(\chi_n) \subset \mathcal{E} \times \mathbb{R}$, $[\chi_n](Y,\tau) \in \mathcal{V}$, and $(\nu^{(3)}(Y,\tau), \nu^{(1)}(Y,\tau)) \in \mathcal{V}\times\mathbb{R}$ denote, respectively, the jump set of the piecewise smooth motion χ_n, the jump of χ_n at a point $(Y,\tau) \in \Gamma(\chi_n)$, and the unit normal to the jump set $\Gamma(\chi_n)$ at the point (Y,τ). The integrand in (14.4) is the tensor product of $[\chi_n](Y,\tau)$ and $\nu^{(3)}(Y,\tau)$, both vectors in $\mathcal{V}$, while the integrand in (14.5) is the scalar $\nu^{(1)}(Y,\tau)$ multiplied by the vector $[\chi_n](Y,\tau) \in \mathcal{V}$.

Relations (14.4) and (14.5) tell us that the tensor $M(X,t)$ and the vector $\dot{\chi}_d(X,t)$ both are measures of the jumps $[\chi_n]$ of the approximating piecewise-smooth motions χ in the Approximation Theorem. Because $\nu^{(3)}(Y,\tau) \in \mathcal{V}$ and $\nu^{(1)}(Y,\tau) \in \mathbb{R}$ are, respectively, the space-like and time-like components of the unit normal $(\nu^{(3)}(Y,\tau), \nu^{(1)}(Y,\tau)) \in \mathcal{V}\times\mathbb{R}$ to the jump set $\Gamma(\chi_n)$ at the point (Y,τ), the tensor $M(X,t)$ reflects only the "space-like directed jumps" $[\chi_n] \otimes \nu^{(3)}$, while the vector $\dot{\chi}_d(X,t)$ reflects only the "time-like directed jumps" $\nu^{(1)}[\chi_n]$. It is worth keeping in mind that in (2.11), the analogue of (14.4), the region of integration is a ball in the three-dimensional space $\mathcal{E}$, centered at X. By contrast, the region of integration in (14.4) is a ball in the four-dimensional space $\mathcal{E} \times \mathbb{R}$, centered at (X,t).

In view of these remarks, the identification relations (14.4) and (14.5) provide a precise distinction between space-like and time-like disarrangements, and the interpretations now available for G, $M = \nabla\chi - G$, $\dot{\chi}_\backslash$, and $\dot{\chi}_d = \dot{\chi} - \dot{\chi}_\backslash$ permit us to understand and interpret various examples of structured motions.

14.2 Rigid motions without disarrangements

If $r : \mathcal{A}\times(0,t_f) \to \mathcal{E}$ is a smooth mapping that at each time preserves distances, then $\nabla r : \mathcal{A}\times(0,t_f) \to \mathrm{Lin}\mathcal{V}$ has orthogonal values and $\nabla\dot{r}\,\nabla r^T : \mathcal{A}\times(0,t_f) \to \mathrm{Lin}\mathcal{V}$ has skew values. The structured motion

$$\chi := r, \quad G := \nabla r, \quad \dot{\chi}_\backslash := \dot{r} \quad (14.6)$$

satisfies $M = \nabla\chi - G = 0$ and $\dot{\chi}_d = \dot{\chi} - \dot{\chi}_\backslash = 0$. Here, there is no deformation due to disarrangements and no velocity due to disarrangements, and the identification relations (14.4) and (14.5) tell us that no space-like or time-like jumps in approximating motions contribute to the macroscopic deformation and velocity. Accordingly, the triple $(r, \nabla r, \dot{r})$ will be called a *classical rigid motion* or a *rigid motion without disarrangements*.

14.3 Rigid motions caused by time-like disarrangements

If r again is smooth and distance-preserving, the structured motion

$$\chi := r, \quad G := \nabla r, \ \dot{\chi}_{\backslash} := 0 \tag{14.7}$$

satisfies $M = \nabla\chi - G = 0$ and $\dot{\chi}_d = \dot{\chi} - \dot{\chi}_{\backslash} = \dot{r}$. Here, there is no deformation due to disarrangements, so that no space-like jumps in approximating motions contribute to the macroscopic deformation. However, the relation $\dot{r} = \dot{\chi}_d$ and the identification relation (14.5) tell us that the macroscopic velocity field $\dot{\chi} = \dot{\chi}_d$ arises entirely from time-like jumps in approximating piecewise smooth motions. The triple $(r, \nabla r, 0)$ will be called a *rigid motion caused by (time-like) disarrangements.* An example in [8] (Part Two, Section 6.3: the "blinking motion") of approximating piecewise smooth motions provides instances of the time-like disarrangements required for a rigid motion caused by disarrangements, and the "movie motion" discussed in Section 14.5 provides more general motions caused by time-like disarrangements.

14.4 Superposed rigid motions

The definition of composition of two structured deformations in (2.6), (2.7) was shown in [8], Part Two, Section 6.5 to lead to the following notion of composition for structured motions:

$$(\eta, H, \dot{\eta}_{\backslash}) \circ (\chi, G, \dot{\chi}_{\backslash}) = (\eta \diamond \chi, (H \diamond \chi)G, (H \diamond \chi)\dot{\chi}_{\backslash} + \dot{\eta}_{\backslash} \diamond \chi)\,. \tag{14.8}$$

Here, the symbol $\diamond$ denotes the "composition in the first variable", for example: $(\eta \diamond \chi)(X,t) := \eta(\chi(X,t),t)$ for all $(X,t) \in \mathcal{A}{\times}(0,t_f)$. This formula provides the following formulas for a structured motion $(\chi, G, \dot{\chi}_{\backslash})$ followed by rigid motions of the types just considered:

$$(r, \nabla r, \dot{r}) \circ (\chi, G, \dot{\chi}_{\backslash}) = (r \diamond \chi, (\nabla r \diamond \chi)G, (\nabla r \diamond \chi)\dot{\chi}_{\backslash} + \dot{r} \diamond \chi) \tag{14.9}$$

$$(r, \nabla r, 0) \circ (\chi, G, \dot{\chi}_{\backslash}) = (r \diamond \chi, (\nabla r \diamond \chi)G, (\nabla r \diamond \chi)\dot{\chi}_{\backslash})\,. \tag{14.10}$$

We interpret the compositions in (14.9) and (14.10) as *superpositions* of a rigid motion (without disarrangements or caused by disarrangements) on the given structured motion $(\chi, G, \dot{\chi}_{\backslash})$. The superpositions result in new structured motions given by the right-hand sides of (14.9) and (14.10), and the velocity without disarrangements for $(r, \nabla r, \dot{r}) \circ (\chi, G, \dot{\chi}_{\backslash})$ is the third entry on the right-hand side of (14.9): $(\nabla r \diamond \chi)\dot{\chi}_{\backslash} + \dot{r} \diamond \chi$, while the velocity without disarrangements for $(r, \nabla r, 0) \circ (\chi, G, \dot{\chi}_{\backslash})$ is the third entry of the right-hand side of (14.10): $(\nabla r \diamond \chi)\dot{\chi}_{\backslash}$. Moreover, we may compute the velocity due to disarrangements for each of the superpositions by subtracting the third entry from the time derivative of the first. For the superposition $(r, \nabla r, \dot{r}) \circ (\chi, G, \dot{\chi}_{\backslash})$ the velocity due to disarrangements is

$$\begin{aligned}(r \diamond \chi)^{\cdot} - ((\nabla r \diamond \chi)\dot{\chi}_{\backslash} + \dot{r} \diamond \chi) &= \dot{r} \diamond \chi + (\nabla r \diamond \chi)(\dot{\chi} - \dot{\chi}_{\backslash}) - \dot{r} \diamond \chi \\ &= (\nabla r \diamond \chi)\dot{\chi}_d \end{aligned} \tag{14.11}$$

while for $(r, \nabla r, 0) \circ (\chi, G, \dot{\chi}_\backslash)$ the velocity due to disarrangements is

$$\begin{aligned}(r \diamond \chi)^{\cdot} - (\nabla r \diamond \chi)\dot{\chi}_\backslash &= \dot{r} \diamond \chi + (\nabla r \diamond \chi)(\dot{\chi} - \dot{\chi}_\backslash) \\ &= \dot{r} \diamond \chi + (\nabla r \diamond \chi)\dot{\chi}_d. \end{aligned} \tag{14.12}$$

If r and $t_o \in (0, t_f)$ are such that $\nabla r(x, t_o) = I$ for all $X \in \mathcal{A}$, then the formulas just obtained may be simplified as follows when evaluated at t_o :

structured motion	**velocity without disarrangements**	**velocity due to disarrangements**
$(r, \nabla r, \dot{r}) \circ (\chi, G, \dot{\chi}_\backslash)$	$\dot{\chi}_\backslash(\cdot, t_o) + \dot{r}(\chi(\cdot, t_o), t_o)$	$\dot{\chi}_d(\cdot, t_o)$
$(r, \nabla r, 0) \circ (\chi, G, \dot{\chi}_\backslash)$	$\dot{\chi}_\backslash(\cdot, t_o)$	$\dot{\chi}_d(\cdot, t_o) + \dot{r}(\chi(\cdot, t_o), t_o)$.

In particular, if for every time t the mapping $r(\cdot, t)$ is a translation of $\mathcal{E}$, then $\nabla r = I$ holds at all times and $\dot{r}$ is a constant. The deformation without disarrangements for the superpositions in (14.9) and (14.10) are the second entries in each triple of the right-hand side, and the deformation due to disarrangements is obtained by subtracting the second entry on the right-hand sides from the gradient of the first. At a time $t_o \in (0, t_f)$ such that $\nabla r(x, t_o) = I$ for all $X \in \mathcal{A}$, we find:

structured motion	**deformation without disarrangements**	**deformation due to disarrangements**
$(r, \nabla r, \dot{r}) \circ (\chi, G, \dot{\chi}_\backslash)$	$G(\cdot, t_o)$	$M(\cdot, t_o)$
$(r, \nabla r, 0) \circ (\chi, G, \dot{\chi}_\backslash)$	$G(\cdot, t_o)$	$M(\cdot, t_o)$.

14.5 Further examples of structured motions

A generalization of the rigid motion caused by time-like disarrangements $(r, \nabla r, 0)$ is the *movie motion* $(\chi, \nabla \chi, 0)$, in which the macroscopic velocity $\dot{\chi}$ equals the velocity due to disarrangements $\dot{\chi}_d$, because $\dot{\chi}_\backslash = 0$. An approximating piecewise smooth motion χ_n may be chosen to be piecewise constant in time, corresponding to the succession of frames of a movie - - each frame depicting a static picture of the body over a short time interval. With this choice of approximating motion, $\dot{\chi}_n$ vanishes for all n. As n increases, the duration of the time-intervals of constancy decreases, the number of intervals of constancy increases, χ_n converges to χ, $\nabla \chi_n$ converges to $\nabla \chi$, and $\dot{\chi}_n$ converges to 0. Because $G = \nabla \chi$, we have $M = 0$, and all of the disarrangements are time-like. In the movie motion, we are justified in saying that the macroscopic motion χ is due entirely to the instantaneous changes in frame embodied in the time-like jumps of χ_n.

For the movie motion, it is possible to replace the piecewise constant approximating motions, for which the space-time disarrangement sites are horizontal lines $t = const.$, by approximating motions for which the disarrangement sites are nearly horizontal lines that tend to horizontal as n tends to infinity. One obtains the movie motion in the limit, as before, but the new approximating motions change a given frame by rapidly sweeping through the modifications needed to reach the next. These approximations are reminiscent of a series of dislocations sweeping through a crystal: each dislocation makes a small, rapid change in the crystal, and the entire series makes a macroscopically detectable change.

An example in which space-like disarrangements predominate is a structured motion of the form $(\chi, I, \dot{\chi})$, a *submacroscopic translation.* Here, the deformation gradient $\nabla\chi$ and the deformation without disarrangements I generally are different, so that $M = \nabla\chi - I$ does not vanish, while the velocity due to disarrangements $\dot{\chi}_d$ vanishes, because $\dot{\chi}_\backslash = \dot{\chi}$. Approximating piecewise smooth motions χ_n can be chosen so that the body initially breaks up into a finite number of pieces, each of which undergoes a succession of translational motions on short time-intervals, and, hence, $\nabla\chi_n = I$. The translational motions can be chosen so that $\dot{\chi}_n$ is close to the macroscopic velocity $\dot{\chi}$ and so that $\chi_n(X, \cdot)$ is continuous for each point X not in the disarrangement site. As n increases, the number of initial pieces increases, their size decreases, the duration of the time-intervals of translation decreases, χ_n converges to χ, $\nabla\chi_n$ converges to I, and $\dot{\chi}_n$ converges to $\dot{\chi}$. While the classical motion $(\chi, \nabla\chi, \dot{\chi})$ corresponds to having the same smooth spatial deformations at both the macroscopic and submacroscopic scales, the submacroscopic translation $(\chi, I, \dot{\chi})$ corresponds to an arbitrary smooth geometrical change at the macrolevel and to rigid translations at the submacrosopic level. One instance of such an approximating motion χ_n corresponds to the motion of a deck of very thin, rigid cards in which each card undergoes its own succesion of translational motions.

As a final example, we wish to describe a *jostling motion* in which a body that appears on a long time scale not to move at all, is observed on a short time scale to translate in one direction at constant velocity. If w_0 denotes a fixed, non-zero vector representing the short-term translational velocity, and if π denotes the trivial motion $(X, t) \mapsto \pi(X, t) = X$ that leaves each point X fixed for all times, the structured motion (π, I, w_0) has the desired properties. Here, an approximating motion χ_n begins with a translation with velocity w_0 of the entire body for a short time interval, at the end of which the body instantaneously returns to its initial position, followed by repetitions of the translation and instantaeous return, so that $\dot{\chi}_n = w_0$ and $\nabla\chi_n = I$. As n increases, the lengths of the time-intervals decrease, χ_n converges to π, $\nabla\chi_n$ converges to I, and $\dot{\chi}_n$ converges to w_0. All of the disarrangements are time-like, so that $M = 0$. Moreover, $\dot{\chi}_\backslash = w_0$ and $\dot{\chi}_d = -w_0$, because $\dot{\chi} = \dot{\pi} = 0$.

15 Contact and Body Forces

15.1 Decompositions

We recall from Section 3.1 that we may write a refined law of balance of forces as:

$$div(SK^*) + div((\det K)S - SK^*) - S\,\nabla \det K + (\det K)\, b_{ref} = 0 \qquad (15.1)$$

in which we called SK^* the stress without disarrangements and $(\det K)S - SK^*$ the stress due to disarrangements; we now call $(\det K)\, b_{ref}$ the *body force without disarrangements* and $-S\,\nabla \det K$ the *body force due to disarrangements.* It is immediate that when $M = 0$ or, equivalently, when $K = I$, the stress and the body force due to disarrangements vanish, the body force without disarrangements reduces to b_{ref}, and the stress without disarrangements reduces to S. More generally, the body force due to disarrangements vanishes whenever the volume fraction $\det K$ does not vary with position.

If we define the body force per unit volume in the virgin configuration b_v to be the sum of the body force without disarrangements and the body force due to disarrangements, the considerations above lead us to the following decompositions of stress and body force:

$$(\det K)S = SK^* + ((\det K)S - SK^*) \tag{15.2}$$

$$b_v = (\det K)\, b_{ref} - S\,\nabla \det K. \tag{15.3}$$

As in Section 3.1, the stress tensor $(\det K)S$ is an analogue of the "weighted Cauchy tensor" $(\det \nabla\chi)T \diamond \chi$. Of course, the consistency relation (3.6), as well as its equivalent form (3.7) and special form (3.8), remain valid here.

For the general case of structured motions, not only do we employ these measures of stress and body force and their decompositions, but we also must study the same issues for inertial forces. In the context of "second-order structured deformations" [21], [2] it will be possible to decompose additively the acceleration into a part due to disarrangements and a part without disarrangements, so that an analogous decomposition of inertial forces is possible. Instead, in the present context we proceed simply by differentiating with respect to time the decomposition $(14.3)_2$ to obtain the following decomposition of the inertial force $b_v^{in} := -(\det K)\rho_{ref}\ddot{\chi}$ (measured per unit volume in the virgin configuration):

$$b_v^{in} = (b_v^{in})_\backslash + (b_v^{in})_d \tag{15.4}$$

into an inertial force without disarrangements

$$(b_v^{in})_\backslash := -(\det K)\rho_{ref}(\dot{\chi}_\backslash)^{\cdot} \tag{15.5}$$

plus an inertial force due disarrangements

$$(b_v^{in})_d := -(\det K)\rho_{ref}(\dot{\chi}_d)^{\cdot}. \tag{15.6}$$

15.2 Transformation laws

Consider now a rigid motion r and a time t_o such that $\nabla r(\cdot, t_o) = I$. From the analysis in Section 14.4, both $\nabla\chi$ and G are unchanged at time t_o, and it follows that the tensor field $K = (\nabla\chi)^{-1}G$ is unchanged at time t_o under superposition of $(r, \nabla r, \dot{r})$ and $(r, \nabla r, 0)$. Moreover, we assume that the Piola-Kirchhoff stress S at time t_o also is unchanged under these superpositions, and we conclude that the stress without disarrangements SK^*, the stress due to disarrangements $(\det K)S - SK^*$, and the body force due to disarrangements $-S\,\nabla \det K$ in a given structured motion all remain unchanged at time t_o under superposition of $(r, \nabla r, \dot{r})$ and $(r, \nabla r, 0)$.

We specify the transformation law for the *total body force* $b^* := b_v + b_v^{in}$ first by noting that the decompositions (15.3) and (15.4) yield

$$\begin{aligned} b^* &= b_v + b_v^{in} \\ &= (\det K)\, b_{ref} - S\,\nabla \det K + (b_v^{in})_\backslash + (b_v^{in})_d, \end{aligned} \tag{15.7}$$

then by defining $b_\backslash^*$, the *total body force without disarrangements*, and b_d^*, the *total body force due to disarrangements*, through the formulas

$$b_\backslash^* := (\det K)\, b_{ref} + (b_v^{in})_\backslash = (\det K)\, b_{ref} - (\det K)\rho_{ref}(\dot{\chi}_\backslash)^{\cdot} \tag{15.8}$$

$$b_d^* := -S\,\nabla \det K + (b_v^{in})_d = -S\,\nabla \det K - (\det K)\rho_{ref}(\dot{\chi}_d)^{\cdot}, \tag{15.9}$$

and finally by requiring that *both* $b^*_\backslash$ *and* b^*_d *are unchanged at time* t_o (i) *under superposition of* $(r, \nabla r, \dot{r})$ *and,* (ii) *when* r *is a* **translation** *under superposition of* $(r, \nabla r, 0)$. The motivation for limiting the requirement (ii) of invariance under superposition of $(r, \nabla r, 0)$ to the case of translations will be discussed in the next section, where a similar postulate of invariance is set forth for the power expended.

16 Power Expended

Following the idea proposed in [22], we postulate that the power in a structured motion $(\chi, G, \dot{\chi}_\backslash)$ expended at time $t \in (0, T)$ on a subbody $\mathcal{S} \subset \mathcal{A}$ by its exterior is given by the formula

$$\begin{aligned} P(\mathcal{S}, t) \quad = \quad & \textstyle\int_{\mathrm{bdy}\mathcal{S}} \left(S_\backslash(X,t)\nu(X) \cdot \dot{\chi}_\backslash(X,t) + S_d(X,t)\nu(X) \cdot \dot{\chi}_d(X,t)\right) dA_X \\ & + \textstyle\int_{\mathcal{S}} \left(b^*_\backslash(X,t) \cdot \dot{\chi}_\backslash(X,t) + b^*_d(X,t) \cdot \dot{\chi}_d(X,t)\right) dV_X. \end{aligned} \tag{16.1}$$

In this formula, $\nu(X)$ denotes the unit outer normal to bdy$\mathcal{S}$ at X, and

$$S_\backslash := SK^*, \quad S_d := (\det K)S - SK^* \tag{16.2}$$

are the stress without disarrangements and the stress due to disarrangements introduced in Section 3.1 and reviewed in Section 15.1. If the structured motion is a classical motion, i.e., if $G = \nabla\chi$ and $\dot{\chi}_\backslash = \dot{\chi}$, then by (15.4) - (15.6), $S_\backslash = S$ and $(b^{in}_v)_\backslash = b^{in}_v$, and (16.1) becomes

$$P(\mathcal{S}, t) = \int_{\mathrm{bdy}\mathcal{S}} S(X,t)\nu(X) \cdot \dot{\chi}(X,t) dA_X + \int_{\mathcal{S}} b^*(X,t) \cdot \dot{\chi}(X,t) dV_X, \tag{16.3}$$

which is (4.1), the classical formula for the power expended.

In each integral on the right-hand side of the defining formula (16.1) for the power, the velocity without disarrangements is paired with a density of force without disarrangements, and there is a corresponding pairing for the velocity due to disarrangements. This choice for the power expended exploits the distinction provided only here in Part II between velocities arising away from disarrangement sites, as measured by $\dot{\chi}_\backslash$, and velocities associated with disarrangement sites as measured by $\dot{\chi}_d$. In Part I, this refinement and corresponding decomposition $\dot{\chi} = \dot{\chi}_\backslash + \dot{\chi}_d$ for velocities was not available for the purpose of defining the power. Only in the *reduced form of the power* (4.4) in Part I, obtained after invoking invariance of the power under classical, superposed rigid motions, did the kinematical quantity $\nabla\dot{\chi}$ appear that decomposes into parts with and without disarrangements, $\nabla\dot{\chi} = \dot{G} + \dot{M}$. In Section 20.1, further comparisons of and contrasts between the treatments of the stress-power in Parts I and II are provided.

We now strengthen the classical invariance assumptions affecting the power [11],[12] by requiring: *for every rigid motion* r *and time* t_o satisfying $\nabla r(\cdot, t_o) = I$, *for every part* $\mathcal{S}$, *and for every structured motion* $(\chi, G, \dot{\chi}_\backslash)$, *the power expended at time* t_o *in the superposition* $(r, \nabla r, \dot{r}) \circ (\chi, G, \dot{\chi}_\backslash)$ *equals the the power expended in* $(\chi, G, \dot{\chi}_\backslash)$ *at time*

t_o; *moreover, in the case where* r is a **translation**, *the power expended at time* t_o *in the superposition* $(r, \nabla r, 0) \circ (\chi, G, \dot{\chi}_\backslash)$ *equals the power expended at time* t_o *in* $(\chi, G, \dot{\chi}_\backslash)$.

We remark that this postulate of invariance of the power, while stronger than the standard invariance requirement under superposed classical rigid motions $(r, \nabla r, \dot{r})$ [11],[12], is weaker than the non-standard requirement made in [22], where invariance under the superposition of $(r, \nabla r, 0)$ was required for *all* rigid motions r (and times t_o satisfying $\nabla r(\cdot, t_o) = I$). The present limitation of invariance under superposed non-classical motions $(r, \nabla r, 0)$ to the case of *translations* r is made in order to broaden the range of applicability of the resulting field theory. In fact, were the stronger requirement of invariance [22] to be imposed here, an additional field relation (the "balance of moments due to disarrangements" - $(31)_2$ in [22]) would have to be satisfied, one that limits the range of applicability of the field theory. We also note that the invariance requirement on total body forces, imposed at the end of the previous section, would not be satisfied by b_d^* if, in part (ii) of that requirement, the collection of rotations r were admitted. This provides another reason for the weakening here of the invariance requirements imposed in [22].

17 Balance Laws

Let $a \in \mathcal{V}$ be given, and put $r(x,t) := x + at$ for all $x \in \mathcal{E}$ and $t \in (0, t_f)$. We then can use the results in the table below (14.12), with $t_o := t$ an arbitrary time, and apply the invariance of the power, of the stresses with and without disarrangements, and of the total body forces with and without disarrangements first for the superposition of the classical translation $(r, \nabla r, \dot{r})$ on the motion $(\chi, G, \dot{\chi}_\backslash)$ to obtain the relation

$$\int_{\text{bdy}\mathcal{S}} S_\backslash(X,t)\nu(X) \cdot a\, dA_X + \int_{\mathcal{S}} b_\backslash^*(X,t) \cdot a\, dV_X = 0 \tag{17.1}$$

and next for the superposition of the non-classical translation $(r, \nabla r, 0)$ on the same motion $(\chi, G, \dot{\chi}_\backslash)$ to obtain the relation

$$\int_{\text{bdy}\mathcal{S}} S_d(X,t)\nu(X) \cdot a\, dA_X + \int_{\mathcal{S}} b_d^*(X,t) \cdot a\, dV_X = 0. \tag{17.2}$$

Using the arbitrariness of a and of the part $\mathcal{S}$ as well as the Divergence Theorem, we obtain from (17.1) and (17.2) the *twin balance laws for linear momentum* [22]:

$$\text{div} S_\backslash + b_\backslash^* = 0; \qquad \text{div} S_d + b_d^* = 0. \tag{17.3}$$

Addition of $(17.3)_1$ and $(17.3)_2$ yields after elementary calculations the *overall balance law for linear momentum:*

$$\text{div} S + b_{ref} = \rho_{ref} \ddot{\chi}, \tag{17.4}$$

which is the relation (4.2) derived in Part I. Subtraction of either of the two equations in (17.3) from (17.4) yields the other equation in (17.3). Consequently, by successive omission of one of the three relations (17.3) and (17.4), we obtain three equivalent pairs of balance laws for linear momentum.

In view of the twin balance laws for linear momentum, the power (16.1) reduces to:

$$\begin{aligned} P(\mathcal{S},t) &= \int_{\mathrm{bdy}\mathcal{S}} \left(S_\backslash(X,t)\nu(X)\cdot\dot{\chi}_\backslash(X,t) + S_d(X,t)\nu(X)\cdot\dot{\chi}_d(X,t)\right) dA_X \\ &\quad - \int_{\mathcal{S}} \left(\mathrm{div}S_\backslash(X,t)\cdot\dot{\chi}_\backslash(X,t) + \mathrm{div}S_d(X,t)\cdot\dot{\chi}_d(X,t)\right) dV_X. \end{aligned}$$

Applying the Divergence Theorem to the surface integral and the product rule

$$\mathrm{div}(A^T w) = \mathrm{div}A\cdot w + A\cdot\nabla w$$

to the integrand in the resulting volume integral, we obtain the *reduced formula for the power expended*:

$$P(\mathcal{S},t) = \int_{\mathcal{S}} \left(S_\backslash(X,t)\cdot\nabla\dot{\chi}_\backslash(X,t) + S_d(X,t)\cdot\nabla\dot{\chi}_d(X,t)\right) dV_X. \qquad (17.5)$$

The integrand is the *total stress-power* $S_\backslash\cdot\nabla\dot{\chi}_\backslash + S_d\cdot\nabla\dot{\chi}_d$, decomposed as the sum of the *stress-power without disarrangements* $S_\backslash\cdot\nabla\dot{\chi}_\backslash$ and the *stress-power due to disarrangements* $S_d\cdot\nabla\dot{\chi}_d$.

For each $t\in(0,t_f)$, $x_o\in\mathcal{E}$, and skew tensor $W\in\mathrm{Lin}\mathcal{V}$, we put

$$r(y,\tau) := x_o + e^{W(\tau-t)}(y-x_o) \qquad \text{for all } y\in\mathcal{E},\ \tau\in(0,t_f), \qquad (17.6)$$

and note that $\dot{r}(\chi(X,t),t) = W(\chi(X,t)-x_o)$ and $\nabla(\dot{r}\circ\chi)(X,t) = W\,\nabla\chi(X,t)$ for all $X\in\mathcal{A}$. We again can use the results below (14.12), with t_o the given time t, and apply the invariance of the power, in the reduced form (17.5), as well as the invariance of the stresses with and without disarrangements for the superposition $(r,\nabla r,\dot{r})\circ(\chi,G,\dot{\chi}_\backslash)$ to obtain the relation

$$\int_{\mathcal{S}} S_\backslash(X,t)\cdot W\,\nabla\chi(X,t)\,dV_X = 0. \qquad (17.7)$$

The arbitrariness of the skew tensor W and of the part $\mathcal{S}$ imply $S_\backslash(X,t)\nabla\chi(X,t)^T$ is symmetric for each $(X,t)\in\mathcal{A}\times(0,t_f)$ and yields the *balance law for angular momentum without disarrangements*:

$$\mathrm{sk}(S_\backslash F^T) = 0. \qquad (17.8)$$

An equivalent form of the balance law (17.8) follows immediately from the definition $T_\backslash := TH^*$ of the stress in the current configuration without disarrangements, introduced in Section 10, and from the relations $K = F^{-1}G$ and $H = GF^{-1}$:

$$\mathrm{sk}T_\backslash = 0. \qquad (17.9)$$

This expression of the balance of angular momentum affords an easy comparison with the *(overall) balance law for angular momentum* (4.3) derived in Part I, which is equivalent to the relation

$$\mathrm{sk}T = 0. \qquad (17.10)$$

Thus, in the standard balance laws of Part I, the symmetry of the Cauchy stress T expresses the content of the law of balance of angular momentum, whereas in the new balance laws of Part II, the symmetry of the stress in the current configuration without disarrangements $T_\backslash$ expresses the content of the law of balance of angular momentum. Actually, as in Part I, we shall not impose balance of angular momentum directly, but instead invoke an appropriate expression of material frame-indifference and show that (17.8) follows.

18 Dynamical Processes, Dissipation, and a Constitutive Class

We augment the list χ, G, S, ψ that defined a dynamical process in Section 6 by including here information about the velocity without disarrangements $\dot{\chi}_\backslash$. In doing so, we note that the present theory contains two balance laws for linear momentum (17.3) and only one body force b_{ref} in (15.8), (15.9) not already determined by χ, G, S, ψ, and $\dot{\chi}_\backslash$. Therefore, given the density field ρ_{ref}, the balance law $(17.3)_1$ is automatically satisfied for given fields χ and S by appropriate choice of b_{ref}. In view of (15.9), the remaining balance law for linear momentum $(17.3)_2$ is a first-order differential equation for $\dot{\chi}_d = \dot{\chi} - \dot{\chi}_\backslash$ that determines $\dot{\chi}_d$ in terms of given fields χ, G, and S, provided only that we specify $\dot{\chi}_d(\cdot, t_o)$, the velocity due to disarrangements at a fixed but arbitrary time t_o. Equivalently, we may specify $\dot{\chi}_\backslash(\cdot, t_o)$ in place of $\dot{\chi}_d(\cdot, t_o)$, and we conclude that the augmented list χ, G, S, ψ, $\dot{\chi}_\backslash(\cdot, t_o)$ provides the information required to determine the remaining quantities $\dot{\chi}_\backslash$, b_{ref}, $S_\backslash$, and S_d that enter our kinematical and dynamical description of a body undergoing disarrangements, while satisfying the laws of balance of linear momentum (17.3). As in Section 6, we here do not impose the balance law for angular momentum without disarrangements (17.8) on dynamical processes, because imposing the requirement of material frame-indifference will turn out to guarantee satisfaction of (17.8). Accordingly, *our notion of dynamical process from now on will correspond to the augmented list* χ , G, S, ψ, $\dot{\chi}_\backslash(\cdot, t_o)$.

In view of the reduced expression for the power (17.5), the *dissipation inequality* takes the form (in the virgin configuration):

$$(\det K)\dot{\psi}(X,t) \le S_\backslash(X,t) \cdot \nabla\dot{\chi}_\backslash(X,t) + S_d(X,t) \cdot \nabla\dot{\chi}_d(X,t). \tag{18.1}$$

We wish to identify a collection of dynamical processes χ, G, S, ψ, $\dot{\chi}_\backslash(\cdot, t_o)$, i.e., a constitutive class, that reflects the ability of a body to store energy in a manner compatible with the dissipation inequality in the form (18.1). To this end, we assume given a smooth free energy response function $(M, G) \mapsto \tilde{\Psi}(M, G)$ and rewrite the dissipation inequality in the equivalent form

$$\begin{aligned} &((\det K)D_M\tilde{\Psi}(M,G) - S_d) \cdot \dot{M} + ((\det K)D_G\tilde{\Psi}(M,G) - S_\backslash) \cdot \dot{G} \\ \le\ & (S_\backslash - S_d) \cdot (\nabla\dot{\chi}_\backslash - \dot{G}), \end{aligned} \tag{18.2}$$

where we have omitted the argument (X, t) and have used the relation

$$\nabla\dot{\chi}_\backslash + \nabla\dot{\chi}_d = \nabla\dot{\chi} = \dot{F} = \dot{G} + \dot{M}. \tag{18.3}$$

We permit the free energy to depend also on the material point X, but we shall not make this dependence explicit in what follows. The inequality (18.2) now suggests, by analogy with the choice made in Section 7, the following choice of constitutive class $\mathcal{E}_{td}$, namely, the collection of all dynamical processes χ, G, S, ψ, $\dot{\chi}_\backslash(\cdot, t_o)$ satisfying

$$\psi(X,t) = \tilde{\Psi}(M(X,t), G(X,t)) \tag{18.4}$$

$$S_d(X,t) = (\det K) D_M \tilde{\Psi}(M(X,t), G(X,t)) \tag{18.5}$$

$$S_\backslash(X,t) = (\det K) D_G \tilde{\Psi}(M(X,t), G(X,t)) \tag{18.6}$$

and

$$(S_\backslash(X,t) - S_d(X,t)) \cdot (\nabla \dot{\chi}_\backslash(X,t) - \dot{G}(X,t)) \geq 0 \tag{18.7}$$

for all X and t. It is convenient to call $(S_\backslash - S_d) \cdot (\nabla \dot{\chi}_\backslash - \dot{G})$ the *driving power* and to call (18.7) the *driving power inequality*. It follows immediately from (18.2), (18.5), (18.6), and (18.7) that *every dynamical process in* $\mathcal{E}_{td}$ *satisfies the dissipation inequality, and the internal dissipation equals the driving power*:

$$\begin{aligned} S_\backslash \cdot \nabla \dot{\chi}_\backslash + S_d \cdot \nabla \dot{\chi}_d - (\det K)\dot{\psi} &= (S_\backslash - S_d) \cdot (\nabla \dot{\chi}_\backslash - \dot{G}) \\ &= (\det K)(D_G \tilde{\Psi} - D_M \tilde{\Psi}) \cdot (\nabla \dot{\chi}_\backslash - \dot{G}) \geq 0. \end{aligned} \tag{18.8}$$

The constitutive relations (18.5), (18.6) and the decompositon (3.5) imply the stress relation

$$S(X,t) = D_M \tilde{\Psi}(M(X,t), G(X,t)) + D_G \tilde{\Psi}(M(X,t), G(X,t)), \tag{18.9}$$

and the constitutive relations (18.4) - (18.7), (18.9) all represent restrictions on dynamical processes for the body. They may be rewritten for the choice of variables F and G and free energy response $(F,G) \mapsto \Psi(F,G)$ as:

$$\psi(X,t) = \Psi(F(X,t), G(X,t)) \tag{18.10}$$

$$S_d(X,t) = (\det K(X,t)) D_F \Psi(F(X,t), G(X,t)) \tag{18.11}$$

$$S_\backslash(X,t) = (\det K(X,t)) \left(D_F \Psi(F(X,t), G(X,t)) + D_G \Psi(F(X,t), G(X,t)) \right) \tag{18.12}$$

$$D_G \Psi(F(X,t), G(X,t)) \cdot (\nabla \dot{\chi}_\backslash(X,t) - \dot{G}(X,t)) \geq 0 \tag{18.13}$$

and

$$S(X,t) = 2 D_F \Psi(F(X,t), G(X,t)) + D_G \Psi(F(X,t), G(X,t)). \tag{18.14}$$

19 Material Frame-Indifference

The response functions $(M,G) \mapsto \tilde{\Psi}(M,G)$ and $(F,G) \mapsto \Psi(F,G)$ are assumed to satisfy the conditions of material frame-indifference (8.1) and (8.2), respectively, so that the relations (8.7) - (8.9) characterizing their frame-indifference (stated in terms of the response function $\tilde{\Psi}$) are valid here. In place of the notion that the mixed power is frame-indifferent, defined and analyzed in Section 8, we here shall define and study frame-indifference of the driving power $(D_G\tilde{\Psi} - D_M\tilde{\Psi})\cdot(\nabla\dot{\chi}_\backslash - \dot{G})$. To this end, we must append the following transformation laws to the list at the beginning of Section 8:

$$\dot{\chi}_\backslash \quad\rightarrow\quad Q\dot{\chi}_\backslash + \dot{Q}(\chi - o) \tag{19.1}$$

$$\dot{\chi}_d \rightarrow Q\dot{\chi}_d, \tag{19.2}$$

$$\nabla\dot{\chi}_\backslash \rightarrow Q\nabla\dot{\chi}_\backslash + \dot{Q}F \tag{19.3}$$

$$\nabla\dot{\chi}_d \rightarrow Q\nabla\dot{\chi}_d, \tag{19.4}$$

which follow from the transformation laws

$$\begin{array}{lcc} \chi_n - o & \rightarrow & Q(\chi_n - o) \\ \dot{\chi}_n & \rightarrow & Q\dot{\chi}_n + \dot{Q}(\chi_n - o). \end{array} \tag{19.5}$$

Here, o denotes a fixed point in space. It is interesting to keep in mind the fact that, unlike $\dot{\chi}_\backslash$ and $\dot{\chi}$, the velocity due to disarrangements $\dot{\chi}_d$ transforms objectively under change of observer, according to (19.2). This fact also is evident in view of the identification relation (14.5), which shows that the velocity due to disarrangements $\dot{\chi}_d$ transforms in the same manner as the jumps $[\chi_n]$.

We define *frame-indifference of the driving power* to mean that for all smooth time-parameterized families $t \mapsto Q(t)$ and for all families

$$t \mapsto (M(X,t), G(X,t), \nabla\dot{\chi}_\backslash(X,t))$$

arising from dynamical processes in $\mathcal{E}_{td}$, there holds

$$\begin{aligned} & (D_G\tilde{\Psi}(M,G) - D_M\tilde{\Psi}(M,G))\cdot(\nabla\dot{\chi}_\backslash - \dot{G}) \\ = \; & (D_G\tilde{\Psi}(QM,QG) - D_M\tilde{\Psi}(QM,QG)\cdot((Q\nabla\dot{\chi}_\backslash + \dot{Q}F) - (Q\dot{G} + \dot{Q}G)). \end{aligned} \tag{19.6}$$

As in Section 8 in the case of material frame-indifference of the mixed power, we now obtain easily the following characterization: *given the frame-indifference of $\tilde{\Psi}$, the driving power is frame-indifferent if and only if there holds*

$$\mathrm{sk}\left(D_G\tilde{\Psi}(M(X,t),G(X,t))M^T(X,t) - D_M\tilde{\Psi}(M(X,t),G(X,t))M^T(X,t)\right) = 0 \tag{19.7}$$

for all dynamical processes in $\mathcal{E}_{td}$.

The symmetry conditions (8.9) and (19.7) on the partial derivatives of $\tilde{\Psi}$ that follow from the frame-indifference of $\tilde{\Psi}$ and of the driving power have the concise forms:

$$\mathrm{sk}(D_M\tilde{\Psi}(M,G)M^T + D_G\tilde{\Psi}(M,G)G^T) = 0 \tag{19.8}$$

$$\mathrm{sk}(D_G\tilde{\Psi}(M,G)M^T - D_M\tilde{\Psi}(M,G)M^T) = 0, \tag{19.9}$$

the first holding throughout the domain of $\tilde{\Psi}$ and the second on all dynamical processes in $\mathcal{E}_{td}$. Adding these two relations yields, in view of (18.6), the relation

$$(detK)^{-1}\mathrm{sk}(S_\backslash F^T) = \mathrm{sk}(D_G\tilde{\Psi}(M,G)(G^T + M^T)) = 0,$$

which is the balance law for angular momentum without disarrangements (17.8). Moreover, (19.9), (18.5), and (18.6) tell us that

$$\mathrm{sk}(S_\backslash M^T) = \mathrm{sk}(S_d M^T), \tag{19.10}$$

i.e., in the terminology of Section 5, the offset moments with and without disarrangements are equal. In summary, *material frame-indifference of $\tilde{\Psi}$ and of the driving power imply the balance law for angular momentum without disarrangements (17.8) and the equality of the offset moments (19.10).*

We note that frame-indifference does not yield the overall balance balance law for angular momentum (17.10), i.e., the symmetry of the Cauchy stress. In fact, *granted the material frame-indifference of $\tilde{\Psi}$ (8.1) and of the driving power (19.6), the following are equivalent:*

1. *the symmetry of the Cauchy stress (17.10),*
2. *the balance law for angular momentum due to disarrangements:* $\mathrm{sk}(S_d F^T) = 0$,
3. *the frame-indifference of the mixed power (8.10).*

Indeed, from the analysis above, the decomposition $(\det K)S = S_\backslash + S_d$ and relations (19.8), (19.9) tell us that

$$\begin{aligned} (\det K)\mathrm{sk}(SF^T) &= \mathrm{sk}(S_\backslash F^T + S_d F^T) = \mathrm{sk}(S_d F^T) \\ &= \mathrm{sk}(S_d M^T + S_d G^T) = \mathrm{sk}(S_\backslash M^T + S_d G^T), \end{aligned} \tag{19.11}$$

which yields the desired equivalence.

20 Field Relations

Given a free-energy response function $\tilde{\Psi}$ satisfying frame-indifference (8.1), given the body force field b_{ref} and the density field ρ_{ref} in the reference configuration, the field relations that govern the dynamical processes in $\mathcal{E}_{td}$ that can occur are determined by the twin balance laws for linear momentum (17.3), the frame-indifference of the driving power (19.6), the constitutive relations (18.5), (18.6), the consistency relation (3.6), and the driving power inequality (18.7). The field relations, expressed in terms of $\tilde{\Psi}$, are equivalent to the relations

$$\mathrm{div}(D_M\tilde{\Psi} + D_G\tilde{\Psi}) + b_{ref} = \rho_{ref}\ddot{\chi} \tag{20.1}$$

$$\mathrm{div}((\det K)D_G\tilde{\Psi}) + (\det K)b_{ref} = (\det K)\rho_{ref}(\dot{\chi}_\backslash)^{\cdot} \tag{20.2}$$

$$D_G\tilde{\Psi}(K^{-T} - I) + D_M\tilde{\Psi}K^{-T} = 0 \tag{20.3}$$

$$\mathrm{sk}(D_G\tilde{\Psi}M^T - D_M\tilde{\Psi}M^T) = 0 \tag{20.4}$$

$$(D_G\tilde{\Psi} - D_M\tilde{\Psi}) \cdot (\nabla\dot{\chi}_\backslash - \dot{G}) \geq 0. \tag{20.5}$$

The equations (20.1) - (20.4) amount to $3+3+9+3=18$ scalar equations for the unknowns χ, G, and $\dot{\chi}_\backslash$, having 15 components.

The free energy response alternatively may be expressed in terms of the pair (F,G) or in terms of the pair (F,K), leading as in Section 9 to equivalent versions of the field relations. In this connection, we recall the form (18.13) that the driving power inequality (20.5) assumes in terms of the pair (F,G):

$$D_G\Psi(F(X,t),G(X,t))\cdot((\nabla\dot{\chi}_\backslash(X,t)-\dot{G}(X,t))\geq 0. \tag{20.6}$$

Because the driving power equals the internal dissipation, relation (20.6) shows that dynamical processes through pairs $(F(X,t),G(X,t))$ corresponding to stationary values of the partial free energy response $G\mapsto\Psi(F,G)$ proceed without dissipation. This partial response corresponds to holding the macroscopic deformation F fixed and varying the deformation without disarrangements G. Therefore, since local minima of this partial response would be among the pairs (F,G) where stationary values are encounterd, we can say that *submacroscopic instabilities are necessary for there to be positive internal dissipation.* Taking into account the fact that $\nabla\dot{\chi}_\backslash$ is the gradient of a limit of time-derivatives and $\dot{G}$ is the time-derivative of a limit of gradients, we find that the difference $\nabla\dot{\chi}_\backslash-\dot{G}$ is a measure of submacroscopic incompatibility that, according to (20.6), also is necessary for there to be positive internal dissipation. Consequently, *the driving power inequality reveals two submacroscopic mechanisms that must be present when the body is dissipating energy: submacroscopic instability and submacroscopic incompatibility.*

We note also that the difference $\nabla\dot{\chi}_\backslash-\dot{G}$ may be decomposed as

$$\nabla\dot{\chi}_\backslash-\dot{G}=(\nabla\dot{\chi}_\backslash-\nabla\dot{\chi})+(\dot{F}-\dot{G}), \tag{20.7}$$

in which the term $\nabla\dot{\chi}_\backslash-\nabla\dot{\chi}=\nabla(\dot{\chi}_\backslash-\dot{\chi})$ accounts for the presence of time-like disarrangements and the term $\dot{F}-\dot{G}=\dot{M}$ accounts for space-like disarrangements. Thus, *both space-like and time-like disarrangements may contribute to the internal dissipation.*

We remark that (20.2) may be replaced by the remaining twin balance law

$$\operatorname{div}((\det K)D_M\tilde{\Psi})-(D_M\tilde{\Psi}+D_G\tilde{\Psi})\nabla\det K=(\det K)\rho_{ref}(\dot{\chi}_d)^{\cdot} \tag{20.8}$$

and (20.5) may be replaced by

$$(D_G\tilde{\Psi}-D_M\tilde{\Psi})\cdot(\dot{M}-\nabla\dot{\chi}_d)\geq 0. \tag{20.9}$$

The resulting system (20.1), (20.8), (20.3), (20.4), (20.9) is a system with unknowns χ, G, $\dot{\chi}_d$ that is equivalent to (20.1) - (20.5).

20.1 Comparison with the field relations in Part I

We now compare the field relations (20.1) - (20.5), that take into account time-like disarrangements through inclusion of $\dot{\chi}_d=\dot{\chi}-\dot{\chi}_\backslash$, with the field relations (9.1) - (9.4) obtained in Part I, that do not. Relation (20.1) is the overall law of balance for linear momentum (17.4), obtained by adding the twin balance laws for linear momentum

(17.3), and is identical to relation (9.1). Relation (20.2) is the law of balance for linear momentum without disarrangements and does not have a counterpart in the field relations in Section 9. The consistency relation (20.3) is identical to (9.2), but the fact that the symmetry condition on stress (17.9) holds here, in place of the symmetry of the Cauchy stress (4.3) in Part I, means that the consistency relation provides here nine scalar relations, instead of the six provided in Part I. Relation (20.4), the condition of frame-indifference of the driving power, replaces (9.3), the condition of frame-indifference of the mixed power. Finally, the driving power inequality (20.5) is the counterpart of the mixed power inequality (9.4) in Section 9. The field relations in Section 9 amount to 12 scalar relations in 12 unknowns, while the present version has 18 scalar relations in 15 unknowns: the additional 6 scalar relations in Part II correspond to the balance law for linear momentum without disarrangements and the replacement of (4.3) by (17.9), as explained just above; the additional 3 scalar unknowns correspond to the field $\dot{\chi}_\backslash$, the velocity without disarrangements that appears in Part II but not in Part I.

Each of the expressions for the internal dissipation, the mixed power in case of the field relations in Section 9 and the driving power in the field relations in this section, arises from a particular choice of the form of the power, combined with the requirement that the stresses with and without disarrangements are determined through the partial derivatives of the free energy response. It is useful to compare the stress-power $S_\backslash \cdot \nabla\dot{\chi}_\backslash + S_d \cdot \nabla\dot{\chi}_d$, derived while taking into account time-like disarrangements, with the stress-power $(\det K)S \cdot \nabla\dot{\chi}$, obtained in Part I without taking into account time-like disarrangements. An easy calculation shows that the difference

$$(\det K)S \cdot \nabla\dot{\chi} - \left(S_\backslash \cdot \nabla\dot{\chi}_\backslash + S_d \cdot \nabla\dot{\chi}_d\right)$$

between the forms of the stress-power in Parts I and II is the same as the difference

$$S_\backslash \cdot \dot{M} + S_d \cdot \dot{G} - (S_\backslash - S_d) \cdot (\nabla\dot{\chi}_\backslash - \dot{G})$$

between the mixed power and the driving power. At an instant when time-like disarrangements do not occur within a given part of the body, we have $\nabla\dot{\chi}_\backslash = \nabla\dot{\chi} = \dot{F}$, and the difference between the mixed power and the driving power becomes

$$\begin{aligned} S_\backslash \cdot \dot{M} + S_d \cdot \dot{G} - (S_\backslash - S_d) \cdot (\dot{F} - \dot{G}) &= \\ S_d \cdot \dot{F} &= O(H - I). \end{aligned} \tag{20.10}$$

References

[1] Del Piero, G., Foundations of the theory of structured deformations; in this volume.

[2] Paroni, R., Second-order structured deformations: approximation theorems and energetics; in this volume.

[3] Deseri, L., Crystalline plasticity and structured deformations; in this volume.

[4] Deseri, L., and Owen, D. R., Toward a field theory for elastic bodies undergoing disarrangements, *Journal of Elasticity* **70**, 197-236, 2003. Research Report No. 02-CNA-017, September, 2002, Center for Nonlinear Analysis, Department of Mathematical Sciences, Carnegie Mellon University

[5] Del Piero, G., and Owen, D.R., Structured deformations of continua, *Archive for Rational Mechanics and Analysis* **124**, 99-155, 1993.

[6] Del Piero, G., and Owen, D.R., Integral-gradient formulae for structured deformations, *Archive for Rational Mechanics and Analysis* **131**, 121-138, 1995.

[7] Deseri, L., and Owen, D.R., Invertible Structured Deformations and the Geometry of Multiple Slip in Single Crystals, *International Journal of Plasticity* **18**, 833-849, 2002.

[8] Del Piero, G., and Owen, D.R., *Structured Deformations*, Quaderni dell' Istituto Nazionale di Alta Matematica, Gruppo Nazionale di Fisica Matematica. No. 58, 2000.

[9] Owen, D.R., Structured deformations and the refinements of balance laws induced by microslip, *International Journal of Plasticity* **14**, 289-299, 1998.

[10] Haupt, P., *Continuum Mechanics and Theory of Materials*, Springer Verlag, Berlin, etc., 2000.

[11] Noll, W., La mécanique classique, basée sur un axiome d'objectivité, pp. 47-56 of *La Méthode Axiomatique dans les Mécaniques Classiques and Nouvelles* (Colloque International, Paris, 1959), Paris, Gauthier-Villars, 1963.

[12] Green, A.E., and Rivlin, R.S., On Cauchy's equations of motion, *Journal of Applied Mathematics and Physics* **15**, 290-292, 1964.

[13] Gurtin, M.E., *An Introduction to Continuum Mechanics* , Academic Press, New York, etc., 1981.

[14] Coleman, B.D., and Noll, W., The thermodynamics of elastic materials with heat conduction and viscosity, *Archive for Rational Mechanics and Analysis* **13**, 167-178, 1963.

[15] Dafermos, C.M., Quasilinear hyperbolic systems with involutions, *Archive for Rational Mechanics and Analysis* **94** , 373-389, 1986.

[16] Choksi, R., Del Piero, G., Fonseca, I., and Owen, D.R., Structured deformations as energy minimizers in models of fracture and hysteresis, *Mathematics and Mechanics of Solids* **4**, 321-356, 1999.

[17] Deseri, L., and Owen, D.R., Energetics of two-level shears and hardening of single crystals, *Mathematics and Mechanics of Solids* **7**, 113-147, 2002.

[18] Noll, W., On the continuity of solid and fluid states, *Journal of Rational Mechanics and Analysis*, **4**, 3-91, 1955.

[19] Šilhavý, M., and Kratochvíl, J., A theory of inelastic behavior of materials, Part I, *Archive for Rational Mechanics and Analysis*, **65**, 97-129, 1977; Part II, *Archive for Rational Mechanics and Analysis*, **65**, 131-152, 1977.

[20] Bertram, A., An alternative approach to finite plasticity based on material isomorphisms, *International Journal of Plasticity* , **14**, 353-374, 1999.

[21] Owen, D.R., and Paroni, R., Second-order structured deformations, *Archive for Rational Mechanics and Analysis*, **155**, 215-235, 2000.

[22] Owen, D.R., Twin balance laws for bodies undergoing structured motions, in *Rational Continua: Classical and New*, P. Podio-Guidugli and M. Brocato, eds., Springer Verlag, New York, etc., 2002.

MIX
Papier aus verantwortungsvollen Quellen
Paper from responsible sources
FSC® C105338

If you have any concerns about our products,
you can contact us on
ProductSafety@springernature.com

In case Publisher is established outside the EU,
the EU authorized representative is:
Springer Nature Customer Service Center GmbH
Europaplatz 3, 69115 Heidelberg, Germany

Printed by Libri Plureos GmbH
in Hamburg, Germany